Second Edition

Applications of
Environmental Aquatic Chemistry

A Practical Guide

Eugene R. Weiner

CRC Press
Taylor & Francis Group
Boca Raton   London   New York

CRC Press is an imprint of the
Taylor & Francis Group, an **informa** business

CRC Press
Taylor & Francis Group
6000 Broken Sound Parkway NW, Suite 300
Boca Raton, FL 33487-2742

© 2008 by Taylor & Francis Group, LLC
CRC Press is an imprint of Taylor & Francis Group, an Informa business

No claim to original U.S. Government works
Printed in the United States of America on acid-free paper
10 9 8 7 6 5 4 3 2 1

International Standard Book Number-13: 978-0-8493-9066-1 (Hardcover)

Library of Congress Cataloging-in-Publication Data

Weiner, Eugene R.
    Applications of environmental aquatic chemistry : a practical guide / Eugene R. Weiner. -- 2nd ed.
        p. cm.
    Rev. ed. of: Applications of environmental chemistry / Eugene R. Weiner. 2000.
    Includes bibliographical references and index.
    ISBN 978-0-8493-9066-1 (alk. paper)
    1. Environmental chemistry. 2. Water quality. I. Weiner, Eugene R. Applications of environmental chemistry. II. Title.

TD193.W45 2007
628.1'68--dc22   2007048068

Visit the Taylor & Francis Web site at
http://www.taylorandfrancis.com

and the CRC Press Web site at
http://www.crcpress.com

Contents

Preface to the Second Edition

Much new material has been added to this second edition. Besides a totally new chapter on radionuclides, the text has been reorganized and updated with separate chapters on metals, light nonaqueous phase liquids (LNAPLs), dense nonaqueous phase liquids (DNAPLs), and biodegradation. Also, some end-of-chapter exercises have been added. The dictionary of inorganic pollutants has been enlarged and some important organic pollutants added. The former appendices listing drinking water standards, water quality criteria, and sample collection protocols have been omitted because these data are continually changing and are readily available on the Internet. However, the goals of the book remain the same—to help non-chemist environmental professionals and students work with chemical information in their work and studies.

Preface to the First Edition

By sensible definition, any by-product of a chemical operation for which there is no profitable use is a waste. The most convenient, least expensive way of disposing of said waste – up the chimney or down the river – is the best.

From *American Chemical Industry – A History*, by W. Haynes,
Van Nostrand Publishers, 1954

The quote above describes the usual approach to waste disposal as it was practiced in the first half of the 1900's. Current disposal and cleanup regulations are aimed at correcting problems caused by such misguided advice and go further toward trying to maintain a non-degrading environment. Regulations such as federal and state Clean Water Acts have set in motion a great effort to identify the chemical components and other characteristics which influence the quality of surface and groundwaters and the soils through which they flow. The number of drinking water contaminants regulated by the United States Government has increased from about five in 1940 to over 150 in 1999.

There are two distinct spheres of interest for an environmental professional, the ever-changing, constructed sphere of regulations and the comparatively stable sphere of the natural environment. Much of the regulatory sphere is bounded by classifications and numerical standards for waters, soils, and wastes. The environmental sphere is bounded by the innate behavior of chemicals of concern. While this book focuses on the environmental sphere, it makes an excursion into a small part of the regulatory sphere in Chapter 1, where the rationale for stream classifications and standards and the regulatory definition of water quality are discussed.

This book is intended to be a guide and reference for professionals and students. It is structured to be especially useful for those who must use the concepts of environmental chemistry but are not chemists and do not have the time and/or inclination to learn all the relevant background material. Chemistry topics that are most important in environmental applications are succinctly summarized, with a genuine effort to walk the middle ground between too much and too little information. Frequently used reference materials are also included, such as water solubilities, partition coefficients, natural abundance of trace metals in soil, and federal drinking water standards. Particularly useful are the frequent "rules-of-thumb" lists, which conveniently offer ways to quickly estimate important aspects of the topic being discussed.

Although it is often true that "a little knowledge can be dangerous", it is also true that a little chemical knowledge of the "right sort" can be a great help to the busy non-chemist. Although no "practical guide" will please everyone with its choice of inclusions and omissions, I have based my choices on the most frequently asked questions from my colleagues and the material I find myself looking-up over and over again. The main goal of this book is to offer non-chemist readers enough chemical insight to help them contend with those environmental chemistry problems that seem to arise most frequently in the work of an environmental professional. Environmental chemists and students of environmental chemistry should also find the book valuable as a general-purpose reference.

Author

Eugene R. Weiner, PhD is professor emeritus of chemistry at the University of Denver, Colorado. In 1965 he joined the University of Denver's faculty. From 1967 to 1992, he was a consultant with the U.S. Geological Survey, Water Resources Division in Denver, and has consulted on environmental issues for many other private, state, and federal entities. After 27 years of research and teaching environmental and physical chemistry, he joined Wright Water Engineers Inc., an environmental and water resources engineering firm in Denver, as a senior scientist.

He received a BS in mathematics from Ohio University, an MS in physics from the University of Illinois, and a PhD in chemistry from Johns Hopkins University. He has authored and coauthored more than 400 research articles, books, and technical reports. In recent years, he has conducted 16 short courses, dealing with the movement and fate of contaminants in the environment, in major cities around the United States for the continuing education program of the American Society of Civil Engineers.

1 Water Quality

1.1 DEFINING ENVIRONMENTAL WATER QUALITY

Water quality means different things to different people, depending on their goals for the water. A chemist in the laboratory will regard high-quality laboratory water as water free from chemical impurities or suspended solids. High-quality environmental water has different criteria. The same chemist on a wilderness backpack trip might identify high-quality water as water in a pristine environment unaltered by human activity. If the chemist is also a fisherman, she or he might regard high-quality water as a good habitat for fish and other aquatic organisms. A drinking water treatment plant manager will define high-quality water as water with a minimum amount of substances that have to be removed or treated to produce safe and palatable drinking water. A broad view of high-quality water will take into consideration its suitability for particular uses.

The U.S. Congress recognized this when they enacted the Federal Water Pollution Control Act Amendments of 1972 (Public Law 92500, also known as the Clean Water Act) where it is stated, "The objective of this act is to restore and maintain the chemical, physical, and biological integrity of the Nation's waters." In the law, water quality is a measure of its suitability for particular designated uses. Implementing the law entails identifying these uses, setting standards that are protective of the designated uses, and providing enforcement procedures that require compliance with the standards.

1.1.1 WATER-USE CLASSIFICATIONS AND WATER QUALITY STANDARDS

In most parts of the world, the days are long gone when rivers, lakes, springs, and wells from which one can directly drink could readily meet almost all needs for high-quality water. Where such water remains, mostly in high mountain regions untouched by mining, grazing, or industrial fallout, it must be protected by strict regulations. In the United States, many states seek to preserve high-quality waters with antidegradation policies. But most of the water that is used for drinking water supplies, irrigation, and industry, not to mention supplying a supporting habitat for natural flora and fauna, is much-reused water that often needs treatment to become acceptable.

Whenever it is recognized that water treatment is required, new issues arise concerning the degree of water quality sought, the costs involved, and, perhaps, restrictions imposed on the uses of the water. Since it is economically impossible to make all waters suitable for all purposes, it becomes necessary to designate for which uses various waters are suitable.

In this context, a practical evaluation of water quality depends on how the water is used, as well as its chemical makeup. The quality of water in a stream might be considered good if the water is used for irrigation but poor if it is used as a drinking water supply. To determine water quality, one must first identify the ways in which the water will be used and only then determine appropriate numerical standards for important water quality parameters that will support and protect the designated water uses.

Strictly speaking, a water impurity is any substance in the water that is not a part of a water molecule, and absolutely pure water is unattainable in any realistic water sample. High-quality water is not pure; it just contains amounts of impurities too small to be harmful to its intended uses. Many impurities in water are beneficial. For example, carbonate (CO_3^{2-}) and bicarbonate (HCO_3^{2-}) make water less sensitive to acid rain and acid mine drainage; hardness and alkalinity decrease the solubility and toxicity of metals; nutrients, dissolved carbon dioxide (CO_2), and dissolved oxygen (O_2) are essential for aquatic life. Outside a chemical laboratory, extremely pure water generally is not desirable. Water with very low concentrations of dissolved impurities is more corrosive (aggressive) to metal pipes than water containing a measure of hardness, it cannot sustain aquatic life, and it certainly does not taste as good as natural water saturated with dissolved oxygen and containing a healthy mix of minerals.

In this book, the following definitions are used:

- A water impurity is any substance in water that is not derived from a water molecule only,* regardless of whether it is considered harmful, beneficial, or neutral to the intended uses of the water.
- A water pollutant or contaminant is any substance in water that is not derived from a water molecule only, and is considered, when present in sufficient concentration, to be harmful to the water's intended uses.

For most purposes, the quality of water is not judged by its purity but rather by its suitability for the different uses intended for it. The water contaminant nitrate (NO_3^-) illustrates this point. In drinking water supplies, nitrate concentrations greater than 10 mg/L are considered a potential health hazard, particularly to young children. On the other hand, nitrate is a beneficial plant nutrient in agricultural water and is added as a fertilizer. Water containing more than 10 mg/L of nitrate is of poor quality if it is used for potable water but may be of good quality for agricultural use.

Thus, water uses must be identified before water quality can be judged. Once the water uses are defined, numerical water quality standards for harmful impurities can be established to protect each use.

There are three different types of water quality standards set by state and federal regulations:

---

* The species H^+, OH^-, H_3O^+, $H_5O_2^+$ can all be derived from water molecules only. Species such as Cl^-, Na^+, MnO_4^-, $Ca(OH)_2$, etc., clearly cannot.

1. Surface and groundwater standards, for the ambient quality of natural waters (rivers, lakes, reservoirs, wetlands, and groundwater). These standards are chosen to protect the current and intended uses of natural waters, as discussed later.
2. Effluent standards, controlled by discharge permits under the National Pollutant Discharge Elimination System (NPDES). Effluent standards are chosen so that wastewaters discharging into natural waters do not cause the receiving waters to exceed their surface and groundwater standards. Effluent standards are affected by the ambient standards for the receiving water, the assimilation capacity of the receiving water, the total pollutant load contributed by all dischargers into that water, etc.
3. Drinking water standards, which apply both to groundwater used as a public water supply and to water delivered to the public from drinking water treatment plants. Drinking water standards are chosen to protect the public health.

1.1.2 WATER QUALITY CLASSIFICATIONS AND STANDARDS FOR NATURAL WATERS

The following preliminary steps, taken by a state or federal agency, are a common approach to evaluating water quality in natural waters:

1. Define in general the basic purposes for which natural waters will potentially be used (water supply, aquatic life, recreation, agriculture, etc.). These will be the categories used for classifying uses for existing bodies of water.
2. Set numerical water quality standards for physical and chemical characteristics that will support and protect the different water-use categories.
3. Compare the water quality standards with field measurements of existing bodies of water, and then assign appropriate use classifications to the water bodies according to whether their present or potential quality is suitable for the assigned water uses.

After a natural body of water is classified for one or more uses, compile an appropriate set of numerical standards to protect its assigned use classifications. Where different assigned classifications have different standards for the same parameter, the more stringent standard will apply.

It is clear that measuring the chemical composition of a water sample collected in the field is just one step in determining water quality. The sample data must then be compared with the standards assigned to that water body. If no standards are exceeded, the water quality is defined as good within its classified uses. As new information is collected about environmental and health effects of individual water constituents, it may be necessary to revise the standards for different water uses. Federal and state regulations require that water quality standards be reviewed periodically and modified when appropriate.

In addition to the review process for existing standards, the Safe Drinking Water Act requires the Environmental Protection Agency (EPA) to periodically publish a

Drinking Water Contaminant Candidate List (CCL), a list of contaminants that, "at the time of publication, are not subject to any proposed or promulgated national primary drinking water regulations, are known or anticipated to occur in public water systems, and may require regulations."* Contaminants on the list are studied until EPA concludes that there are sufficient data and information to either propose appropriate regulations or conclude that no action is currently necessary.

1.1.3 Setting Numerical Water Quality Standards

Numerical water quality standards are chosen to protect the current and intended uses (classifications) for environmental waters. The water quality standards for each water body are based on all the uses for which it is classified. In addition, site-specific standards may be established where special conditions exist, such as where aquatic life has become acclimated to high levels of dissolved metals. Each state has tables of water quality standards for each classified water body. In addition to standards for environmental waters, there are separate human health-based standards for ground-water used for public drinking water supplies, and for treated drinking water as delivered from a water treatment plant or, for some parameters such as lead and copper, as delivered at the tap.

The states, not the U.S. EPA, have the primary responsibility for setting water quality standards. However, EPA sets baseline standards for different use classifications that serve as minimum requirements for the state standards. In addition, EPA issues guidance and model regulations regarding standards, and EPA approval is required before standards can be adopted or changed by states.

Water quality standards are defined in terms of

- Chemical composition: Concentrations of metals, organic compounds, chlorine, nitrates, ammonia, phosphorus, sulfate, etc.
- General physical and chemical properties: Temperature, alkalinity, conductivity, pH, dissolved oxygen, hardness, total dissolved solids (TDS), chemical oxygen demand, etc.
- Biological characteristics: Biological oxygen demand, *Escherichia coli*, fecal coliforms, whole effluent toxicity (WET), etc.
- Radionuclides: Radium-226, radium-228, uranium, radon, gross alpha and gross beta emissions, etc.

1.1.4 Typical Water-Use Classifications

All states classify surface waters and groundwater according to their current and intended uses. Typical classifications are described in the following sections.

1.1.4.1 Recreational

Class 1—Primary contact: These surface waters are suitable or intended to become suitable for recreational activities in or on the water when the ingestion of small

---

* More information may be found at: www.epa.gov/safewater/ccl/index.html.

quantities of water is likely to occur and prolonged and intimate contact with the body is expected, e.g., swimming, rafting, kayaking, tubing, windsurfing, waterskiing, etc. The Clean Water Act requires that waters shall be presumed, by default, to be suitable or potentially suitable for class 1 uses, unless a use attainability analysis demonstrates that there is not a reasonable potential for primary contact uses to occur in the water segments in question within the next 20 year period. For waters that are not currently suitable for class 1 status but could be restored within 20 years, states may subdivide the primary contact recreation classification into class 1a and class 1b:

> *Class 1a—Existing primary contact*: Class 1a waters are those in which primary contact uses have been documented or are presumed to be present. Waters for which no use attainability analysis has been performed demonstrating that a recreation class 2 classification is appropriate shall be assigned a class 1a classification, unless a reasonable level of inquiry has failed to identify any existing class 1 uses of the water segment.
>
> *Class 1b—Potential primary contact*: This classification shall be assigned to water segments for which no use attainability analysis has been performed demonstrating that a recreation class 2 classification is appropriate, and a reasonable level of inquiry has failed to identify any existing class 1 uses of the water segment.

Class 2—Secondary contact: Surface waters not suitable for a primary contact classification but are suitable or intended to become suitable for recreational uses in or around the water, which are not included in the primary contact categories, e.g., wading, fishing, motor yachting, and other streamside or lakeside recreation activities.

1.1.4.2 Aquatic Life

Surface waters that presently support aquatic life uses as described below, or may reasonably be expected to do so in the future due to the suitability of present conditions. Aquatic life classifications also apply to waters that are intended to become suitable for such uses as a goal. Separate standards should be applied to protect:

Class 1—Cold water aquatic life: These are waters that (1) are currently capable of sustaining a wide variety of cold water biota (considered to be the inhabitants of water in which temperatures normally do not exceed 20°C), including sensitive species or (2) could sustain such biota but for correctable water quality conditions. Waters shall be considered capable of sustaining such biota where physical habitat, water flows or levels, and water quality conditions result in no substantial impairment of the abundance and diversity of species.

Class 1—Warm water aquatic life: These are waters that (1) are currently capable of sustaining a wide variety of warm water biota (considered to be the inhabitants of water in which temperatures normally exceed 20°C), including sensitive species or (2) could sustain such biota but for correctable water quality conditions. Waters shall be considered capable of sustaining such biota where physical habitat, water flows or levels, and water quality conditions result in no substantial impairment of the abundance and diversity of species.

Class 2—Cold and warm water aquatic life: These are waters that are not capable of sustaining a wide variety of cold or warm water biota, including sensitive species, due to conditions of physical habitat, water flows and levels, or uncorrectable water quality, which result in substantial impairment of the abundance and diversity of species.

1.1.4.3 Agriculture

Surface waters that are suitable or intended to become suitable for irrigation of crops and that are not hazardous as drinking water for livestock.

1.1.4.4 Domestic Water Supply

Surface waters that are suitable or intended to become suitable for potable water supplies. After receiving standard treatment—defined as coagulation, flocculation, sedimentation, filtration, and disinfection with chlorine or its equivalent—these waters will meet federal and state drinking water standards.

1.1.4.5 Wetlands

Wetlands may be defined as areas that are inundated or saturated by surface or groundwater at a frequency and duration sufficient to support, and under normal circumstances do support, a prevalence of vegetation and organisms typically adapted for life under saturated soil conditions. Surface water and groundwater that supply wetlands may be subject to the same standards applied to wetlands.

A state may adopt a wetlands classification based on the functions of the wetlands in question. Wetland functions that may warrant site-specific protection include groundwater recharge or discharge, flood flow alteration, sediment stabilization, sediment or other pollutant retention, nutrient removal or transformation, biological diversity or uniqueness, wildlife diversity or abundance, aquatic life diversity or abundance, and recreation.

1.1.4.6 Groundwater

Subsurface waters in a zone of saturation that are at the ground surface or can be brought to the ground surface or brought to surface waters through wells, springs, seeps, or other discharge areas. Separate standards are applied to groundwater used for

Domestic use: Groundwaters that are used or are suitable for a potable water supply.

Agricultural use: Groundwaters that are used or are suitable for irrigating crops and livestock water supply.

Surface water quality protection: This classification is used for groundwaters that feed surface waters. It places restrictions on proposed or existing activities that could impact groundwaters in a way that water quality standards of classified surface water bodies could be exceeded.

Potentially usable: Groundwaters that are not used for domestic or agricultural purposes, where background levels are not known or do not meet human health and agricultural standards, where TDS levels are less than 10,000 mg/L, and where domestic or agricultural use can be reasonably expected in the future.

Limited use: Groundwaters where TDS levels are equal to or greater than 10,000 mg/L, where the groundwater has been specifically exempted by regulations of the state, or where the criteria for any of the above classifications are not met.

RULE OF THUMB

Generally, the most stringent standards are for drinking water and aquatic life classifications.

1.1.5 STAYING UP-TO-DATE WITH STANDARDS AND OTHER REGULATIONS

This is a daunting challenge and, in the opinion of some, an impossible one. Not only are the federal regulations continually changing but also individual states may promulgate different rules because of local needs. The usual approach is to obtain the latest regulatory information as the need arises, always recognizing that your current knowledge may be outdated. Part of the problem is that few environmental professionals can find time to regularly read the *Federal Register*, where the EPA first publishes all proposed and final regulations.

Fortunately, most trade magazines and professional journals highlight important changes in standards and regulations that are of interest to their readers. If you stay abreast of this literature, you will be aware of the regulatory changes and their implications. For the greatest level of security, one has to often contact state and federal information centers to ensure you are working with the regulations that are currently being enforced. Among the most useful sources for staying abreast of the latest information is the EPA Web site on the Internet (www.epa.gov). This Web site has links to information hotlines, laws and regulations, databases and software, available publications, and other information sources. Each state environmental agency also has its own Web site.

1.2 SOURCES OF WATER IMPURITIES

As discussed in Section 1.1.1, a water impurity is any substance other than water (H_2O) that is found in the water sample, whether harmful or beneficial.

Thus, calcium carbonate ($CaCO_3$) is a water impurity even though it is not considered hazardous and is not regulated. Impurities can be divided into three classes: (1) regulated impurities (pollutants) considered harmful or aesthetically objectionable, (2) unregulated impurities not considered harmful, and (3) unregulated impurities not yet evaluated for their potential health risks.*

In water quality analysis, unregulated as well as regulated impurities are measured. For example, hardness is a water quality parameter that results mainly from the presence of dissolved calcium and magnesium ions, which are unregulated impurities. However, high hardness levels can partially mitigate the toxicity of many dissolved metals to aquatic life. Hence, it is important to measure water hardness in order to evaluate the hazards of dissolved metals.

Data concerning unregulated water impurities are also helpful for anticipating certain non-health-related potential problems, such as a tendency for the water to form deposits in pipes and boilers, to cause metal corrosion, and for irrigation water to cause soils to swell and diminish their permeability. Unregulated impurities can also help to identify the recharge sources of wells and springs, identify the mineral formations through which surface water or groundwater passes, and age-date water samples.

1.2.1 NATURAL SOURCES

Snow and rainwater contain dissolved and particulate minerals collected from atmospheric particulate matter, and small amounts of gases dissolved from atmospheric gases. Snow and rainwater have virtually no bacterial content until they reach the surface of the earth.

After precipitation reaches the surface of the earth and flows over and through the soil, there are innumerable opportunities for introduction into the water of mineral, organic, and biological substances. Water can dissolve at least a little of nearly anything it contacts. Because of its relatively high density, water can also carry suspended solids. Even under pristine conditions, surface and groundwater will usually contain various dissolved and suspended chemical substances.

1.2.2 HUMAN-CAUSED SOURCES

Many human activities cause additional possibilities for water contamination. Some important sources are

- Construction and mining where freshly exposed soils and minerals can contact flowing water
- Industrial waste discharges and spills
- Petroleum leaks and spills from storage tanks, pipelines, tankers, and trucks
- Agricultural applications of chemical fertilizers, herbicides, and pesticides

---

* The 1996 Amendments to the Safe Drinking Water Act created a Contaminant Candidate List and a process to determine if new regulations are needed to protect drinking water safety. EPA is required to periodically publish a list of potential contaminants that, "at the time of publication, are not subject to any proposed or promulgated national primary drinking water regulation, which are known or anticipated to occur in public water systems, and which may require regulation."

- Urban storm water runoff, which may contact all the debris of a city, including spilled fuels, animal feces, dissolved metals, organic scraps, road salt, tire and brake particles, construction rubble, etc.
- Effluents from industries and waste treatment plants
- Leachate from landfills, septic tanks, treatment lagoons, and mine tailings
- Fallout from atmospheric pollution

Environmental professionals must remain alert to the possibility that natural impurity sources also may be contributing to problems that at first appear to be solely the result of human-caused sources. Whenever possible, one should obtain background measurements that demonstrate what impurities are present in the absence of known human-caused contaminant sources. For instance, groundwater in an area impacted by mining often contains relatively high concentrations of dissolved metals. Before any remediation programs are initiated, it is important to determine what the groundwater quality would have been if the mines had not been there. This generally requires finding, if possible, a location upgradient of the area influenced by mining, where the groundwater encounters subsurface mineral structures similar to those in the mined area.

1.3 MEASURING IMPURITIES

There are four characteristics of water impurities that are important for an initial assessment of water quality:

1. What kinds of impurities are present? Are they regulated compounds?
2. How much of each impurity is present? Are any standards exceeded for the water body being sampled?
3. How do the impurities influence water quality? Are they hazardous? Beneficial? Unaesthetic? Corrosive?
4. What is the fate of the impurities? How will their location, quantity, and chemical form change with time?

1.3.1 WHAT IMPURITIES ARE PRESENT?

The chemical content of a water sample is found by qualitative chemical analysis of collected environmental samples. Qualitative analysis identifies the chemical species present but not the quantity (although qualitative and quantitative analyses are often combined in a single measurement). Some of the analytical methods used are gas and ion chromatography, mass spectroscopy, optical emission and absorption spectroscopy, electrochemical probes, and immunoassay testing.

1.3.2 HOW MUCH OF EACH IMPURITY IS PRESENT?

The amount of impurity is found by quantitative chemical analysis of the water sample. The amount of impurity can be expressed in terms of total mass (e.g., "There are 15 tons of nitrate in the lake."), or in terms of concentration (e.g., "Nitrate is present at a concentration of 12 mg/L."). Concentration is usually the measure of interest for predicting the effect of an impurity on the environment. It is used for defining environmental standards, and is reported in most laboratory analyses. In addition to concentration standards, an additional limit of total mass may be applied to some rivers in the form of total maximum daily loads (TMDLs) of certain pollutants. Total maximum daily loads are used in setting standards for waste effluent discharges so that allowable total loads are not exceeded.

1.3.3 WORKING WITH CONCENTRATIONS

Unfortunately, there is not one all-purpose method for expressing concentration. The best choice of concentration units depends in part on the medium (liquid, solid, or gas), and in part on the purpose of the measurement. The example problems in this and the following sections illustrate some applications of concentration calculations.

For regulatory compliance purposes, concentration is usually expressed as mass of impurity per unit volume or unit mass of sample.

- In water samples, impurity concentrations are typically reported as milli-grams (mg), micrograms (μg), or nanograms (ng) of impurity per liter (L) of water sample. Although this is actually comparing a weight to a volume, it is generally assumed that the liter of water sample weighs exactly 1000 grams, so that an impurity concentration of 1 mg/L is equivalent to 1 gram of impurity in 1 million grams of water, or one part per million (1 ppm).*
  $1 \text{ mg/L} = 10^{-3} \text{ g/L} = 1$ part per million (ppm)
  $1 \text{ μg/L} = 10^{-6} \text{ g/L} = 1$ part per billion (ppb)
  $1 \text{ ng/L} = 10^{-9} \text{ g/L} = 1$ part per trillion (ppt)
- In soil samples, impurity concentrations are typically reported as milli-grams, micrograms, or nanograms of impurity per kilogram of soil sample.
  $1 \text{ mg/kg} = 10^{-3} \text{ g/kg} = 1$ part per million (ppm)
  $1 \text{ μg/kg} = 10^{-6} \text{ g/kg} = 1$ part per billion (ppb)
  $1 \text{ ng/kg} = 10^{-9} \text{ g/kg} = 1$ part per trillion (ppt)
- In gas samples (normally air samples), concentrations cannot be expressed as simply as in water or soils, because gas volumes and densities are strongly dependent on temperature and pressure. In addition, the amount of some air pollutants (such as carbon monoxide or organic vapors) can be as large as or as greater than the oxygen and nitrogen levels in severely polluted air; thus, the approximation of the footnote below, used for water concentrations, may not apply to air concentrations.

For these reasons, parts per million for gases is different from parts per million for liquids and solids. For liquids and solids, ppm is a ratio of two masses

---

* Note that the actual mass of the water sample includes the mass of the water plus the mass of the impurity. Since 1 L of pure water at 4°C and 1 atm pressure weigh 1000 g, there is an inherent assumption when equating 1 mg/L to 1 ppm, that the mass of the impurity in the sample is negligible compared to the mass of water and that the density of the sample does not change significantly over the temperature and pressure ranges encountered in environmental sampling.

(sometimes written as ppm (w/w)), whereas for gases, ppm means a ratio of two volumes* (ppm (v/v) or ppmv).

There is no consensus regarding the appropriate units by which to express concentrations of substances in air. Air pollution standards are usually promulgated as ppm, whereas air pollutant concentrations in reports and other literature may be expressed as

- Percent (parts per hundred) by volume (%v)
- Parts per million by volume (ppmv)
- Mass of pollutant per cubic meter of air (mg/m^3, $\mu g/m^3$, or ng/m^3)
- Molecules of a pollutant gas per cubic centimeter of air (molecules/cc or molecules/cc^3)
- Moles of a pollutant gas per liter of air (mol/L)
- Partial pressure of pollutant gas
- Mole fraction of pollutant gas

With so many definitions of gas concentrations in common use, it clearly is useful to be able to convert gas concentrations from one set of units to another. The rules of thumb box below illustrate the principles for converting between the two different units of air pollutant concentrations most commonly used, ppmv and mg/m^3.

RULES OF THUMB

1. To convert a gaseous pollutant concentration (C_{pol}) from ppbv to mg/m^3 use

$$C_{pol}\,(mg/m^3) = \left(\frac{C_{pol}\,(ppbv) \times MW_{pol} \times P\,(atm)}{T\,(^\circ K)}\right) \times \left(\frac{1}{0.08205}\right) \quad (1.1)$$

2. To convert a gaseous pollutant concentration (C_{pol}) from mg/m^3 to ppbv use

$$C_{pol}\,(ppbv) = \left(\frac{T\,(^\circ K)}{C_{pol}\,(ppbv) \times MW_{pol} \times P\,(atm)}\right) \times \left(\frac{0.08205}{1}\right) \quad (1.2)$$

where
C_{pol} = pollutant concentration in the desired units
MW_{pol} = molecular weight of the pollutant in g/mole
P = pressure of air (atm)
T = temperature of air (K)
0.08205 (L atm)/(K mol) = R, the ideal gas constant

Note: $T(K) = T(^\circ C) + 273.15$;
$P(atm) = P(mm\,Hg)/760 = P(torr)/760 = P(hPa)/1013$
$= P(mbar)/1013 = P(dyne/cm^2)/1.013 \times 106$

---

* The ratio v/v means the ratio of the volume that a gaseous pollutant would have if it were isolated from the air, to the volume of air without the pollutant, both volumes having the same temperature and pressure. According to the ideal gas law, a v/v ratio is equivalent to a ratio of the number of pollutant gas molecules (n_{pol}) to the number of normal air molecules (n_{air}), or n_{pol}/n_{air}.

1.3.4 MOLES AND EQUIVALENTS

For chemical calculations (as opposed to regulatory compliance calculations), concentrations in any phase are usually expressed either as moles* of impurity per liter of sample (abbreviated as mol/L), moles of impurity per kilogram of sample (mol/kg), or as equivalents of impurity per liter (eq/L) or kilogram (eq/kg) of sample. Moles per liter are related to the number of impurity molecules, rather than the mass of impurity. Because chemical reactions involve one-on-one molecular interactions, regardless of the mass of the reacting molecules, moles are best for chemical calculations, such as balancing chemical reactions and calculating reaction rates. A common chemical notation for expressing a concentration in mol/L is to enclose the constituent in square brackets. Thus, writing $[Na^+] = 16.4$, is the same as writing $Na^+ = 16.4$ mol/L.

EXAMPLE 1

CALCULATING A CONCENTRATION IN WATER

A 45.6 mL water sample was found to contain 0.173 mg of sodium. What is the concentration in mg/L of sodium in the sample?

Answer:

$$\frac{0.173 \text{ mg}}{45.6 \text{ mL}} \times \frac{1000 \text{ mL}}{L} = 3.79 \text{ mg/L or } 3.79 \text{ ppm}$$

(Note that 3.79 ppm means all of the following: 3.79 g of sodium in 10^6 g of solution, or 3.79×10^{-3} g (3.79 mg) of sodium in 10^3 g (1 L) of solution, or 3.79×10^{-6} g (3.79 μg) of sodium per gram of solution.)

---

* Operationally, 1 mole of any pure element or compound is that quantity of the substance that has a mass equal to the atomic or molecular weight, in grams, of that substance. Thus, 1 mole of pure sodium (Na) metal is the amount that weighs 23.00 g; 1 mole of sodium chloride (NaCl) is the amount that weighs 58.45 g. This arises from the definition of a mole (abbreviated mol in chemical notation, as in mol/L): the term mole indicates a particular number of things, just as a dozen indicates 12 things and a pair indicates 2 things. The number of things indicated by a mole is defined to be the number of carbon atoms found in exactly 12 g of the ^{12}C isotope. The number of atoms present in 12 g of ^{12}C has been determined experimentally to be 6.022×10^{23} atoms (given here to 4 significant figures). This large number is called Avogadro's number, after the first scientist to deduce its value.

   The molecular (or atomic) weight of any molecule (or atom), expressed in grams, contains one mole, or 6.022×10^{23} molecules or atoms.

   Thus, 1 mole of pure calcium metal (weighing 40.08 grams) contains 6.022×10^{23} calcium atoms. As an example of a molecule, 1 mole of pure calcium chloride ($CaCl_2$; weighing $40.08 + 2 \times 35.45 = 110.98$ g) contains 6.022×10^{23} calcium chloride molecules. When 1 mole of calcium chloride dissolves, it dissociates into 1 mole of Ca^{2+} ions and 2 moles of Cl^- ions.

EXAMPLE 2

CALCULATING A CONCENTRATION IN AIR

The equations used in these calculations are based on the ideal gas laws, which are discussed in introductory general chemistry textbooks. The equations are used here without derivation.

The ozone (O_3) level in the Denver, Colorado, atmosphere was reported to be 2.50 ppmv (2.50 μL/L). Express this in mg/m^3 at ambient conditions of 37°C and 722 mm Hg.

Answer:

Using Equation 1.1

$$2.5 \text{ ppmv} = 2500 \text{ ppbv}$$
$$MW(O_3) = 3 \times 16.0 \text{ g/mol} = 48 \text{ g/mol}$$
$$P(\text{atm}) = 722 \text{ mm Hg} \times \frac{1 \text{ atm}}{760 \text{ mm Hg}} = 0.950 \text{ atm}$$
$$T(\text{K}) = 37°\text{C} + 273 = 310 \text{ K}$$
$$C_{O_3}(\text{mg/m}^3) = \left(\frac{2500 \text{ ppbv} \times 48 \text{ g/mol} \times 0.950 \text{ atm}}{310 \text{ K}} \right) \left(\frac{1}{0.08205} \right)$$
$$C_{O_3}(\text{mg/m}^3) = 4.482 \times 10^3 \text{ mg/m}^3$$

EXAMPLE 3

CONVERTING MG/L TO MOLES/L

Benzene in a water sample was reported as 0.017 mg/L. Express this concentration as mol/L. (To convert mg/L to mol/L, divide by 1000 mg/g and then divide by the molecular weight, in grams (g) of the impurity. Obtain the molecular weight by adding the atomic weights of all the atoms in the molecule. Use the periodic table inside the front cover of this book to find the atomic weights.)

Answer:

The chemical formula for benzene is C_6H_6 (meaning that one molecule of benzene contains six atoms of carbon and six atoms of hydrogen). Therefore, its molecular weight is $(6 \times 12 + 6 \times 1) = 78$ g/mol. The concentration of benzene in the sample can be expressed as

$$C_{\text{benzene}} = \frac{0.017 \text{ mg/L}}{(1000 \text{ mg/g})(78 \text{ g/mol})} = 2.18 \times 10^{-7} \text{ mol/L}$$

EXAMPLE 4

USING MOLES, PPM, AND MG/L TOGETHER

The federal primary drinking water standard for nitrate is 10 mg of nitrate–nitrogen per liter of water (written as: 10 mg NO_3–N/L). It is defined in terms of the nitrogen content of the nitrate ion present in the sample, neglecting the mass of the oxygen atoms in the molecule.

If a laboratory analysis includes the mass of the oxygen atoms and reports the nitrate concentration in a water sample as 33 ppm NO_3/L (not NO_3–N/L), does the analysis indicate that the water source is in compliance with the federal drinking water standard?

Answer:

$$33 \text{ ppm} = 33 \text{ mg/L} = 33 \times 10^{-3} \text{ g/L}$$

$$\text{Moles of } NO_3 \text{ in 1 L of sample} = \frac{\text{weight of } NO_3 \text{ in 1 L of sample}}{\text{molecular weight of } NO_3}$$

$$= \frac{33 \times 10^{-3} \text{ g/L}}{62.0 \text{ g/mol}} = 0.53 \times 10^{-3} \text{ mol/L or } 0.53 \text{ mmol/L}$$

Each mole of NO_3 contains 1 mole of N and 3 moles of O. Therefore, 0.53 mmol/L of NO_3 contains 0.53 mmol/L of N.

$$0.53 \times 10^{-3} \text{ mol N/L} \times 14 \text{ g N/mol} = 7.4 \times 10^{-3} \text{ g } NO_3\text{–N/L}$$

This sample does not exceed the federal standard of 10 mg NO_3–N/L and the source is in compliance.

EXAMPLE 5

USING MOLES, PPM, AND MG/L TOGETHER

2.00 g of the disinfectant calcium hypochlorite, $Ca(OCl)_2$, is added to a hot tub containing 1050 L of water. $Ca(OCl)_2$ dissociates in water by the reaction

$$Ca(OCl)_2 + 2H_2O \rightarrow 2HOCl + Ca^{2+} + 2OH^- \tag{1.3}$$

HOCl partially dissociates further by

$$HOCl + H_2O \leftrightarrow OCl^- + H_3O^+ \tag{1.4}$$

a. What would be the concentration of $Ca(OCl)_2$ in the hot tub water if $Ca(OCl)_2$ did not dissociate? In other words, what concentration of $Ca(OCl)_2$ is initially added to the hot tub water?

Answer:

$$\text{Concentration of } Ca(OCl)_2 = \frac{2.00 \text{ g}}{1050 \text{ L}} = 0.0019 \text{ g/L} = 1.9 \text{ mg/L} = 1.9 \text{ ppm}$$

b. The active disinfecting species are HOCl and OCl^-, which together are called the available free chlorine. What is the concentration of the available free chlorine in ppm after the reaction (in Equation 1.4) is complete? Assume that the $Ca(OCl)_2$ dissociates in water completely, and that the available free chlorine is 50% HOCl and 50% OCl$^-$ (in terms of number of molecules, or moles, not by weight), as is the case at pH $= 7.5$.

Procedure: (Use the periodic table for atomic weights.)

(1) Determine the number of moles in 2.00 g of Ca(OCl)$_2$.
(2) Determine the moles of HOCl formed by 2.00 g of Ca(OCl)$_2$, if no further dissociation occurred.
(3) Determine the moles of HOCl and OCl$^-$ in the water after complete dissociation (Equation 1.4).
(4) Determine the weight of HOCl and OCl$^-$ in the water.
(5) Calculate the ppm of HOCl + OCl$^-$ in 1050 L of water.

Answer:

(1) MW of Ca(OCl)$_2 = 40.1 + 2\times16.0 + 2\times35.4 = 142.9$ g/mol

$$\text{Moles of Ca(OCl)}_2 \text{ in 2.00 g} = \frac{2.00 \text{ g}}{142.9 \text{ g/mol}} = 0.014 \text{ mol}$$

(2) Equation 1.3 indicates that 2 moles of HOCl are formed from 1 mole of Ca(OCl)$_2$. Therefore

2×0.014 mol $= 0.028$ moles of HOCl are formed from 2.00 g of Ca(OCl)$_2$.

(3) Half of the HOCl dissociates to OCl$^-$, resulting in 0.014 moles of HOCl and 0.014 moles of OCl$^-$.

(4) MW of HOCl $= 1.0 + 16.0 + 35.4 = 52.4$ g/mol
MW of OCl$^- = 16.0 + 35.4 = 51.4$ g/mol
Weight of HOCl in the water $= 0.014$ mol $\times$ 52.4 g/mol $= 0.73$ g
Weight of OCl$^-$ in the water $= 0.014$ mol $\times$ 51.4 g/mol $= 0.72$ g
Total weight of HOCl + OCl$^- =$ weight of free chlorine $= 1.45$ g

(5) Concentration of free chlorine in hot tub $= \dfrac{1.45 \text{ g}}{1050 \text{ L}} = 0.0014$ g/L $= 1.4$ ppm

EXAMPLE 6

USING CONCENTRATION CALCULATIONS TO PREDICT A PRECIPITATE

In this example, the result of a chemical reaction must be determined and it is necessary to use concentration units of moles/L (molarity).

A mole is the amount of a compound that has a weight in grams equal to its molecular weight. Molecular weight is the sum of the atomic weights of all the atoms in the molecule. (See periodic table inside front cover for atomic weights.)

For example, the atomic weight of oxygen is 16. An oxygen molecule (O$_2$) contains two oxygen atoms and, thus, has a molecular weight of 32. A mole of O$_2$ is the amount, or number of molecules, that weighs 32 g. Another example: the atomic weights of carbon and calcium are 12 and 40, respectively. The molecular weight of CaCO$_3$ is

$$40 + 12 + 3 \times 16 = 100 \text{ g}$$

One mole of CaCO$_3$ is the quantity that weighs 100 g.

A water sample is taken from a stream that passes through soils containing gypsum ($CaSO_4$), some of which dissolves. The stream already carries some Ca^{2+} and SO_4^{2-} dissolved from other mineral sources. Laboratory analysis of the water shows $SO_4^{2-} = 576$ mg/L; $Ca^{2+} = 244$ mg/L.

Will a precipitate of $CaSO_4$ develop in the stream?

Calculation:

The precipitation reaction is $Ca^{2+} + SO_4^{2-} \rightarrow CaSO_4(\text{solid})$.

The solubility product is $K_{eq} = [Ca^{2+}][SO_4^{2-}] = 2.4 \times 10^{-5}$.

This means that if the product of the calcium and sulfate stream concentrations (in moles/L) exceeds 2.4×10^{-5}, then $CaSO_4$ will precipitate when the system is at equilibrium. The quantity $[Ca^{2+}][SO_4^{2-}]$, using experimental concentrations, is called the reaction quotient.

1. Convert mg/L to mol/L:

$$Ca^{2+}: \frac{244 \times 10^{-3} \text{ g/L}}{40 \text{ g/mol}} = 6.1 \times 10^{-3} \text{ mol/L}$$

$$SO_4^{2-}: \frac{576 \times 10^{-3} \text{ g/L}}{96 \text{ g/mol}} = 6.0 \times 10^{-3} \text{ mol/L}$$

2. Insert environmental concentrations into the reaction quotient:

$$[Ca^{2+}][SO_4^{2-}] = (6.1 \times 10^{-3})(6.0 \times 10^{-3}) = 3.7 \times 10^{-5}$$

The reaction quotient is larger than K_{eq}. Therefore, $CaSO_4$ may precipitate downstream when equilibrium is reached, if the stream is not diluted by additional water carrying less calcium or sulfate.

1.3.4.1   Working with Equivalent Weights

Equivalents per liter (eq/L) express the moles of ionic charge per liter of sample. This is useful for chemical calculations involving ions, because ionic reactions must always balance electrically, i.e., with respect to ionic charge. Since environmental waters normally contain many ionic species, equivalent weights are often useful in water quality calculations.

The equivalent weight of an ion is its molecular weight (for molecular ions such as HCO_3^-) or atomic weight (for single atom ions such as Na^+ or Cl^-) divided by its magnitude of charge (without regard for the sign of the charge). For nonionic species (such as $CaSO_4$), divide the molecular weight by what the charge would be if the molecules were dissolved (also called the oxidation number).

Another way of stating this is to define the equivalent weight of an ion as the weight that would carry 1 mole of charge. Thus,

- One equivalent weight of a singly charged ion is equal to its molecular or atomic weight, because 1 mole of the ions carries 1 mole of charge.
- One equivalent weight of a doubly charged ion is equal to one-half of its molecular or atomic weight, because 1 mole of the ions carries 2 moles of charge.

• One equivalent weight of a triply charged ion is equal to one-third of its molecular or atomic weight, because 1 mole of the ions carries 3 moles of charge.

For example, the equivalent weight of Ca^{2+} is ½ its atomic weight, because each calcium ion carries 2 positive charges, and a ½ mole of Ca^{2+} contains 1 mole of positive charge:

$$\text{eq. wt. of } Ca^{2+} = \frac{1}{2} \times 40.08 = 20.04 \text{ g/eq (20.04 grams/equivalent)}$$

To determine the equivalent weight of the neutral molecule $CaSO_4$, you must recognize that it dissolves in water to form Ca^{2+} and SO_4^{2-}.

$$CaSO_4 \xrightarrow{H_2O} Ca^{2+} + SO_4^{2-}$$

Since each ion formed carries a charge of magnitude 2, the equivalent weight of $CaSO_4$ is its molecular weight divided by 2, or

$$\text{eq. wt. of } CaSO_4 = \frac{(40.08 + 32.07 + 4 \times 16.00)}{2} = 68.08 \text{ g/eq}$$

Equivalents per liter of an impurity are equal to the moles per liter multiplied by the ionic charge or oxidation number, because, for example, 1 mole of Ca^{2+} contains 2 moles of charge (or two equivalents of charge). That this is consistent with the fact that the equivalent weight of a substance is its molecular weight divided by the charge or oxidation number is shown by Example 2.

EXAMPLE 7

EQUIVALENT WEIGHT OF AN ION

What is the equivalent weight of Cr^{3+}?

Answer:

The equivalent weight of Cr^{3+} is the mass that contains 1 mole of charge. Since each ion of Cr^{3+} contains 3 units of charge, the moles of charge in a given amount of chromium are 3 times the moles of ions. Thus, 1 mole of Cr^{3+}, or 52 grams, contains 3 moles of charge, or 3 equivalent weights. Therefore,

$$\text{eq. wt. of } Cr^{3+} = (\text{at. wt. of } Cr^{3+})/3 = 52.0/3 = 17.3 \text{ g/eq}$$

If a water sample contains 1 mol/L (52 g/L) of Cr^{3+}, it contains 3×17.3 g/L or 3 eq/L of Cr^{3+}.

Working with equivalents is useful for comparing the balance of positive and negative ions in a water sample or making cation exchange calculations. To convert mol/L to eq/L, multiply by the ionic charge or oxidation number of the impurity. Use the absolute value of the charge or oxidation number, i.e., multiply by +2 for a charge of either +2 or −2.

TABLE 1.1

Molecular Weights and Equivalent Weights of Some Common Water Species

Species	Atomic Weight	Absolute Charge	Equivalent Weight	Species	Atomic Weight	Absolute Charge	Equivalent Weight
Na^+	23.0	1	23.0	Cl^-	35.4	1	35.4
K^+	39.1	1	39.1	F^-	19.0	1	19.0
Li^+	6.9	1	6.9	Br^-	79.9	1	79.9
Ca^{2+}	40.1	2	20.04	NO^{3-}	62.0	1	62.0
Mg^{2+}	24.3	2	12.2	NO^{2-}	46.0	1	46.0
Sr^{2+}	87.6	2	43.8	HCO^{3-}	61.0	1	61.0
Ba^{2+}	137.3	2	68.7	CO_3^{2-}	60.0	2	30.0
Fe^{2+}	55.8	2	27.9	CrO_4^{2-}	116.0	2	58.0
Mn^{2+}	54.9	2	27.5	SO_4^{2-}	96.1	2	48.03
Zn^{2+}	65.4	2	32.7	S^{2-}	32.1	2	16.0
Al^{3+}	27.0	3	9.0	PO_4^{3-}	95.0	3	31.7
Cr^{3+}	52.0	3	17.3	$CaCO_3$	100.1	2	50.04
NH_4^+	18.0	1	18.0	$CaSO_4$	136.2	2	68.1

EXAMPLE 8

EQUIVALENT WEIGHT OF A COMPOUND

Alkalinity in a water sample is reported as 450 mg/L of $CaCO_3$. Using Table 1.1, convert this result to eq/L of $CaCO_3$. Alkalinity is a water quality parameter that results from more than one constituent. It is expressed as the amount of $CaCO_3$ that would produce the same analytical result as the actual sample (see Chapter 2).

Answer:

The molecular weight of $CaCO_3$ is $(1 \times 40 + 1 \times 12 + 3 \times 16) = 100$ g/mol. The dissolution reaction of $CaCO_3$ is

$$CaCO_3 \xrightarrow{H_2O} Ca^{2+} + CO_3^{2-}$$

Since the absolute value of charge for either the positive or negative species equals 2, eq/L = mol/L × 2.

$$450 \text{ mg/L} = \frac{450 \text{ mg/L}}{(100 \text{ g/mol})(1000 \text{ mg/g})} = 4.5 \times 10^{-3} \text{ mol/L or } 4.5 \text{ mmol/L}$$

$$4.5 \times 10^{-3} \text{ mol/L} \times 2 \text{ eq/mol} = 9.0 \times 10^{-3} \text{ eq/L or } 9.0 \text{ meq/L}$$

EXAMPLE 9

USING EQUIVALENT WEIGHT

Chromium(III) in a water sample is reported as 0.15 mg/L. Express the concentration as eq/L. (The Roman numeral III indicates that the oxidation number of chromium in the sample is +3. It also indicates that the dissolved ionic form would have a charge of +3.)

Answer:

The atomic weight of chromium is 52.0 g/mol. Chromium(III) ionizes as Cr^{3+}, so its concentration in mol/L is multiplied by 3 to obtain its equivalent weight.

$$mol/L \text{ of } Cr^{3+} = [Cr^3] = \frac{0.15 \text{ mg/L}}{(52.0 \text{ g/mol})(1000 \text{ mg/g})}$$
$$= 2.8810^{-6} \text{ mol/L (or 2.88 } \mu mol/L)$$

Concentration of Cr^{3+} in equivalents $= [Cr^3]_{eq}$

$$[Cr^3]_{eq} = 2.88 \times 10^{-6} \text{ mol/L} \times 3 \text{ eq/mol} = 8.64 \times 10^{-6} \text{ eq/L} = 8.64 \text{ } \mu eq/L$$

EXAMPLE 10

USING MG/L, MOLES/L, AND EQUIVALENTS TOGETHER

Using the dissolution reaction $CaCl_2 \rightarrow Ca^{2+} + 2Cl^-$

a. Calculate how many moles/L of $CaCl_2$ are needed to produce a solution with 500 mg/L of Ca^{2+}, with 500 mg/L of Cl^-.
b. Show that dissolving in water one equivalent weight of calcium chloride ($CaCl_2$), results in an electrically neutral solution containing 1 mole of positive ions and 1 mole of negative ions.

Answer:

a. Calculate how many moles/L of Ca are in 500 mg/L of Ca^{2+}.

$$\frac{500 \times 10^{-3} \text{ g/L}}{40.1 \text{ g/mol}} = 0.0125 \text{ mol/L}$$

The dissolution reaction shows that the moles of Ca^{2+} produced are equal to the moles of $CaCl_2$ dissolved. Therefore, 0.0125 mol/L of $CaCl_2$ are needed. Calculate how many moles/L of Cl are in 500 mg/L of Cl.

$$\frac{500 \times 10^{-3} \text{ g/L}}{35.5 \text{ g/mol}} = 0.0141 \text{ mol/L}$$

The dissolution reaction shows that the moles of Cl^- produced are two times the moles of $CaCl_2$ dissolved. Therefore, 0.0141/2 = 0.00704 mol/L of $CaCl_2$ are needed.

b. The exact amount of $CaCl_2$ dissolved is irrelevant. However, many moles or equivalents are dissolved in any amount of water, the concentration of Cl^- produced is always twice the concentration of Ca^{2+} produced and the total negative charge is always equivalent to the total positive charge.

Given that

- The total negative charge in the solution is $1 \times [Cl^-]$
- The total positive charge in the solution is $2 \times [Ca^{2+}]$
- Total negative charge must equal total positive charge: $[Cl^-] = 2 \times [Ca^{2+}]$

Then, negative charge $= 1 \times [Cl^-] = 1 \times (2 \times [Ca^{2+}]) =$ positive charge.

This result is general. Dissolving electrically neutral compounds in water always results in an electrically neutral solution.

1.3.5 CASE HISTORY EXAMPLE

A shallow aquifer below an industrial park was contaminated with toxic halogenated hydrocarbons (in this case, hydrocarbons containing chlorine and bromine). Although other pollutants were present, only the halogenated hydrocarbons were found to threaten a municipal drinking water supply. The state environmental authorities mandated a remediation program to be paid for by the responsible parties, which were several industrial facilities in the park. In order to allocate an appropriate share of the cleanup expenses to each responsible party, it was necessary to estimate what percentage of the total pollution was caused by each party.

An automobile rental agency was cited as one of the responsible parties, even though they did not use halogenated chemicals in their business, because they had had a leaking underground gasoline storage tank that released approximately 2500 gal of leaded gasoline to the subsurface above the aquifer. The gasoline contained additives with chlorine and bromine compounds.

During the time period between the late 1920s and the early 1990s, lead compounds, particularly tetraethyl lead (also called TEL), were added to automotive and aviation gasoline as an octane enhancer. During that time, it was common practice to also add halogenated organic compounds, particularly 1,2-dichloroethane (also called ethylene dichloride* or EDC) and 1,2-dibromoethane (also called ethylene dibromide[†] or EDB), to leaded gasoline to serve as lead scavengers, helping to prevent lead deposits from accumulating in gasoline engines.

a. Use the data below to calculate the mass in grams of EDC and EDB that was potentially added to the aquifer from the spill of 2500 gal of leaded gasoline.
b. Measurements of the contaminant plume indicated that the gasoline spill impacted about 9×10^6 ft^3 of aquifer volume. Assume that 25% of this volume was occupied by water in the soil pore space and calculate the potential average concentrations of EDC and EDB in the aquifer water (ignoring biodegradation, evaporation, and other loss mechanisms).

---

*,[†] The use of these common trade names may be confusing because the name ethylene normally means that there is a double bond between two carbons, whereas the compounds 1,2-dichloroethane and 1,2-dibromoethane contain only single bonds. The use of ethylene dichloride and ethylene dibromide as trade names arose because 1,2-dichloroethane and 1,2-dibromoethane were often manufactured from ethylene with chlorine or bromine.

Data:

Weight percentages in the gasoline additive package were 62% TEL, 18% EDC, 18% EDB, and 2% other nonrelevant compounds. The amount of additive used was sufficient to yield 2.0 g of lead per gallon of gasoline. Assume both EDC and EDB are fully dissolved in water.

Chemical formulas: TEL $= C_8H_{20}Pb$, EDC $= C_2H_4Cl_2$, and EDB $= C_2H_4Br_2$. Use the periodic table on the inside front cover to calculate molecular weights.

Conversion factors: 1 gal $= 3.785412$ L; 1 ft$^3 = 28.31685$ L

Calculation:

a. The formula for tetraethyl lead (TEL) is $Pb(C_2H_5)_4$ or PbC_8H_{20}:

MW TEL $= 207.2 + (8 \times 12.0) + (20 \times 1.0) = 323.2$ g/mol

Wt% lead in TEL $= \%Pb = \dfrac{207.2}{323.2} \times 100\% = 64\%$, to two significant figures

2.0 g Pb/gal gasoline $= 0.64 \times$ (g of TEL per gallon of gasoline)

g of TEL/gal $= \dfrac{2.0}{0.64} = 3.1$ g TEL/gal

TEL is 62% of additive package.

Therefore, total grams of additive $= \dfrac{3.1 \text{ g TEL}}{0.62} = 5.0$ g additive/gal

EDC and EDB each equal 18% of the additive package
EDC and EDB each $= 0.18$ (5.0 g additive/gal) $= 0.9$ g/gal
Therefore, the 2500 gal spill contained about

0.9 g/gal $\times$ 2500 gal $= 2250$ g each of EDC and EDB

b. Volume of aquifer water receiving 2250 g each of EDC and EDB:

$(9 \times 10^6$ ft^3 of aquifer$) \times (0.25) = 2.2 \times 10^6$ ft^3 of water $= 6.4 \times 10^7$ L

The potential concentration of each pollutant in mg/L is

$\dfrac{2250 \text{ g}}{6.4 \times 10^7 \text{ L}} \times 1 \times 10^3$ mg/g $= 0.035$ mg/L, or 35 ppb

For comparison, the drinking water MCLs (EPA maximum contaminant levels) are EDC $= 0.005$ ppm or 5 ppb and EDB $= 0.00005$ ppm or 0.05 ppb. Therefore, the gasoline releases were likely to have polluted the groundwater aquifer with EDC and EDB in excess of the drinking water standards.

1.3.6 HOW DO IMPURITIES INFLUENCE WATER QUALITY?

The effects of different impurities on water quality are found by research and experience. For example, concentrations of arsenic in drinking water greater than

0.01 mg/L are deemed hazardous to human health. This judgment is based on research and epidemiological studies. Frequently, regulations have to be based on an interpretation of studies that are not rigorously conclusive. Such regulations may be controversial, but until they are revised due to the emergence of new information, they serve as the legal definition of the concentration above which an impurity is deemed to have a harmful effect on water quality.

The EPA has a policy of publishing newly proposed regulatory rules before the rules are finalized, and explaining the rationale used to justify the rules, in order to receive feedback from interested parties. During the period dedicated for public comment, interested parties can support or take issue with the EPA's position. The public input is then added to the database used for establishing a final regulation. Examples of such regulations may be a numerical standard for a chemical not previously regulated, a revised standard for a chemical already regulated, or a new procedure for the analysis of a pollutant. The EPA publishes extensive documentation for all their standards, describing the data on which the numerical values are based.

EXERCISES

1. 475 mL of a water sample was evaporated to determine the amount of dissolved salts contained in it. After evaporation, the dried precipitated salts weighed 1475 mg. What was the concentration in ppm of dissolved salts (also called total dissolved solids or TDS)?
2. The annual arithmetic mean ambient air quality standard for sulfur dioxide (SO_2) is 0.03 ppmv. What is this standard in $\mu g/m^3$?
3. The primary drinking water MCL (maximum contaminant level) for barium (Ba) is 2.0 mg/L. If the sole source of barium is barium sulfate ($BaSO_4$), what weight of $BaSO_4$ salt is present in 1 L of water that contains 2.0 mg/L of Ba? (Hint: The moles of Ba in 2.0 mg equal the moles of $BaSO_4$ in one liter of sample.)
4. Most people can detect the odor of ozone in concentrations as low as 10 ppb. Could they detect the odor of ozone in samples with an ozone level of (a) 0.118 ppm, (b) 25 ppm, and (c) 0.001 ppm?
5. Determine the percentage by volume of the different gases in a mixture containing 0.3 L O_2, 1.6 L N_2, and 0.1 L CO_2. Assume that each separate gas is at 1 atm pressure before mixing and that the pressure of the combined gases after mixing is also at 1 atm.
6. What is the significance of the fact that the percentage of oxygen in air is 21% by numbers of molecules and 23% by mass?
7. Express the 0.9% argon content of air in parts per million (ppm).
8. Express 400 ppm of CO_2 in cigarette smoke as a percentage of the smoke inhaled.
9. The permissible limit for ozone for a 1 h average is 0.12 ppm. If Little Rock, Arkansas registers a reading of 0.15 ppm for 1 h, by what percent is Little Rock over the limit for atmospheric ozone?
10. A certain water-soluble pesticide is fatal to fish at 0.5 mg/L (ppm). Five kilograms (kg) of the pesticide are spilled into a stream. The stream flow rate was 10 L/s at 1 km/h. For what approximate distance downstream could fish potentially be killed?

2 Contaminant Behavior in the Environment: Basic Principles

2.1 BEHAVIOR OF CONTAMINANTS IN NATURAL WATERS

Every part of our world is continually changing, essential ecosystems as well as unwelcome contaminants. Some changes occur imperceptibly on a geological time-scale; others are rapid, occurring within days, minutes, or less. Oil and coal are formed from animal and vegetable matter over millions of years. When oil and coal are burned, they can release their stored energy in fractions of a second.

Control of environmental contamination depends on learning how to bring about desired changes within a useful timescale, a task that requires an understanding of how pollutants are affected by environmental conditions. For example, metals that are dangerous to our health, such as lead, are often more soluble in water under acidic conditions than under basic conditions. Knowing this, one can plan to remove dissolved lead from drinking water by raising the pH and making the water basic. Under basic conditions, a large part of dissolved lead can be made to precipitate as a solid and can be removed from drinking water by settling out or filtering.

Contaminants in the environment are driven to change by

- Physical forces that move contaminants to new locations, often without significant change in their chemical properties. Contaminants released into the soil and water can move into regions far from their origin under the forces of wind, gravity, and water flow. An increase in temperature will cause an increase in the rate at which gases and volatile substances evaporate from water or soil into the atmosphere. Electrostatic attractions can cause dissolved substances and small particles to adsorb to solid surfaces, where they may leave the water flow and become immobilized in soils or filters. Water flow can erode soils and transport sediments carrying sorbed pollutants over long distances.
- Chemical changes, such as oxidation and reduction, which break and make chemical bonds, allowing atoms to rearrange into new compounds with different properties. Chemical change often has the potential to destroy pollutants by converting them into less undesirable substances.

- Biological activity, whereby microbes, in their constant search for survival energy, break down many kinds of contaminant molecules and return their atoms to the environmental cycles that circulate carbon, oxygen, nitrogen, sulfur, phosphorus, and other elements repeatedly through our ecosystems. Biological processes are a special kind of chemical change.

We are particularly interested in processes that move pollutants to less hazardous locations or change the nature of a pollutant to a less harmful form, because these processes are the tools of environmental protection. The effectiveness of these processes depends on properties of the pollutant and its water and soil environment. It is often said that every remediation project is unique and site specific. The reason for this is that, although each pollutant has its predictable and, generally, tabulated chemical and physical properties, each project site has properties that are always different from others to some extent, depending on its long-term geologic history and its more recent anthropomorphic disturbances.

Important properties of pollutants can usually be found in handbooks or chemistry references. However, the important properties of the water and soil in which the pollutant resides are always unique to the particular site and must be measured or estimated anew for every project.

2.1.1 Important Properties of Pollutants

The six properties of pollutants listed below are the most important for predicting the environmental behavior of a pollutant. They are usually tabulated in handbooks and other chemistry references, to the extent of current knowledge:

- Solubility in water
- Volatility
- Density
- Chemical reactivity
- Biodegradability
- Strength of sorption* to solids

If not readily found, these properties can often be estimated from the chemical structure of the pollutant. Whenever possible, this book will offer "rules of thumb" for estimating pollutant properties. The ability to guesstimate the environmental behavior of a pollutant is often an important first step in developing a remediation strategy.

---

* Sorption is a general term that includes all the possible processes by which a molecule originally in a gas or liquid phase becomes bound to a solid. Sorption includes both adsorption (becoming bound to a solid surface) and absorption (becoming bound within pores and passages in the interior of a solid). It also includes all the variants of binding mechanisms, such as chemisorption (where chemical bonds are formed between a molecule and the surface), and physisorption (where physical attractions such as van der Waals and London forces hold a molecule to a surface).

2.1.2 IMPORTANT PROPERTIES OF WATER AND SOIL

The properties of water and soil that influence pollutant behavior can be expected to differ at every location and must be measured or estimated for each project. Since environmental conditions are so varied, it is difficult to generate a simple set of water and soil properties that should always be measured. The lists below include the most commonly needed properties. Discussions and examples throughout this book will illustrate how knowledge of important soil and water properties are used in protecting and restoring the natural environment.

Water properties

- Temperature
- Water quality (chemical composition, pH, oxidation–reduction potential, alkalinity, hardness, turbidity, dissolved oxygen, biological oxygen demand, fecal coliforms, etc.)
- Flow rate and flow pattern

Properties of solids and soils in contact with water

- Mineral composition
- Percentage of organic matter
- Sorption coefficients for contaminants (attractive forces between solids and contaminants)
- Mobility of solids (colloid and particulate movement)
- Porosity
- Particle size distribution
- Hydraulic conductivity

The properties of environmental waters and soils are always site specific and must be estimated or measured in the field.

2.2 WHAT ARE THE FATES OF DIFFERENT POLLUTANTS?

There are three possible naturally occurring (rather than engineered) fates of pollutants:

1. All or a portion might remain unchanged in their present location.
2. All or a portion might be carried elsewhere by transport processes.
   a. Movement to other phases (air, water, or soil) by volatilization, dissolution, adsorption, and precipitation.
   b. Movement within a phase under gravity, diffusion, and advection.
3. All or a portion might be transformed into other chemical species by natural chemical and biological processes.
   a. Biodegradation (aerobic and anaerobic): Pollutants are altered structurally by biological processes, mainly the metabolism of microorganisms present in aquatic and soil environments.

b. Bioaccumulation: Pollutants accumulate in plant and animal tissues to higher concentrations than in their original environmental locations.
c. Weathering: Pollutants undergo a series of environmentally induced non-biological chemical changes, by processes such as oxidation–reduction, acid-base, hydration, hydrolysis, complexation, and photolysis reactions.

2.3 PROCESSES THAT REMOVE POLLUTANTS FROM WATER

2.3.1 NATURAL ATTENUATION

Even without human intervention, pollutant concentrations in the environment have a tendency to diminish with time due to natural causes. The rate of attenuation, however, depends strongly on the chemical and physical properties of the pollutants (e.g., solubility; biodegradability; chemical stability; whether solid, liquid or gas; etc.) and on many characteristics of the polluted site (soil permeability, average precipitation and temperature, geologic features, etc.). Where natural processes are fast enough, the simplest approach to remediation is to wait until pollutant levels are no longer deemed hazardous. The case study in Section 5.10 is an example of when this approach may be the best choice.

Because every case is different and highly site specific, the progress of natural attenuation generally must be closely monitored before considering it as the preferred remediation option. Monitored natural attenuation is a recognized approach to pollution remediation, (OSWER Directive 9200.4-17P, Use of Monitored Natural Attenuation at Superfund, RCRA Corrective Action, and Underground Storage Tank Sites, April 21, 1999; http://www.epa.gov/swerust1/directiv/d9200417.htm), defined by EPA as the

> ... reliance on natural attenuation processes (within the context of a carefully controlled and monitored site cleanup approach) to achieve site-specific remediation objectives within a time frame that is reasonable compared to that offered by other more active methods. The 'natural attenuation processes' that are at work in such a remediation approach include a variety of physical, chemical, or biological processes that, under favorable conditions, act without human intervention to reduce the mass, toxicity, mobility, volume, or concentration of contaminants in soil or groundwater. These in-situ processes include biodegradation; dispersion; dilution; sorption; volatilization; radioactive decay; and chemical or biological stabilization, transformation, or destruction of contaminants.

Natural attenuation processes are described more fully below and in later chapters.

2.3.2 TRANSPORT PROCESSES

Contaminants that are dissolved or suspended in water can move to other phases by the following processes:

1. Volatilization: Dissolved and sorbed contaminants move from water and soil into air, in the form of gases or vapors.

2. Sorption: Dissolved contaminants become bound to solids by attractive chemical, physical, and electrostatic forces.
3. Sedimentation: Small suspended solids in water grow large enough to settle out of water under gravity. There are two stages to sedimentation:
   a. Coagulation: Suspended solids generally carry an electrostatic charge that keeps them apart. Chemicals may be added to lower the repulsive electrostatic energy barrier between the particles (destabilization), allowing thermal energy to bring them closer together.
   b. Flocculation: Lowering the repulsive energy barrier by coagulation allows suspended solids to collide and clump together because of short-range attractive forces, to form a floc. When floc particles aggregate, they can become heavy enough to settle out of the water.

2.3.3 ENVIRONMENTAL CHEMICAL REACTIONS

The following are brief descriptions of some important environmental chemical reactions that can remove pollutants from water. More detailed discussions are given throughout this book.

- Photolysis: In molecules that absorb solar radiation, exposure to sunlight can break chemical bonds and start chemical breakdown. Many natural and synthetic organic compounds are susceptible to photolysis.
- Complexation and chelation: Polar or charged dissolved species (such as metal ions) bind to electron-donor ligands* to form complex or coordination compounds. Complex compounds are often soluble and resist removal by precipitation because the ligands must be displaced by other anions (such as sulfide) before an insoluble species can be formed. Common ligands include hydroxyl, carbonate, carboxylate, phosphate, and cyanide anions, as well as water molecules, humic acids, and synthetic chelating agents such as nitrilotriacetate (NTA) and ethylenediaminetetraacetate (EDTA).
- Acid-base: Protons (H^+ ions) are transferred between chemical species. Acid-base reactions are part of many environmental processes and influence the reactions of many pollutants.
- Oxidation–reduction (OR or redox): Electrons are transferred between chemical species, changing the oxidation states and the chemical properties of the electron donor and the electron acceptor. Water disinfection, electrochemical reactions such as metal corrosion, and most microbial reactions such as biodegradation are oxidation–reduction reactions.
- Hydrolysis and hydration: A compound forms chemical bonds to water molecules or hydroxyl anions (OH^-). In water, all ions and polar compounds develop a hydration shell of water molecules. When the attraction to water is strong enough, a chemical bond can result. Hydrolysis reactions cause many

* Ligands are polyatomic chemical species that contain nonbonding (within the ligand) electron pairs that are used to bond the ligand to a central atom. The ligand contributes both of the electrons that forms the bond, instead of the more common case where each bonded atom contributes one electron.

metal ions to form hydroxides of low solubility. With organic compounds, a water molecule may replace an atom or group, a step that often breaks the organic compound into smaller fragments. Even dissolved gases can undergo hydration. Hydration of dissolved carbon dioxide (CO_2) and sulfur dioxide (SO_2) forms carbonic acid (H_2CO_3) and sulfurous acid (H_2SO_3), respectively.

- Chemical precipitation: Two or more dissolved species react to form an insoluble solid compound, or there is a change in pH, redox potential, or concentrations, resulting in the formation of a solid from dissolved species. For example, precipitation can occur if a solution of a salt becomes over-saturated, (when the concentration of a salt is greater than its solubility limit). For the salt calcium carbonate ($CaCO_3$) its solubility at 25°C is about 10 mg/L. In a water solution containing 5 mg/L of $CaCO_3$, all the calcium carbonate will be dissolved. If more $CaCO_3$ is added or water is evaporated, the concentration of dissolved $CaCO_3$ can increase only to 10 mg/L. Any $CaCO_3$ in excess of the solubility limit will precipitate as solid $CaCO_3$.

Chemical precipitation can also occur if two soluble salts react to form a different salt of low solubility. For example, silver nitrate ($AgNO_3$) and sodium chloride (NaCl) are both highly soluble. They react in solution to form the insoluble salt silver chloride (AgCl) and the soluble salt sodium nitrate ($NaNO_3$). The insoluble silver chloride precipitates as a solid, while the soluble silver nitrate remains dissolved.

Breaking the reaction into two separate conceptual steps (Equations 2.1 and 2.2) helps to visualize what happens. Refer to the solubility table inside the back cover, which gives qualitative solubilities for ionic compounds in water.

In the first conceptual step, the soluble salts silver nitrate and sodium chloride are added to water and dissolve as ions:

$$AgNO_3(s) \xrightarrow{H_2O} Ag^+(aq) + NO_3^-(aq) \qquad (2.1)^*$$

$$NaCl(s) \xrightarrow{H_2O} Na^+(aq) + Cl^-(aq) \qquad (2.2)$$

After the dissolution step and before any further reaction, the solution contains Ag^+, Na^+, Cl^-, and NO_3^- ions.

While in solution, all ions move about freely. Ions with charges having opposite signs (positive/negative) are attracted to one another while ions with charges having the same sign (positive/positive and negative/negative) are repelled from one another. Charged ions with opposite signs tend to pair up randomly, regardless of their chemical identity.

Therefore, in the second conceptual step, the ions can combine in all possible ways that pair a positive ion with a negative ion. Besides the original Ag^+/NO_3^- and

---

* Placing the chemical formula for water, H_2O, above the reaction arrows means that the reaction requires the presence of water, even though water does not react chemically with the other reagents and does not appear in the overall reaction. The suffix (aq), abbreviation for aqueous, following a chemical species means that the species are dissolved in water, see also Chapter 4, Section 4.11. The suffix (s), abbreviation for solid, following a chemical species means that the species is in solid form.

Na^+/Cl^- pairs, both of which are soluble, Ag^+/Cl^- and Na^+/NO_3^- are also possible pairs. Since $NaNO_3$ is a soluble ionic compound, which dissolves to form Na^+ and NO_3^-, the Na^+ and NO_3^- ions simply remain in solution. However, $AgCl$ is insoluble and will precipitate as a solid. The overall reaction is written as

$$AgNO_3(aq) + NaCl(aq) \xrightarrow{H_2O} Na^+(aq) + NO_3^-(aq) + AgCl(s) \qquad (2.3)$$

Thus, adding the two soluble salts, $AgNO_3$ and $NaCl$ to water results in a solution containing Na^+ and NO_3^- ions and the precipitated solid compound $AgCl$. If equal moles of the two salts, $AgNO_3$ and $NaCl$, were mixed initially, only very small amounts of Ag^+ and Cl^- will remain unprecipitated, because the solubility of $AgCl$ is very small.

2.3.4 BIOLOGICAL PROCESSES

Microbes can degrade organic pollutants by facilitating oxidation–reduction reactions. During microbial metabolism (the biological reactions that convert organic compounds into energy and carbon for microbe growth), there may be a transfer of electrons from a pollutant molecule to other compounds present in the soil or water environment. It is necessary that compounds are present that can serve as electron acceptors. The electron acceptors most commonly available in the environment are molecular oxygen (O_2), carbon dioxide (CO_2), nitrate (NO_3^-), sulfate (SO_4^{2-}), manganese (Mn^{2+}), and iron (Fe^{3+}). When molecular oxygen (O_2) is available, it is always the preferred electron acceptor and the process is called aerobic biodegradation. In the absence of O_2, it is called anaerobic biodegradation. Aerobic and anaerobic biodegradation are examples of oxidation–reduction reactions, discussed in Chapter 3, Section 3.3.

Organic pollutants are generally toxic because of their chemical structure. Changing their structure in any way will change their properties and may make them innocuous or, in a few cases, more toxic. Eventually, usually after many reaction steps, in a process called mineralization, biodegradation converts organic pollutants into carbon dioxide, water, and mineral salts. Although these final products represent the destruction of the original pollutant, some of the intermediate steps may temporarily produce compounds that are also pollutants, sometimes more toxic than the original. Biodegradation is discussed in more detail in Chapter 8.

2.4 MAJOR CONTAMINANT GROUPS AND NATURAL PATHWAYS FOR THEIR REMOVAL FROM WATER

Only brief and general introductory descriptions of major contaminant groups and natural removal processes are given here, as an introduction to the discussion of intermolecular forces that are the basis for their removal processes. There are less common removal pathways not discussed here, such as photolysis and radiolysis, which can become important or even dominant under special conditions.

2.4.1 METALS

Dissolved metals such as iron, lead, copper, cadmium, mercury, etc., are removed from water mainly by sorption and precipitation processes. Some metals—particularly

As, Cd, Hg, Ni, Pb, Se, Te, Sn, and Zn—can form volatile metal-organic compounds in the natural environment by microbial reactions. For these, volatilization can be an important removal mechanism. Bioaccumulation of metals in animals usually is not very significant as a removal process, although it can have very toxic effects. Bioaccumulation in plants on the other hand, has been developed into a useful remediation technique called phytoremediation. Biotransformation of metals, in which redox reactions involving bacteria can cause some metals to precipitate, has also shown promise as a removal method. The aqueous chemistry of metals is discussed in Chapter 4.

2.4.2 CHLORINATED PESTICIDES

Chlorinated pesticides, such as atrazine, chlordane, DDT, dicamba, endrin, heptachlor, lindane, 2,4-D, etc., are removed from water mainly by sorption, volatilization, and biotransformation. Chemical processes like oxidation, hydrolysis, and photolysis usually play a minor role.

2.4.3 HALOGENATED ALIPHATIC HYDROCARBONS

Halogenated hydrocarbons in the environment arise mostly from industrial and household solvents. Compounds such as 1,2-dichloropropane, 1,1,2-trichlorethane, tetrachlorethylene, carbon tetrachloride, chloroform, etc., are removed mainly by volatilization. Under natural conditions, aerobic biotransformation and biodegradation processes are usually very slow, with half-lives of tens to hundreds of years. However, natural and engineered anaerobic biodegradation processes have been identified that have short enough half-lives to be useful remediation techniques.

2.4.4 FUEL HYDROCARBONS

Gasoline, diesel fuel, and heating oils are mixtures of hundreds of different organic hydrocarbons. The lighter weight compounds such as benzene, toluene, ethylbenzene, xylenes, naphthalene, trimethylbenzenes, and the smaller alkanes, etc., are removed mainly by sorption, volatilization, and biotransformation. The heavier compounds including polycyclic aromatic hydrocarbons (PAHs) such as fluorene, benzo(a)pyrene, anthracene, phenanthrene, etc., are not volatile and are removed mainly by sorption, sedimentation, and biodegradation.

2.4.5 INORGANIC NONMETAL SPECIES

These include ammonia, chloride, bromide, fluoride, cyanide, nitrite, nitrate, phosphate, sulfate, sulfide, etc. They are removed mainly by sorption, volatilization, chemical reactions, and biotransformation.

It should be noted that many normally minor removal pathways, such as photolysis, can become important, or even dominant, in special circumstances. For example, low volatility pesticides in a clear, shallow stream with little organic matter might be degraded primarily by photolysis.

2.5   CHEMICAL AND PHYSICAL REACTIONS IN THE WATER ENVIRONMENT

Chemical and physical reactions in water can be

- Homogeneous—occurring entirely among dissolved species
- Heterogeneous—occurring at the liquid–solid–gas interfaces

Most environmental water reactions are heterogeneous. Purely homogeneous reactions are rare in natural waters and wastewaters. Among the most important heterogeneous reactions are those that move pollutants from one phase to another: volatilization, dissolution, and sorption:*

- Volatilization: At the liquid–air and solid–air interfaces, volatilization transfers volatile contaminants from water and solid surfaces into the atmosphere and into air in soil pore spaces. Volatilization is most important for compounds with high vapor pressures. Contaminants in the vapor phase are the most mobile in the environment.
- Dissolution: At the solid–liquid and air–liquid interfaces, dissolution transfers contaminants from air and solids to water. It is most important for contaminants of significant water solubility. The environmental mobility of contaminants dissolved in water is generally intermediate between volatilized and sorbed contaminants.
- Sorption: At the liquid–solid and air–solid interfaces, sorption transfers contaminants from water and air to soils and sediments. It is most important for compounds of low solubility and low volatility. Sorbed compounds undergo chemical and biological transformations at different rates and by different pathways than dissolved compounds. The binding strength with which different contaminants become sorbed depends on the nature of the solid surface (sand, clays, organic particles, etc.) and on the properties of the contaminant. Contaminants sorbed to solids (except for solid colloids, see Chapter 5, Section 5.8) are the least mobile in the environment.

Eventually, as described in the next section using diesel oil as an example, a portion of every pollutant released to the environment becomes distributed by heterogeneous reactions into all the liquid, gas, and solid phases with which it comes into contact, as diagramed in Figure 2.1. Predicting the amount of pollutant that will enter different phases is an important subject that is treated later in this text.

2.6   PARTITIONING BEHAVIOR OF POLLUTANTS

A pollutant in contact with water, soil, and air will partially dissolve into the water, partially volatilize into the air, and partially sorb to the soil surfaces, as illustrated in Figure 2.1. The relative amounts of pollutant that are found in each phase with

---

* See footnote on page 24.

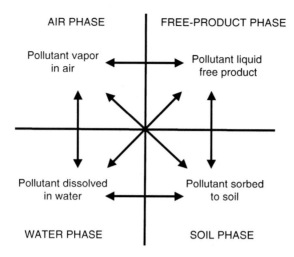

FIGURE 2.1 Partitioning of a pollutant among air, water, soil, and free-product phases. Arrows indicate all possible phase change pathways.

which it is in contact depend on intermolecular attractive forces existing between pollutant, water, and soil molecules. The most important factor for predicting the partitioning behavior of contaminants in the environment is an understanding of the intermolecular attractive forces between contaminants and the water and soil materials in which they are found.

2.6.1 PARTITIONING FROM A DIESEL OIL SPILL

Consider, for example, what happens when diesel oil is spilled at the soil's surface. Some of the liquid diesel oil (commonly called free product) flows downward under gravity through the soil toward the groundwater table. Before the spill, the soil pore spaces above the water table (called the soil unsaturated zone or the vadose zone) were filled with air and water, and the soil surfaces were partially covered with adsorbed water.

As diesel oil, which is a mixture of many different compounds, passes downward through the soil, its different components become partitioned in the air and water within the soil pore spaces, on the soil particle surfaces, and some remain within the oil-free product. After the spill, the pore spaces are filled with air containing diesel vapors, water carrying dissolved diesel components, and diesel free-product that has changed in composition by losing some of its components to other phases. The soil surfaces are partially covered with diesel-free product and adsorbed water containing dissolved diesel components.

Diesel oil is a mixture of hundreds of different compounds each having a unique partitioning, or distribution pattern. The pore space air will contain mainly the most volatile components, the pore space water will contain mainly the most soluble components, and the soil particles will sorb mainly the least volatile and least soluble components. The quantity of free product diminishes continually as it moves downward through the soil because a significant portion is lost to other phases. The composition of the free product also changes continually because the most volatile, soluble, and strongly sorbed compounds are lost preferentially.

The distribution of the various diesel compounds among different phases attains quasi-equilibrium, with compounds continually passing back and forth across each phase interface, as indicated in Figure 2.1. As the remaining free product continues to change by losing components to other phases (part of the "weathering" process), it increasingly resists further change. Since the lightest weight components tend to be the most volatile and soluble, they are the first to be lost to other phases, and the remaining free product becomes increasingly more viscous and less mobile. Severely weathered free product is very resistant to further change, and can persist in the soil for decades. It only disappears by biodegradation or by actively engineered removal.

Depending on the amount of diesel oil spilled, it is possible that all of the diesel free-product becomes "immobilized" in the soil before it can reach the water table. This occurs when the mass of free product diminishes and its viscosity increases to the point where capillary forces in the soil pore spaces can hold the remaining free product in place against the force of gravity. There is still pollutant movement, however, mainly in the non-free-product phases. The volatile components in the vapor state usually diffuse rapidly through the soil, moving mostly upward toward the soil surface and along any high permeability pathways through the soil, such as a sewer line backfill. New water percolating downward, from precipitation or other sources, can dissolve additional diesel compounds from the sorbed phase and carry it downward. Downward percolating water can also displace some free product held by capillary forces and soil pore water already containing dissolved pollutants, forcing them to move farther downward. Although the diesel free-product is not truly immobilized, its downward movement can become imperceptible.

However, if the spill is large enough, diesel free-product may reach the water table before becoming immobilized. If this occurs, liquid diesel free-product, being less dense than water, cannot enter the water-saturated zone but remains above it, effectively floating on top of the water table. There, the free product spreads horizontally on the groundwater surface, continuing to partition into groundwater, soil pore space air, and to the surfaces of soil particles. In other words, a portion of the free product will always become distributed among all the solid, liquid, and gas phases with which it comes into contact. This behavior is governed by intermolecular forces that exist between molecules.

RULES OF THUMB

When a pollutant consisting of a mixture of different compounds, such as diesel fuel or gasoline, is released to the environment, its composition and physical properties change as time passes.

1. The most volatile components tend to leave the free product and pass into the atmosphere or into air in the soil pore space.

(Continued)

RULES OF THUMB (Continued)

2. The most water-soluble components tend to dissolve into any surface water or groundwater they contact.
3. The least volatile and soluble components tend to sorb to soil and sediment surfaces as the pollutant is moved by gravity and water flow forces.

The remaining free product progressively becomes denser, more viscous, less mobile, and more resistant to further change.

2.7   INTERMOLECULAR FORCES

Volatility, solubility, and sorption processes all result from the interplay between intermolecular forces. All molecules have attractive forces acting between them. The attractive forces are electrostatic in nature, created by a nonuniform distribution of valence shell electrons around the positively charged nuclei of a molecule. When electrons are not uniformly distributed, the molecule will have regions that carry net positive or negative charges. A charged region on one molecule is attracted to oppositely charged regions on adjacent molecules, resulting in the so-called polar attractive forces. Because of random movement caused by thermal agitation, molecules can experience momentary electrostatic repulsive forces as well. On average, however, molecular arrangements will favor the lower energy attractive positions, and the attractive forces always prevail. The most obvious demonstrations of intermolecular attractive forces are the phase changes of matter that inevitably accompany a sufficient lowering of temperature, where cooling a gas turns it into a liquid and then into a solid, as the temperature becomes low enough.

2.7.1   TEMPERATURE DEPENDENT PHASE CHANGES

Attractive forces always work to bring order to molecular configurations, in opposition to thermal energy, which always works to randomize configurations. Gases are always the higher temperature form of any substance and are the most randomized state of matter. If the temperature of a gas is lowered enough, every gas will condense to a liquid, a more ordered state. Condensation is a manifestation of intermolecular attractive forces. As the temperature falls, the thermal energy of the gas molecules decreases, eventually reaching a point where there is insufficient thermal kinetic energy to keep the molecules separated against the intermolecular attractive forces. The temperature at which condensation occurs is called the boiling point; it is dependent on environmental pressure as well as temperature.

If the temperature of a liquid is lowered further, it eventually freezes to a solid when its thermal energy becomes low enough for intermolecular attractions to pull the molecules into a rigid solid arrangement. Solids are the most highly ordered state of matter. Whenever lowering the temperature causes a change of phase (gas to liquid and liquid to solid) the decrease in thermal energy allows the always-present attractive forces to overcome molecular kinetic energy and pull gas and liquid molecules closer together into more ordered liquid or solid phases.

2.7.2 VOLATILITY, SOLUBILITY, AND SORPTION

The model of attractive forces working to bring increased order, against the randomizing effects of thermal energy, also explains the volatility, solubility, and sorption behavior of molecules.

- Molecules of volatile liquids have relatively weak attractions to one another. Thermal energy at ordinary environmental temperatures is sufficient to allow the most energetic of the weakly held liquid molecules to escape from the electrostatic attractions to their slower liquid neighbors and fly into the gas phase.
- Molecules in water-soluble solids are attracted to water more strongly than they are attracted to themselves. If a water-soluble solid is placed in water, its surface molecules are drawn from the solid phase into the liquid phase by the stronger attractions to water molecules.
- Dissolved molecules that become sorbed to sediment surfaces are held to the sediment particle by attractive forces that pull them away from water molecules.

Understanding intermolecular forces is the key to predicting how contaminants become distributed in the environment.

2.7.3 PREDICTING RELATIVE ATTRACTIVE FORCES

When you can predict relative attractive forces between molecules, you can predict their relative solubility, volatility, and sorption behavior. For example, the volatility of a substance is closely related to its freezing and boiling temperatures, which, in turn are related to the attractive forces between molecules of that substance. The water solubility of a compound is related to the strength of the attractive forces between molecules of the compound and molecules of water. The soil–water partition coefficient of a compound indicates the relative strengths of its attraction to water and soil.

For any compound, the temperature at which it changes phase is an indicator of the intermolecular attractive force existing between its molecules:

- Boiling a liquid means that it has been heated to the point where thermal energy imparts sufficient kinetic energy to the molecules to allow them to overcome their attractive forces and move apart from one another into the gas phase. A higher boiling temperature indicates stronger intermolecular attractive forces between the liquid molecules, because they must attain higher kinetic energy to pull apart. The thermal energy has to be higher in order to overcome the stronger attractions and allow liquid molecules to escape into the gas phase. Thus, the fact that methanol boils at a lower temperature than water means that methanol molecules are attracted to one another more weakly than are water molecules.

- Freezing a liquid means that its thermal energy has been reduced to the point where attractive forces can overcome the randomizing effects of thermal motion and pull freely moving liquid molecules into fixed positions in a solid phase. A lower freezing point indicates weaker attractive forces. The thermal energy has to be reduced to lower values so that the weaker attractive forces can pull the molecules into fixed positions in a solid phase. The fact that methanol freezes at a lower temperature than water is another indicator that attractive forces are weaker between methanol molecules than between water molecules.

- Wax, a mixture of hydrocarbons consisting mainly of carbon and hydrogen atoms, is solid at room temperature (20°C), whereas diesel fuel, also a mixture of hydrocarbons, is liquid. The freezing temperature of diesel fuel is well below room temperature. This indicates that the attractive forces between wax molecules are stronger than between molecules in diesel fuel. At the same temperature where diesel molecules can still move about randomly in the liquid phase, wax molecules are held by their stronger attractive forces in fixed positions in the solid phase. The reasons for differences in attractive forces, discussed below, are important for remediation strategies.

- Compounds that are highly soluble in water have strong attractions to water molecules. When a soluble solid substance is added to water, water molecules are attracted to the solid surface where they literally pull the solid molecules apart from one another and carry them into solution.

- When compounds released to the environment are mostly found sorbed to soils and sediments, rather than dissolved in water or vaporized into the air, it indicates that they have stronger attractions to soil than to water or to their own kind of molecules.

- Compounds found in the environment as a gas, because they volatilize readily at environmental temperatures from water, soil, and their own molecules, must have relatively weak attractions to water, soil, and their own kind of molecules.

2.8 ORIGINS OF INTERMOLECULAR FORCES: ELECTRONEGATIVITIES, CHEMICAL BONDS, AND MOLECULAR GEOMETRY

Intermolecular forces are electrostatic in nature. Molecules are composed of electrically charged particles (electrons and protons), and it is common for there to be regions within a single molecule that are predominantly charged positive or negative. Attractive forces between molecules arise when electrostatic forces attract positive regions on one molecule to negative regions on another. The strength of the attractions between different molecules depends on the polarities of chemical bonds within the molecules and the geometrical shapes of the molecules.

2.8.1 CHEMICAL BONDS

At the simplest level, the chemical bonds that hold atoms together in a molecule are of two types, ionic and covalent.

1. Ionic bonds occur when one atom attracts an electron away from another atom to form a positive and a negative ion. The ions are then bound together by electrostatic attraction. The electron transfer occurs because the electron-receiving atom has a much stronger attraction for electrons in its vicinity than does the electron-losing atom.

2. Covalent bonds are formed when two atoms share electrons, called bonding electrons, in the space between their nuclei. The electron-attracting properties of two covalent bonded atoms are not different enough to allow one atom to pull an electron entirely away from the other. However, unless both atoms attract the bonding electrons equally, the average position of the bonding electrons will be closer to one of the atoms. The atoms are held together because their positive nuclei are attracted to the negative charge of the shared electrons in the space between them.

When two covalent bonded atoms are identical, as in Cl_2, the bonding electrons are always equally attracted to each atom and the electron charge is uniformly distributed between the atoms. Such a bond is called a nonpolar covalent bond, meaning that it has no polarity, i.e., no regions with net positive or negative charge.

When two covalent bonded atoms are of different kinds, as in HCl, one atom can attract the bonding electrons more strongly than the other. This results in a nonuniform distribution of electron charge between the atoms where one end of the bond is more negative than the other, resulting in a polar covalent bond.

Figure 2.2 illustrates the electron distributions in nonpolar and polar covalent bonds. The strength with which an atom attracts bonding electrons to itself is indicated by a quantity called electronegativity, abbreviated EN. Electronegativities

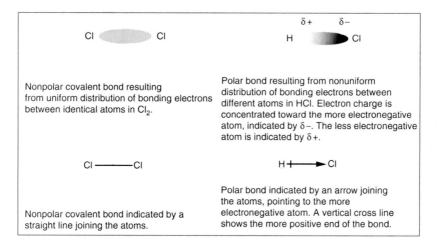

FIGURE 2.2 Uniform and nonuniform electron distributions, resulting in nonpolar and polar covalent chemical bonds. Shading indicates variation in the electron density within the bond; light regions are low density and dark regions are high density. The use of a delta (δ) in front of the + and – signs signifies that the charges are partial, arising from a nonuniform electron charge distribution rather than the transfer of a complete electron.

TABLE 2.1

Electronegativity Values of the Elements (Pauling Scale, to two Significant Figures)

1 1A																
1 **H** 2.2	2 2A											13 3A	14 4A	15 5A	16 6A	17 7A
3 **Li** 1.0	4 **Be** 1.6											5 **B** 2.0	6 **C** 2.5	7 **N** 3.0	8 **O** 3.4	9 **F** 4.0
11 **Na** 0.9	12 **Mg** 1.3	3 3B	4 4B	5 5B	6 6B	7 7B	8 8B	9 8B	10 8B	11 1B	12 2B	13 **Al** 1.6	14 **Si** 1.9	15 **P** 2.2	16 **S** 2.6	17 **Cl** 3.2
19 **K** 0.8	20 **Ca** 1.0	21 **Sc** 1.4	22 **Ti** 1.5	23 **V** 1.6	24 **Cr** 1.7	25 **Mn** 1.6	26 **Fe** 1.8	27 **Co** 1.9	28 **Ni** 1.9	29 **Cu** 1.9	30 **Zn** 1.6	31 **Ga** 1.8	32 **Ge** 2.0	33 **As** 2.2	34 **Se** 2.6	35 **Br** 2.9
37 **Rb** 0.8	38 **Sr** 1.0	39 **Y** 1.2	40 **Zr** 1.3	41 **Nb** 1.6	42 **Mo** 2.2	43 **Tc** 1.9	44 **Ru** 2.2	45 **Rh** 2.3	46 **Pd** 2.2	47 **Ag** 1.9	48 **Cd** 1.7	49 **In** 1.8	50 **Sn** 2.0	51 **Sb** 2.0	52 **Te** 2.1	53 **I** 2.7
55 **Cs** 0.8	56 **Ba** 0.9	57 **'La** 1.1	72 **Hf** 1.3	73 **Ta** 1.5	74 **W** 2.4	75 **Re** 1.9	76 **Os** 2.2	77 **Ir** 2.2	78 **Pt** 1.3	79 **Au** 2.5	80 **Hg** 2.0	81 **Tl** 1.8	82 **Pb** 2.3	83 **Bi** 2.0	84 **Po** 2.0	85 **At** 2.2
87 **Fr** 0.7	88 **Ra** 0.9	89 **#Ac** 1.1	104 **Rf** ?	105 **Db** ?	106 **Sg** ?	107 **Bh** ?	108 **Hs** ?	109 **Mt** ?								

*Lanthanide series	58 **Ce** 1.1	59 **Pr** 1.1	60 **Nd** 1.1	61 **Pm** 1.1	62 **Sm** 1.2	63 **Eu** 1.1	64 **Gd** 1.2	65 **Tb** 1.1	66 **Dy** 1.2	67 **Ho** 1.2	68 **Er** 1.2	69 **Tm** 1.3	70 **Yb** 1.0	71 **Lu** 1.3
# Actinide series	90 **Th** 1.3	91 **Pa** 2.2	92 **U** 1.7	93 **Np** 1.3	94 **Pu** 1.3	95 **Am** 1.3	96 **Cm** 1.3	97 **Bk** 1.3	98 **Cf** 1.3	99 **Es** 1.3	100 **Fm** 1.3	101 **Md** 1.3	102 **No** 1.3	103 **Lr** 1.5

Note: Electronegativity values are below and atomic numbers are above the element symbol.

of the elements, shown in Table 2.1, are relative numbers with an arbitrary maximum value of 4.0 for fluorine, the most electronegative element. Electronegativity values are approximate, to be used primarily for predicting the relative polarities of covalent bonds and relative bond strengths.

The electronegativity difference, ΔEN, between two atoms indicates what kind of bond they will form. The greater the difference in electronegativities of bonded atoms, the more strongly the bonding electrons are attracted to the more electronegative atom, and the more polar is the bond. The following rules of thumb usually apply, with very few exceptions.

RULES OF THUMB (USE TABLE 2.1)

1. If the electronegativity difference between two bonded atoms is zero, they will form a nonpolar covalent bond. Examples are O_2, H_2, and N_2.

RULES OF THUMB (USE TABLE 2.1) (Continued)

2. If the electronegativity difference between two atoms is greater than zero and less than 1.7, they will generally form a polar covalent bond. Examples are HCl, NO, and CO.
3. If the electronegativity difference between two atoms is 1.7 or greater, they will generally form an ionic bond. Examples are NaCl, HF, and KBr.
4. Relative electronegativities of the elements can be predicted by an element's position in the periodic table. Ignoring the inert gases:
   a. The most electronegative element (F) is at the upper right corner of the periodic table.
   b. The least electronegative element (Fr) is at the lower left corner of the periodic table.
   c. In general, electronegativities increase diagonally up and to the right in the periodic table. Within a given period (or row), electronegativities tend to increase in going from left to right; within a given group (or column), electronegativities tend to increase in going from bottom to top.
   d. The farther apart two elements are in the periodic table, the more different are their electronegativities, and the more polar will be a bond between them.
5. The solubility in water of a pure compound is roughly proportional to the polarity of its molecules.
   a. Molecules with no, or small, polarity are generally insoluble or only slightly soluble in water.
   b. Molecules that are ionic or have large polarity are generally soluble in water.*

Because electronegativity differences can vary continuously between zero and four, bond character also can vary continuously between nonpolar covalent and ionic, as illustrated in Figure 2.3.

2.8.2 CHEMICAL BOND DIPOLE MOMENTS

For polar bonds, we can define a quantity called the dipole moment, which serves as a measure of the nonuniform charge separation. Hence, the dipole moment measures the degree of the bond polarity. The more polar the bond, the larger is its dipole moment. In Figure 2.4, the dipole moment, μ, is equal to the magnitude of positive and

---

* There are some exceptions to this principle, such as when a crystalline substance has a high lattice energy, indicating that the atoms are held in place by strong forces. For example, LiF is ionic ($\Delta EN = 3.0$) but is only slightly soluble in water.

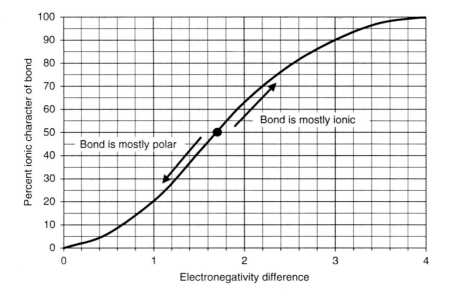

FIGURE 2.3 Bond character as a function of the electronegativity difference.

negative charges at each end of the dipole multiplied by the distance, d, between the charges.

Polarity arrows, as shown in Figure 2.4, are vector quantities. They show both the magnitude and direction of the bond dipole moment. The length of the arrow indicates the magnitude of the dipole moment, and the direction of the arrow points from the positive region toward the negative region of the separated charges.

2.8.3   Molecular Geometry and Molecular Polarity

When a molecule containing polar bonds is itself polar, its polarity will always contribute to its strength of attraction to other molecules. When we know whether a molecule is polar or not, we can estimate its relative water solubility and several other properties. The presence of polar bonds in a molecule is necessary, but not sufficient, for the molecule also to be polar. The geometric symmetry of the molecule also is important.

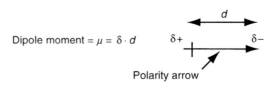

FIGURE 2.4 Bond dipole moment as indicated by a polarity arrow.

Depending on its geometry, a molecule that contains polar bonds may, or may not, be a polar molecule. A molecule with polar bonds will not be a polar molecule if the polar bonds are oriented in a way that the polarity vectors cancel each other (see Section 2.8.4). A molecule with polar bonds will be polar if the polarity vectors of all its bonds add up to give a net polarity vector to the molecule, as in Section 2.8.5. The polarity of a molecule is the vector sum of all its bond polarity vectors. A polar molecule can be experimentally detected by observing whether an electric field exerts a force that makes the molecule align its charged regions with the direction of the field. Polar molecules will point their negative ends toward the positive source of the field, and their positive ends toward the negative source.

RULES OF THUMB

To predict if a molecule is polar, we need to answer two questions:

1. Does the molecule contain polar bonds? If it does, then it might be polar; if it does not, it cannot be polar.
2. If the molecule contains polar bonds, do all the bond polarity vectors add to give a resultant molecular polarity? If the molecule is symmetrical in a way that the bond polarity vectors add to zero, then the molecule is not polar, even though it contains polar bonds. If the molecule is asymmetric and bond polarity vectors add to give a resultant polarity vector, the resultant vector indicates the molecular polarity.

2.8.4 EXAMPLES OF NONPOLAR MOLECULES

Nonpolar molecules invariably have low water solubility. A molecule with no polar bonds cannot be a polar molecule. Thus, all diatomic molecules where both atoms are the same, such as H_2, O_2, N_2, and Cl_2, are nonpolar (and have low water solubility) because there is no electronegativity difference across the bond. On the other hand, a molecule with polar bonds whose dipole moments add to zero because of molecular symmetry is also not a polar molecule. Carbon dioxide, carbon tetrachloride, hexachlorobenzene, p-dichlorobenzene, and boron tribromide are all symmetrical and nonpolar, although all contain polar bonds.

O◄—+C+—►O   Carbon dioxide: Oxygen is more electronegative ($EN(O_2) = 3.5$) than carbon ($EN(C) = 2.5$). Each bond is polar, with the oxygen atom at the negative end of the dipole. Because CO_2 is linear with carbon in the center, the polarity vectors cancel each other and CO_2 is nonpolar.

(Continued)

(Continued)

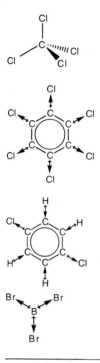

Carbon tetrachloride: EN(C) = 2.5, EN(Cl) = 3.2, C+→Cl. Although each bond is polar, the tetrahedral symmetry of the molecule results in no net dipole moment so that CCl_4 is nonpolar.

Hexachlorobenzene: The bond polarities are the same as in CCl_4 above. C_6Cl_6 is planar with hexagonal symmetry. All the bond polarities cancel one another and the molecule is nonpolar.

p-Dichlorobenzene: This molecule also is planar. It has polar bonds of two magnitudes, the H+→C bond with the smaller polarity and the C+→Cl bond with the larger polarity. The H and Cl atoms are positioned so that all polarity vectors cancel and the molecule is nonpolar. Check the electronegativity values in Table 2.1.

Boron tribromide: EN(B) = 2.0, EN(Br) = 2.8, B+→Br. BBr_3 has trigonal planar symmetry, with 120° between adjacent bonds. All the polarity vectors cancel and the molecule is nonpolar.

2.8.5 EXAMPLES OF POLAR MOLECULES

Polar molecules are generally more water-soluble than nonpolar molecules of similar molecular weight. Any molecule with polar bonds whose dipole moments do not add to zero is a polar molecule. Carbon monoxide, carbon trichloride, pentachlorobenzene, o-dichlorobenzene, boron dibromochloride, and water are all polar.

Carbon monoxide: EN(O) = 3.5, EN(C) = 2.5. Oxygen is more electronegative than carbon. Every diatomic molecule with a polar bond must be a polar molecule.

Carbon trichloride: EN(C) = 2.5, EN(Cl) = 3.2, EN(H) = 2.2. Carbon trichloride has polar bonds of two magnitudes, the smaller polarity H+→C bond and the larger polarity C+→Cl bond. The asymmetry of the molecule results in a net dipole moment, so that $CHCl_3$ is polar.

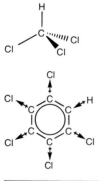

Pentachlorobenzene: The bond polarities are the same as in $CHCl_3$ above. The bond polarities do not cancel one another and the molecule is polar.

(Continued)

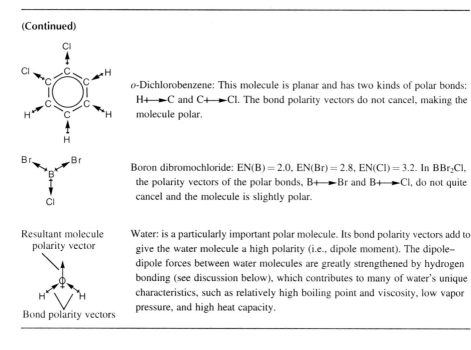

o-Dichlorobenzene: This molecule is planar and has two kinds of polar bonds: H+——►C and C+——►Cl. The bond polarity vectors do not cancel, making the molecule polar.

Boron dibromochloride: EN(B) = 2.0, EN(Br) = 2.8, EN(Cl) = 3.2. In BBr$_2$Cl, the polarity vectors of the polar bonds, B+——►Br and B+——►Cl, do not quite cancel and the molecule is slightly polar.

Resultant molecule polarity vector

Bond polarity vectors

Water: is a particularly important polar molecule. Its bond polarity vectors add to give the water molecule a high polarity (i.e., dipole moment). The dipole–dipole forces between water molecules are greatly strengthened by hydrogen bonding (see discussion below), which contributes to many of water's unique characteristics, such as relatively high boiling point and viscosity, low vapor pressure, and high heat capacity.

2.8.6   THE NATURE OF INTERMOLECULAR ATTRACTIONS

All molecules are attracted to one another because of electrostatic forces. Polar molecules are attracted to one another because the negative end of one molecule is attracted to the positive ends of other molecules, and vice versa. Attractions between polar molecules are called dipole–dipole forces. Similarly, positive ions are attracted to negative ions. Attractions between ions are called ion–ion forces. If ions and polar molecules are present together, as when sodium chloride is dissolved in water, there can be ion–dipole forces, where positive and negative ions (e.g., Na$^+$ and Cl$^-$) are attracted to the oppositely charged ends of polar molecules (e.g., H$_2$O).

However, nonpolar molecules also are attracted to one another although they do not have permanent charges or dipole moments. Evidence of attractions between nonpolar molecules is demonstrated by the fact that nonpolar gases such as methane (CH$_4$), oxygen (O$_2$), nitrogen (N$_2$), ethane (CH$_3$CH$_3$), and carbon tetrachloride (CCl$_4$) condense to liquids and solids when the temperature is lowered sufficiently. Knowing that positive and negative charges attract one another makes it easy to understand the existence of attractive forces among polar molecules and ions. But how can the attractions among nonpolar molecules be explained?

In nonpolar molecules, the valence electrons are distributed about the nuclei so that, on average, there is no net dipole moment. However, molecules are in constant motion, often colliding and approaching one another closely. When two molecules approach closely, their electron clouds interact by electrostatically repelling one another. These repulsive forces momentarily distort the electron distributions within the molecules and create transitory dipole moments in molecules that would be nonpolar if isolated from neighbors. A transitory dipole moment in one molecule induces electron charge distortions and transitory dipole moments in all nearby molecules. At any instant in an assemblage of molecules, nearly every molecule will have a nonuniform charge distribution and an instantaneous dipole moment. An instant later, these dipole moments will have changed direction or disappeared so that, averaged over time, nonpolar molecules have no net dipole moment. However, the effect of these transitory dipole moments is to create a net attraction among nonpolar molecules. Attractions between nonpolar molecules are called dispersion forces or London forces (after Professor Fritz London who gave a theoretical explanation for them in 1928).

Hydrogen bonding: An especially strong type of dipole–dipole attraction, called hydrogen bonding, occurs among molecules containing a hydrogen atom covalently bonded to a small, highly electronegative atom that contains at least one valence shell nonbonding electron pair. An examination of Table 2.1 shows that fluorine, oxygen, and nitrogen are the smallest (implied by their position at the top of their columns in the periodic table)* and the most electronegative elements that also contain nonbonding valence electron pairs. Although chlorine and sulfur have similarly high electronegativities and contain nonbonding valence electron pairs, they are too large to consistently form hydrogen bonds (H-bonds). Because hydrogen bonds are both strong and common, they influence many substances in important ways.

Hydrogen bonds are very strong (10–40 kJ/mole) compared to other dipole–dipole forces (from less than 1 to 5 kJ/mole). The hydrogen atom's very small size makes hydrogen bonding so uniquely strong. Hydrogen has only one electron. When hydrogen is covalently bonded to a small, highly electronegative atom, the shift of bonding electrons toward the more electronegative atom leaves the hydrogen nucleus nearly bare. With no inner core electrons to shield it, the partially positive hydrogen can approach very closely to a nonbonding electron pair on nearby small polar molecules. The very close approach results in stronger attractions than with other dipole–dipole forces.

Because of the strong intermolecular attractions, hydrogen bonds have a strong effect on the properties of the substances in which they occur. Compared with non-hydrogen bonded compounds of similar size, hydrogen bonded substances have relatively high boiling and melting points, low volatilities, high heats of vaporization, and high specific heats. Molecules that can H-bond with water are highly soluble in water; thus, all the substances in Figure 2.5 are water-soluble.

2.8.7 COMPARATIVE STRENGTHS OF INTERMOLECULAR ATTRACTIONS

The strength of dipole–dipole forces depends on the magnitude of the dipole moments. The strength of ion–ion forces depends on the magnitude of the ionic charges. The strength of dispersion forces depends on the polarizability of the nonpolar molecules. Polarizability is a measure of how easily the electron distribution can be distorted by an electric field—that is, how easily a dipole moment can be induced in an atom or a molecule by the electric field carried by a nearby atom or

---

* Atomic size of atoms in the same column of the periodic table tends to increase from the top of the column to the bottom. Thus, for example, the diameter of Li < Na < K < Rb < Cs < Fr and Bi > Sb > As > P > N.

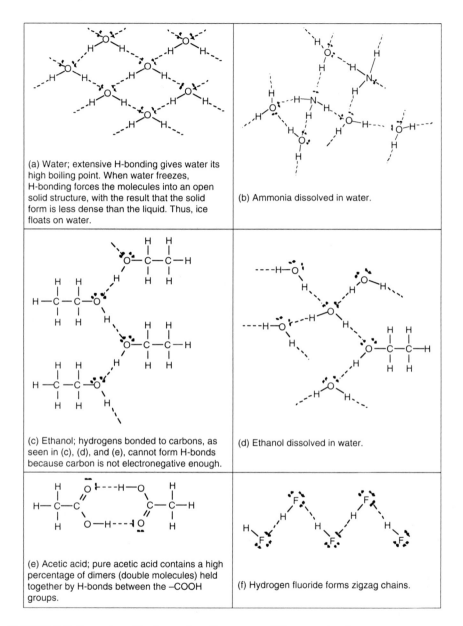

(a) Water; extensive H-bonding gives water its high boiling point. When water freezes, H-bonding forces the molecules into an open solid structure, with the result that the solid form is less dense than the liquid. Thus, ice floats on water.

(b) Ammonia dissolved in water.

(c) Ethanol; hydrogens bonded to carbons, as seen in (c), (d), and (e), cannot form H-bonds because carbon is not electronegative enough.

(d) Ethanol dissolved in water.

(e) Acetic acid; pure acetic acid contains a high percentage of dimers (double molecules) held together by H-bonds between the –COOH groups.

(f) Hydrogen fluoride forms zigzag chains.

FIGURE 2.5 Examples of hydrogen bonding among different molecules.

molecule. Large atoms and molecules have more electrons and larger electron clouds than small ones. In large atoms and molecules, the outer shell electrons are farther from the nuclei and, consequently, are more loosely bound. Their electron distributions can be more easily distorted by external electric fields. In small atoms and molecules, the outer electrons are closer to the nuclei and are more tightly held. Electron charge distributions in small atoms and molecules are less easily distorted.

Therefore, large atoms and molecules are more polarizable than small ones. Since atomic and molecular sizes are closely related to atomic and molecular weights, we can generalize that polarizability increases with increasing atomic and molecular weights. The greater the polarizability of atoms and molecules, the stronger are the intermolecular dispersion forces between them. Molecular shape also affects polarizability. Elongated molecules are more polarizable than compact molecules. Thus, a linear alkane is more polarizable than a branched alkane of the same molecular weight.

All atoms and molecules have some degree of polarizability. Therefore, all atoms and molecules experience attractive dispersion forces, whether or not they also have dipole moments, ionic charges, or can hydrogen bond. Small polar molecules are dominated by dipole–dipole forces since the contribution to attractions from dispersion forces is small. However, dispersion forces may dominate in very large polar molecules.

RULES OF THUMB

1. The higher the atomic or molecular weights of nonpolar molecules, the stronger are the attractive dispersion forces between them.
2. For different nonpolar molecules with the same molecular weight, molecules with a linear shape have stronger attractive dispersion forces than do branched, more compact molecules.
3. For polar and nonpolar molecules alike, the stronger the attractive forces, the higher the boiling point and freezing point, and the lower the volatility of the substance.

EXAMPLE 1

Consider the halogen gases fluorine (F_2, MW = 38), chlorine (Cl_2, MW = 71), bromine (Br_2, MW = 160), and iodine (I_2, MW = 254). All are nonpolar, with progressively greater molecular weights and correspondingly stronger attractive dispersion forces as you go from F_2 to I_2. Accordingly, their boiling and melting points increase with their molecular weights. At room temperature, F_2 is a gas (bp = $-188°C$), Cl_2 is also a gas but with a higher boiling point (bp = $-34°C$), Br_2 is a liquid (bp = $58.8°C$), and I_2 is a solid (mp = $184°C$).

EXAMPLE 2

Alkanes are compounds of carbon and hydrogen only. Although C–H bonds are slightly polar (electronegativity of C = 2.5; electronegativity of H = 2.1), all alkanes are nonpolar because of their bond geometry. In the straight-chain alkanes (called normal-alkanes), as the alkane carbon chain becomes longer, the molecular weights and, consequently, the attractive dispersion forces become greater. Consequently, melting points and boiling points become progressively higher. The physical properties of the normal-alkanes in Table 2.2 reflect this trend.

TABLE 2.2

Some Properties of the First Twelve Straight-Chain Alkanes

Alkane	Formula	Molecular Weight	Melting Point[a] (°C)	Boiling Point (°C)
Methane	CH_4	16	−183	−162
Ethane	C_2H_6	30	−172	−89
Propane	C_3H_8	44	−188	−42
n-Butane	C_4H_{10}	58	−138	0
n-Pentane	C_5H_{12}	72	−130	36
n-Hexane	C_6H_{14}	86	−95	69
n-Heptane	C_7H_{16}	100	−91	98
n-Octane	C_8H_{18}	114	−57	126
n-Nonane	C_9H_{20}	128	−51	151
n-Decane	$C_{10}H_{22}$	142	−29	174
n-Dodecane	$C_{12}H_{26}$	170	−10	216

[a] Deviations from the general trend in melting points occur because melting points for the smallest alkanes are more strongly influenced by differences in crystal structure and lattice energy of the solid.

EXAMPLE 3

Normal-butane (n-C_5H_{12}) and dimethylpropane ($CH_3C(CH_3)_2CH_3$) are both nonpolar and have the same molecular weights (MW = 72). However, n-C_5H_{12} is a straight-chain alkane whereas $CH_3C(CH_3)_2CH_3$ is branched. Thus, n-C_5H_{12} has stronger dispersion attractive forces than $CH_3C(CH_3)_2CH_3$ and a correspondingly higher boiling point.

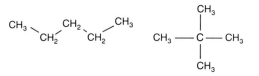

n-Pentane: bp = 36°C   Dimethylpropane: bp = 9.5°C

2.9   SOLUBILITY AND INTERMOLECULAR ATTRACTIONS

In liquids and gases, the molecules are in constant, random, thermal motion, colliding and intermingling with one another. Even in solids, the molecules are in constant, although more limited, motion. If different kinds of molecules are present, random movement tends to mix them uniformly. If there were no other considerations, random motion would cause all substances to dissolve completely into one another. Gases and liquids would dissolve more quickly and solids more slowly.

However, intermolecular attractions must also be considered. Strong attractions between molecules tend to hold them together. Consider two different substances A and B, where A molecules are attracted strongly to other A molecules, B molecules are attracted strongly to other B molecules, but A and B molecules are attracted only weakly to one another. Then, A and B molecules tend to stay separated from each other. *A* molecules try to stay together and *B* molecules try to stay together, each excluding entry from the other. In this case, *A* and *B* are not soluble in one another.

As an example of this situation, let *A* be a nonpolar, straight-chain liquid hydrocarbon such as *n*-octane (C_8H_{18}) and let *B* be water (H_2O). Octane molecules are attracted to one another by strong dispersion forces, and water molecules are attracted strongly to one another by dipole–dipole forces and H-bonding. Dispersion attractions are weak between the small water molecules. Because the small water molecules have low polarizability, octane cannot induce a strong dispersion force attraction to water. Because octane is nonpolar, there are no dipole–dipole attractions to water. When water and octane are placed in the same container, they remain separate forming two layers with the less dense octane floating on top of the water.

However, if there were strong attractive forces between *A* and *B* molecules, it would help them to mix. The solubility of one substance (the solute) in another (the solvent) depends mostly on intermolecular forces and, to a much lesser extent, on conditions such as temperature and pressure. Substances are more soluble in one another when intermolecular attractions between solute and solvent are similar in magnitude to the intermolecular attractions between the pure substances. This principle is the origin of the rules of thumb that say, "like dissolves like" or "oil and water do not mix." "Like" molecules have similar polar or nonpolar properties and, consequently, similar intermolecular attractions. Oil and water do not mix because water molecules are attracted strongly to one another, and oil molecules are attracted strongly to one another; but water molecules and oil molecules are attracted only weakly to one another.

RULES OF THUMB

1. The more symmetrical the structure of a molecule containing polar bonds, the less polar and the less soluble it is in water.
2. Molecules with OH, NO, or NH groups can form hydrogen bonds to water molecules. They are the most water-soluble nonionic compounds, even if they are nonpolar because of geometrical symmetry.
3. The next most water-soluble compounds contain O, N, and F atoms. All have high electronegativities and allow water molecules to H-bond with them.
4. Charged regions in ionic compounds (like sodium chloride) are attracted to polar water molecules. This makes them more soluble.
5. Most compounds in oil and gasoline mixtures are nonpolar. They are attracted to water very weakly and have very low solubilities.
6. All molecules, including nonpolar molecules, are attracted to one another by dispersion forces. The larger the molecule the stronger the dispersion force.
7. Nonpolar molecules, large or small, have low solubilities in water because the small-sized water molecules have weak dispersion forces, and nonpolar molecules have no dipole moments. Thus, there are neither dispersion nor polar attractions to encourage solubility.

EXAMPLE 4

Alcohols of low molecular weight are very soluble in water because of hydrogen bonding. However, their water solubilities decrease as the number of carbons increase. The –OH group on alcohols is hydrophilic (attracted to water by hydrogen bonding), whereas the hydrocarbon part is hydrophobic (repelled from water). If the hydrocarbon part of an alcohol is large enough, the hydrophobic behavior overcomes the hydrophilic behavior of the –OH group and the alcohol has low solubility. Solubilities for alcohols with increasingly larger hydrocarbon chains are given in Table 2.3.

EXAMPLE 5

For alcohols of comparable molecular weight, the more hydrogen bonds a compound can form, the more water-soluble the compound, and the higher the boiling and melting points of the pure compound. In Table 2.3, notice the effect of adding another –OH group to the molecule. The double alcohol 1,5-pentanediol is more water-soluble and has a higher boiling point than single alcohols of comparable molecular weight, because of its two –OH groups capable of hydrogen bonding. This effect is general. Double alcohols (diols) are more water-soluble and have higher boiling and melting points than

TABLE 2.3
Solubilities and Boiling Points of Some Straight-Chain Alcohols

Name	Formula	Molecular Weight	Melting Point[a] (°C)	Boiling Point (°C)	Aqueous Solubility at 25°C (mol/L)
Methanol	CH_3OH	32	−98	65	∞(miscible)
Ethanol	C_2H_5OH	46	−130	78	∞(miscible)
1-Propanol	C_3H_7OH	60	−127	97	∞(miscible)
1-Butanol	C_4H_9OH	74	−90	117	0.95
1-Pentanol	$C_5H_{11}OH$	88	−79	138	0.25
1,5-Pentanediol[b]	$C_5H_{10}(OH)_2$	104	−18	239	∞(miscible)
1-Hexanol	$C_6H_{13}OH$	102	−47	158	0.059
1-Octanol	$C_8H_{17}OH$	130	−17	194	0.0085
1-Nonanol	$C_9H_{19}OH$	144	−6	214	0.00074
1-Decanol	$C_{10}H_{21}OH$	158	+6	233	0.00024
1-Dodecanol	$C_{12}H_{25}OH$	186	+24	259	0.000019

[a] Deviations from the general trend in melting points occur because melting points for the smallest alcohols are more strongly influenced by differences in crystal structure and lattice energy of the solid.

[b] The properties of 1,5-pentanediol deviate from the trends of the other alcohols because it is a diol and has two –OH groups available for hydrogen bonding, see text.

single alcohols of comparable molecular weight. Triple alcohols (triols) are still more water-soluble and have higher boiling and melting points.

EXERCISES

1. a. Describe briefly what physical changes will condense a gas to a liquid and a liquid to a solid. Explain, in terms of forces between molecules, why these changes have the effect described.
   b. In light of your answer to 1a, what conclusions can you draw from the fact that water boils at a higher temperature than does ammonia?
2. Describe briefly the relation between temperature and molecular motion.
3. a. Balance the equation: $H_2 + N_2 \rightarrow NH_3$
   b. When 3 g of hydrogen (H_2) react with 14 g of nitrogen (N_2), 17 g of a compound called ammonia (NH_3) are made: $3.0 \text{ g } H_2 + 14 \text{ g } N_2 \rightarrow 17 \text{ g } NH_3$ How many grams of ammonia will be made if 6.0 g of hydrogen react with 14 g of nitrogen?
4. Balance the following equation: $C_5H_{10} + O_2 \rightarrow CO_2 + H_2O$
5. Various compounds and some of their properties are tabulated below.

Compound	Boiling Point (°C)	Water Solubility (g/100 mL)	Dipole Moment (D)
H_2	−253	2×10^{-4}	0
HCl	−84.9	82	1.08
HBr	−67	221	0.82
CO_2	−78	0.15	0
CH_4	−164	2×10^{-3}	0
NH_3	−33.5	90	1.3
H_2O	100	∞	1.85
HF	19.5	∞	1.82
LiF	1676	0.27	6.33

   a. Which compounds are nonpolar?
   b. Which compounds are polar?
   c. Make a rough plot of boiling point versus dipole moment. What conclusions may be inferred from the graph?
   d. Discuss briefly the trends in water solubility. Why is solubility not related to dipole moment in the same manner as boiling point?
   e. The ionic compound LiF appears to be unique. Try to suggest a reason for its low solubility.

6. The chemical structures of several compounds are shown below. Use the structures to fill in the table with estimated properties. Do not look up reference data for the answers.

Compound	Structure	Physical State at Room Temperature (Gas, Liquid, or Solid)	Water Solubility (Very Low, Low Moderate, High)	Lipid Solubility (Very Low, Low Moderate, High)
Ethane	H_3C-CH_3			
Benzene				
Citric acid				
Anthracene				
Glyphosate				

3 Major Water Quality Parameters and Applications

3.1 INTERACTIONS AMONG WATER QUALITY PARAMETERS

This chapter deals with important water quality parameters that serve as controlling variables, those that strongly influence the behavior of many other constituents present in the water. The major controlling variables are pH, oxidation–reduction (redox) potential, alkalinity and acidity, temperature, and total dissolved solids (TDS). This chapter also discusses several other important parameters, such as ammonia, sulfide, carbonates, dissolved metals, and dissolved oxygen (DO), which are strongly affected by changes in the controlling variables.

It is important to understand that chemical constituents in natural water bodies react in an environment far more complicated than if they were surrounded only by a large number of water molecules. Various impurities in water interact in ways that can affect their chemical behavior markedly. Water quality parameters defined above as controlling variables have an especially strong effect on water chemistry. For example, a pH change from pH 6 to 9 will lower the solubility of Cu^{2+} by four orders of magnitude. The solubility of Cu^{2+} in water at pH 6 is about 40 mg/L, while at pH 9 it is about 4×10^{-3} mg/L—10,000 times smaller. If, for example, a water solution at pH 6 contained 40 mg/L of Cu^{2+} and its pH was raised to 9, all but 4×10^{-3} mg/L of the Cu^{2+} would precipitate as solid $Cu(OH)_2$.

As another example, consider a shallow lake with algae and other vegetation growing in it. Its suspended and bottom sediments contain high concentrations of decaying organic matter. The lake is fed by surface and groundwaters containing high concentration of sulfate. During daytime, photosynthesis can produce enough DO to maintain a positive oxidation–reduction (OR) potential in the water. At night, photosynthesis stops and biodegradation of suspended and lake-bottom organic sediments consumes nearly all of the DO in the lake. This changes the water from oxidizing (aerobic) to reducing (anaerobic) conditions and causes the OR potential to change from positive to negative values. At negative OR potentials (reducing conditions), dissolved sulfate in the lake is reduced to sulfide, producing hydrogen sulfide (H_2S) gas, which smells like rotten eggs. Thus, if there are residences around the lake, there may be an odor problem at night that generally dissipates during the day.

A remedy for this problem entails finding a way to maintain a positive OR potential for longer periods.

RULE OF THUMB

Because they strongly influence other water quality parameters, the controlling variables listed are usually included among the parameters that are routinely measured in water quality sampling programs.

- pH
- Temperature
- Alkalinity and acidity
- Total dissolved solids or conductivity
- Oxidation–reduction (redox) potential

3.2 pH

3.2.1 BACKGROUND

Acidic, basic, and neutral water solutions are measured by a quantity called pH. As shown below, pH is a measure of the hydrogen ion (H^+) concentration in water solutions. Hydrogen ions arise in water from dissociation of the water molecules themselves or from dissociation of other molecules containing hydrogen that are dissolved in water.

Pure water always contains a small number of molecules that self-dissociate because of thermal energy into hydrogen ions (H^+)* and hydroxyl ions (OH^-), as illustrated by Equation 3.1:

$$H_2O \leftrightarrow H^+ + OH^- \tag{3.1}$$

Equation 3.1 is a reversible reaction (indicated by the double arrow); after they are formed, H^+ and OH^- ions can recombine to form uncharged water molecules. At any given temperature, collisions between the more energetic water molecules initiate the forward dissociation reaction of Equation 3.1, and collisions between less energetic H^+ and OH^- ions initiate the reverse recombination reaction. The condition of equilibrium, when the rates of ion formation and recombination are equal, determines the equilibrium concentrations of H^+ and OH^- ions.

---

* Note that a hydrogen atom consists of a nucleus with just one proton, surrounded by an electron cloud with just one electron. Its positive ion is formed by losing its only electron, leaving a bare proton. Without an electron cloud, H^+ has the dimensions of a proton, very much smaller than any other ionic species. Thus, the hydrogen ion is a unique chemical entity, lying between subatomic and atomic domains. It is structurally identical to the proton subatomic positive particle found in all atomic nuclei, but is not within a nucleus because it is formed in a chemical reaction and is available for further chemical interactions. Its very large charge-to-diameter ratio makes it extremely reactive (see discussion of Equation 3.5).

The water dissociation constant, K_w, is defined as the product of the equilibrium concentrations of H^+ and OH^- ions, expressed as moles per liter:

$$K_w = [H^+][OH^-] \tag{3.2}$$

Enclosing the symbol for a chemical species in square brackets is chemical symbolism that represents the species concentration, expressed as moles per liter. For example, if $[Na^+] = 10^{-3}$ mol/L, the concentration of Na^+ in the solution is 0.001 mol/L, and, since the atomic weight of Na is 23, this is equivalent to 0.023 g/L (or 23 ppm) (see footnotes on pages 10 and 12 of Chapter 1).

Because the degree of dissociation increases with temperature, K_w is temperature dependent. At 25°C

$$K_{w,25°C} = [H^+][OH^-] = 1.0 \times 10^{-14} \ (mol/L)^2 \tag{3.3}$$

whereas at 50°C

$$K_{w,50°C} = [H^+][OH^-] = 1.83 \times 10^{-13} \ (mol/L)^2 \tag{3.4}$$

If, for example, an acid is added to water at 25°C (usually considered as the standard default temperature when no other is specified), it dissociates releasing additional H^+ into the water. The H^+ concentration increases but the product, $[H^+] \times [OH^-]$, in Equation 3.3 must remain equal to 1.0×10^{-14} $(mol/L)^2$. This means that if $[H^+]$ increases, $[OH^-]$ must decrease by a corresponding amount. Similarly, adding a base causes $[OH^-]$ to increase and, therefore, $[H^+]$ must decrease correspondingly.

In pure water, or in water with no sources or sinks of H^+ or OH^- other than the dissociation reaction of water, Equation 3.1 predicts that there will be equal numbers of H^+ and OH^- species. By definition, pure water is neither acidic nor basic; it is defined as neutral with respect to its acid-base properties. Thus, the neutral condition in water is when there are equal concentrations of H^+ and OH^- ions. When $[H^+]$ is greater than $[OH^-]$, the water solution is acidic; when $[H^+]$ is less than $[OH^-]$, it is basic.

In pure water at 25°C, the values of $[H^+]$ and $[OH^-]$ must each be equal to 1.0×10^{-7} mol/L, since by Equation 3.3

$$K_{w,25°C} = [H^+][OH^-] = (1.0 \times 10^{-7} \ mol/L)(1.0 \times 10^{-7} \ mol/L)$$
$$= 1.0 \times 10^{-14} \ (mol/L)^2$$

Since pure water defines the condition of acid-base neutrality, neutral water always has equal concentrations of H^+ and OH^-, or $[H^+] = [OH^-]$.

Whatever their separate values, the product of hydrogen ion and hydroxyl ion concentrations must be equal to 1×10^{-14} at 25°C, as in Equation 3.3. If for example $[H^+] = 10^{-5}$ mol/L, then it is necessary that $[OH^-] = 10^{-9}$ mol/L, so that their product is 10^{-14} $(mol/L)^2$. In this example, because $[H^+] > [OH^-]$, the solution is acidic.

Many compounds dissociate in water to form ions. Those that form hydrogen ions (H^+) or consume hydroxyl ions (OH^-) are called acids because, when added to pure water, they produce the condition $[H^+] > [OH^-]$. Compounds that produce the condition $[H^+] < [OH^-]$ when added to pure water are called bases.

RULES OF THUMB

The extent of water self-dissociation increases with temperature. For example

    a. In neutral water at 25°C, $[H^+] = [OH^-] = 1 \times 10^{-7}$ mol/L, and pH 7.0 (by Equation 3.8).
    b. In neutral water at 50°C, $[H^+] = [OH^-] = 4.3 \times 10^{-7}$ mol/L, and pH 6.4 (by Equation 3.8).

    • If $[H^+] > [OH^-]$, the water solution is acidic.
    • If $[H^+] < [OH^-]$, the water solution is basic.
    • If $[H^+] = [OH^-]$, the water solution is neutral.

H^+ is too reactive to exist alone and is always attached to other molecular species. In water solutions, some of the H^+ are always chemically bound to water molecules, most commonly to one or two water molecules. Equations 3.5a and 3.5b are almost instantaneous, so H^+ does not exist long enough to be measured:

$$H^+ + H_2O \rightarrow H_3O^+ \tag{3.5a}$$

$$H_3O^+ + H_2O \rightarrow H_5O_2^+ \rightarrow (H_2O)_nH^+ \tag{3.5b}$$

Further steps in this process to give values for $n > 2$ become less and less likely.

It is the hydrogen ion–water molecule complex $(H_2O)_nH^+$, called the hydrated proton, that gives solutions their acidic characteristics. Since the chemical structure of the hydrated proton is not precisely defined, a convention must be adopted to chemically describe the hydrated proton. Although, H_3O^+ and $H_5O_2^+$ are the two most common varieties of hydrated protons, it does not make any difference to the meaning of a chemical equation whether the presence of an acid is indicated by H^+, H_3O^+, or $H_5O_2^+$. For example, the addition of nitric acid, HNO_3, to water can be written in a manner that emphasizes the hydration process and identifies the hydrated protons:

$$HNO_3 + H_2O \rightarrow H_3O^+ + NO_3^- \tag{3.6a}$$

$$HNO_3 + 2H_2O \rightarrow H_5O_2^+ + NO_3^- \tag{3.6b}$$

Equation 3.7 is an equivalent but more generic equation that emphasizes the dissociation process:

$$HNO_3 \xrightarrow{H_2O} H^+ + NO_3^- \tag{3.7}$$

All the equations are read, "HNO_3 added to water forms H^+(or $H_3O^+/H_5O_2^+$) and NO_3^- ions." In other words, 1 mole of HNO_3 dissociates in water to form 1 mole of hydrated protons, no matter whether we choose to write the hydrated proton as H^+, H_3O^+, or $H_5O_2^+$. For most environmental chemistry purposes, the dissociation process is more important and this text will discuss acidic solutions in terms of their "hydrogen ion (H^+)" concentration.

3.2.2 DEFINING pH

The concentration of H^+ in water solutions commonly ranges from about 1 mol/L (since the molecular weight (MW) of H^+ is 1.0 g/mol,* this is equivalent to 1 g/L or 1000 ppm) for very acidic water, to about 10^{-14} mol/L (10^{-14} g/L or 10^{-11} ppm) for very basic water. Under special circumstances, the range can be even wider.[†]

Rather than work with such a wide numerical range for a measurement that is so common, chemists have developed a way to use logarithmic units for expressing $[H^+]$ as a positive decimal number whose value normally lies between 0 and 14.[†] This number is called the pH, and is defined in Equation 3.8 as the negative of the base 10 logarithm of the hydrogen ion concentration, expressed in moles per liter:

$$pH = -\log_{10} [H^+] \tag{3.8}$$

Note that since logarithms are dimensionless, pH value has no dimensions or units. Frequently, pH is unnecessarily assigned units called *standard units* (SU), even though pH is unitless. This mainly serves to avoid blank spaces in a table that contains a column for units, or to satisfy a computer database that requires an entry in a units field. Also note that if $[H^+] = 10^{-7}$, then $pH = -\log_{10}(10^{-7}) = -(-7) = 7$. A higher concentration of H^+ such as $[H^+] = 10^{-5}$ yields a lower value for pH, i.e., $pH = -\log_{10}(10^{-5}) = 5$. Thus, if pH is less than 7 in a solution at 25°C, the solution contains more H^+ than OH^- and is acidic; if pH is greater than 7, the solution is basic.

3.2.3 ACID-BASE REACTIONS

In acid-base reactions, protons (H^+ ions) are transferred between chemical species, one of which is an acid and the other is a base. The proton donor is the acid and the proton acceptor is the base. For example, if an acid, such as hydrochloric acid (HCl), is dissolved in water, water acts as a base by accepting the proton donated by HCl. The acid-base reaction is written as

$$HCl + H_2O \rightarrow Cl^- + H_3O^+ \tag{3.9}$$

A water molecule that behaved as a base by accepting a proton is turned into an acid, H_3O^+, a species that has a proton available to donate. The species H_3O^+, as noted above, is called a hydronium ion and is the chemical species that gives acid water

---

* In water solutions, the water molecules attached to a hydrated proton readily exchange with other unattached water molecules and are indistinguishable from them. It makes no difference in quantitative calculations when the water molecules bound to hydrated protons are simply included with those of the water solvent. Thus, the molecular weight of the hydrated proton may be taken as 1 g/mol, the same as for H^+.

† It is possible, although not very common, to have H^+ concentrations greater than 1 ppm or less than 10^{-14} ppm. This, of course, would result in pH values less than 0 (negative) or greater than 14. For example, if $[H^+] = 2$ g/L, then $pH = -\log_{10}(2) = -(+0.301) \approx -0.3$. (see Nordstrom, 2000).

solutions their acidic characteristics. The HCl/water solution of Equation 3.9 contains water molecules, hydronium ions, hydroxyl ions (in smaller concentration than H_3O^+), and chloride ions. The solution is termed acidic, with pH < 7. The measurable parameter pH indicates the concentration of protons available for acid-base reactions.

RULES OF THUMB

In an acid-base reaction, H^+ ions are exchanged between chemical species. The specie that donates the H^+ is the acid. The specie that accepts the H^+ is the base.

The concentration of H^+ in water solutions is an indication of how many hydrogen ions are available, at the time of measurement, for exchange between chemical species. The exchange of hydrogen ions changes the chemical properties of the species between which the exchange occurs; the donor, which was an acid, becomes a base because it now can accept a hydrogen ion, and the acceptor, which was a base, becomes an acid, because it now can donate a hydrogen ion.

pH is a measure of $[H^+]$, the hydrogen ion concentration, which determines the acidic or basic quality of water solutions. At 25°C, when

- pH < 7, water solution is acidic
- pH = 7, water solution is neutral
- pH > 7, water solution is basic

EXAMPLE 1

The $[H^+]$ of water in a stream is 3.5×10^{-6} mol/L. What is the pH?

Answer:

$$pH = -\log_{10}[H^+] = -\log_{10}(3.5 \times 10^{-6}) = -(-5.46) = 5.46.$$

An alternate and useful form of Equation 3.8 is

$$[H^+] = 10^{-pH} \tag{3.10}$$

EXAMPLE 2

The pH of water in a stream is 6.65. What is the hydrogen ion concentration?

Answer:

From Equation 3.10, $[H^+] = 10^{-pH} = 10^{-6.65} = 2.24 \times 10^{-7}$ mol/L.

3.2.4 IMPORTANCE OF pH

Measurement of pH is one of the most important and frequently used tests in water chemistry. pH is an important factor in determining the chemical and biological

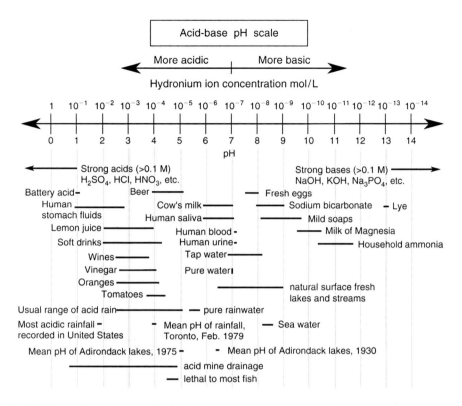

FIGURE 3.1 pH scale and typical pH values of some common substances.

properties of water. It affects the chemical forms and environmental impacts of many chemical substances in water. For example, many metals dissolve as ions at lower pH values, precipitate as hydroxides and oxides at higher pH, and redissolve again at very high pH. Figure 3.1 shows the pH scale and typical pH values of some common substances.

pH also influences the degree of ionization, volatility, and toxicity to aquatic life of certain dissolved substances, such as ammonia, hydrogen sulfide, and hydrogen cyanide. The ionized form of ammonia, which predominates at low pH, is the less toxic ammonium ion (NH_4^+). NH_4^+ transforms to the more toxic unionized form of ammonia, NH_3, at higher pH. Both hydrogen sulfide (H_2S) and hydrogen cyanide (HCN) behave oppositely to ammonia; the less toxic ionized forms, S^{2-} and CN^-, are predominant at high pH, and the more toxic unionized forms, H_2S and HCN, are predominant at low pH. The pH value is an indicator of the chemical state in which these compounds will be found and must be considered when establishing water quality standards.

3.2.5 MEASURING pH

The pH of environmental waters is most commonly measured with electronic pH meters or by wetting special papers impregnated with color-changing dyes with the sample being measured. Battery-operated field meters are common. The pH of a water sample is altered by many processes that can occur after the sample is collected, such as loss or gain of dissolved carbon dioxide (CO_2) or the oxidation of dissolved iron. Therefore, a pH measurement of surface or groundwater is valid only when made in the field or very shortly after sampling. A laboratory determination of pH made hours or days after sampling may be more than one full pH unit (a factor of 10 in H^+ concentration) different from the value at the time of sampling.

Loss or gain of dissolved CO_2 is one of the most common causes for pH changes. When CO_2 dissolves into water, by diffusion from the atmosphere or from microbial activity in water or soil, the pH is lowered. Conversely, when CO_2 is lost, by diffusion to the atmosphere or consumption during photosynthesis of algae or water plants, the pH is raised.

RULES OF THUMB

Under low pH conditions (acidic):

a. Metals tend to dissolve.
b. Cyanide and sulfide are more toxic to fish.
c. Ammonia is less toxic to fish.

Under high pH conditions (basic):

a. Metals tend to precipitate as hydroxides and oxides. However, if the pH gets too high, some precipitates begin to dissolve again because soluble hydroxide complexes are formed (see Chapter 4).
b. Cyanide and sulfide are less toxic to fish.
c. Ammonia is more toxic to fish.

3.2.6 Water Quality Criteria and Standards for pH

The pH of pure water at 25°C is 7.0, but the pH of environmental waters is affected by dissolved CO_2 and exposure to minerals. Most unpolluted groundwater and surface water in the United States have pH values between about 6.0 and 8.5, although higher and lower values can occur because of special conditions, such as sulfide oxidation that lowers the pH, or low CO_2 concentrations which raises the pH. During daylight, photosynthesis in surface waters by aquatic organisms may consume more CO_2 than is dissolved from the atmosphere, causing pH to rise. At night, after photosynthesis has ceased, CO_2 from the atmosphere continues to dissolve and lowers the pH again. In this manner, photosynthesis can cause diurnal pH fluctuations, the magnitude of which depends on the alkalinity buffering capacity of the water. In poorly buffered lakes or rivers, the daytime pH may reach 9.0–12.0.

The permissible pH range for protecting aquatic life depends on factors such as DO, temperature, and concentrations of dissolved anions and cations. A pH range of 6.5–9.0, with no short-term change greater than 0.5 units beyond the normal seasonal maximum or minimum, is deemed protective of freshwater aquatic life and considered harmless to fish. In irrigation waters, the pH should not fall outside a range of 4.5–9.0 to protect plants.

Environmental Protection Agency (EPA) criteria for pH

- Domestic water supplies is 5.0–9.0
- Freshwater aquatic life is 6.5–9.0

RULES OF THUMB

1. The pH of natural unpolluted river water is generally between 6.5 and 8.5.
2. The pH of natural unpolluted groundwater is generally between 6.0 and 8.5.
3. Clean rainwater has a pH of about 5.7 because of dissolved CO_2.
4. After reaching the surface of the earth, rainwater usually acquires alkalinity from carbonate minerals while moving over and through the earth, which may raise the pH and buffer the water against severe pH changes.
5. The pH of drinking water supplies should be between 5.0 and 9.0.
6. Fish acclimate to ambient pH conditions. For aquatic life, pH should be between 6.5 and 9.0 and should not vary more than 0.5 units beyond the normal seasonal maximum or minimum.

3.3 OXIDATION–REDUCTION POTENTIAL

3.3.1 BACKGROUND

Oxidizing or reducing conditions in water are indicated by a measured quantity known as the oxidation potential or, more commonly, redox potential. The oxidation–reduction (redox) potential measures the availability of electrons for exchange between chemical species. This may be viewed as analogous to pH, which measures the availability of protons (H^+ ions) for exchange between chemical species. When H^+ ions are exchanged, the acid or base properties of the species are changed. When electrons are exchanged, the oxidation states of the species and their chemical properties are changed, resulting in OR (oxidation-reduction) reactions. The electron donor is said to be oxidized; the electron acceptor is said to be reduced. For every electron donor, there must be an electron acceptor. For example, whenever one substance is oxidized, another must be reduced.

The most common oxidizing agent in the natural environment is oxygen. When the measured redox potential of a natural water body is positive, indicating oxidizing (or aerobic) conditions, there is sufficient DO (dissolved oxygen) present to allow many dissolved metals and organic species to be oxidized, thereby undergoing chemical changes that influence their toxicity and solubility. Stronger oxidizing agents, such as ozone, chlorine, or permanganate, are those that take electrons from many substances even more readily than oxygen. As noted above, the electron donor is oxidized and, by accepting electrons, the oxidizing agents are themselves reduced. Strong oxidizing agents are those that are easily reduced. Similarly, strong reducing agents are those that are easily oxidized, in other words, they readily give up electrons to other substances that in turn become reduced.

For example, chlorine is a strong oxidizing agent widely used to treat water and sewage. Chlorine oxidizes many pollutants more readily than oxygen to less objectionable forms. When chlorine reacts with hydrogen sulfide (H_2S)—a common sewage pollutant that smells like rotten eggs—it oxidizes the sulfur in H_2S to insoluble elemental sulfur (S_8), which is easily removed by settling or filtering. The reaction is

$$8Cl_2(g) + 8H_2S(aq) \rightarrow S_8(s) + 16HCl(aq) \qquad (3.11)$$

As the sulfur in H_2S is more electronegative than the hydrogens, the sulfur atom in H_2S effectively carries two extra electrons that it has attracted from the hydrogens, giving it an oxidation number of -2 and a redox structure of S^{2-}. Because chlorine is more electronegative than sulfur, it can attract the electrons from S^{2-}, leaving the sulfur with oxidation number zero,* at the same time forming two chloride ions in oxidation state -1, as in Equation 3.12, which shows only the species from Equation 3.11 that undergo a change in oxidation number.

$$Cl_2 + S^{2-} \rightarrow 2Cl^- + S^0 \qquad (3.12)$$

In Equation 3.11, the sulfur in H_2S is oxidized because it donates two electrons that are accepted by the chlorine atoms in Cl_2. Chlorine is reduced because it accepts the electrons. Since chlorine is the agent that causes the oxidation of H_2S, chlorine is called an oxidizing agent. Because H_2S is the agent that causes the reduction of chlorine, H_2S is called a reducing agent.

The class of OR reactions is very large. It includes all combustion processes, such as the burning of gasoline or wood, most microbial reactions, such as those that occur in biodegradation, and all electrochemical reactions, such as those that occur in batteries and metal corrosion. The use of subsurface groundwater treatment using permeable barrier walls containing finely divided iron is based on the reducing properties of iron. Such treatment walls are placed in the path of groundwater contaminant plumes. The iron donates electrons to pollutants as they pass through the permeable barrier. Thus, the iron is oxidized and the pollutant reduced. This often causes the pollutant to decompose into less harmful or inert fragments.

---

* Every atom in its elemental state has the same number of electrons around its nucleus as there are protons in its nucleus. Hence, the net charge on an elemental atom, and its oxidation number, equals zero.

RULES OF THUMB

1. Oxygen gas (O_2) is always an oxidizing agent in its reactions with metals and most nonmetals. If a compound has combined with O_2, it has been oxidized and the O_2 has been reduced. By accepting electrons, O_2 is either changed to the oxide ion (O^{2-}) or combined in compounds such as CO_2 or H_2O.
2. Like O_2, halogen gases (F_2, Cl_2, Br_2, and I_2) are always oxidizing agents in reactions with metals and most nonmetals. They accept electrons to become halide ions (F^-, Cl^-, Br^-, and I^-) or are combined in compounds such as HCl or $CHBrCl_2$.
3. If an elemental metal (Fe, Al, Zn, etc.) reacts with a compound, the metal acts as a reducing agent by donating electrons, usually forming a soluble positive ion such as Fe^{2+}, Al^{3+}, or Zn^{2+}.

3.4 CARBON DIOXIDE, BICARBONATE, AND CARBONATE

3.4.1 BACKGROUND

The reactive inorganic forms of environmental carbon are CO_2, bicarbonate (HCO_3^-), and carbonate (CO_3^{2-}). Organic carbon, such as cellulose and starch, is made by plants from CO_2 and water during photosynthesis. Carbon dioxide is present in the atmosphere and in soil pore space as a gas, and in surface waters and groundwaters as a dissolved gas. The carbon cycle is based on the mobility of CO_2, which is distributed readily through the environment as a gas in the atmosphere and dissolved in rainwater, surface water, and groundwater. Most of the earth's carbon, however, is relatively immobile, being contained in ocean sediments and on continents as minerals. The atmosphere, with about 360 ppmv (parts per million by volume) of mobile CO_2, is the second smallest of the earth's global carbon reservoirs, after life forms which are the smallest.

On land, solid forms of carbon are mobilized as particulates, mainly by weathering of carbonate minerals, biodegradation and burning of organic carbon, and burning of fossil fuels.

3.4.2 SOLUBILITY OF CO_2 IN WATER

Carbon dioxide plays a fundamental role in determining the pH of natural waters. Although CO_2 itself is not acidic, it reacts in water (reversibly) to make an acidic solution by forming carbonic acid (H_2CO_3), as shown in Equation 3.13. Carbonic acid can subsequently dissociate in two steps to release hydrogen ions, as shown in Equations 3.14 and 3.15:

$$CO_2 + H_2O \leftrightarrow H_2CO_3 \qquad (3.13)$$

$$H_2CO_3 \leftrightarrow H^+ + HCO_3^- \qquad (3.14)$$

$$HCO_3^- \leftrightarrow H^+ + CO_3^{2-} \qquad (3.15)$$

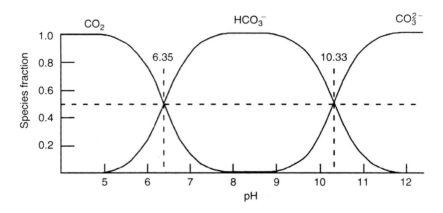

FIGURE 3.2 Distribution diagram showing pH dependence of carbonate species in water.

As a result, pure water exposed to air is not acid-base neutral with a pH near 7.0 because dissolved CO_2 makes it acidic, with a pH around 5.7. The pH dependence of Equations 3.13 through 3.15 is shown in Figure 3.2 and Table 3.1.

Observations from Figure 3.2 and Table 3.1:

- As pH increases, all equilibria in Equations 3.10 through 3.12 shift to the right.
- As pH decreases, all equilibria shift to the left.
- Above pH 10.3, carbonate ion (CO_3^{2-}) is the dominant species.
- Below pH 6.3, dissolved CO_2 is the dominant species.
- Between pH 6.3 and 10.3, a range common to most environmental waters, bicarbonate ion (HCO_3^-) is the dominant species.

The equilibria among only the carbon species (omitting the H^+ species) are

$$CO_2(gas, atm) \leftrightarrow CO_2(aq) \leftrightarrow H_2CO_3(aq) \leftrightarrow HCO_3^-(aq) \leftrightarrow CO_3^{2-}(aq) \qquad (3.16)$$

These dissolved carbon species are sometimes referred to as dissolved inorganic carbon (DIC).

TABLE 3.1
pH Dependence of Carbonate Fractions

pH	Fraction as CO_2	Fraction as HCO_3^-	Fraction as CO_3^{2-}
≪6.35	Essentially 1.00	Essentially 0	Essentially 0
6.35	0.50	0.50	Essentially 0
$\frac{1}{2}(6.35 + 10.33) = 8.34$	0.01	0.98	0.01
10.33	Essentially 0	0.50	0.50
≫10.33	Essentially 0	Essentially 0	Essentially 1.00

Note: Data from Figure 3.2.

3.4.3 Soil CO_2

Processes such as biodegradation of organic matter and respiration of plants and organisms, which commonly occur in the subsurface, consume O_2 and produce CO_2. In the soil subsurface, air in the pore spaces cannot readily equilibrate with the atmosphere and, therefore, pore space air becomes lower in O_2 and higher in CO_2 concentrations.

- Oxygen may decrease from about 21% (210,000 ppmv) in the atmosphere to between 15% and 0% (150,000–0 ppmv) in the soil.
- Carbon dioxide may increase from about 0.04% (~360 ppmv) in the atmosphere to between 0.1% and 10% (1,000–100,000 ppmv) in the soil.

When water moves through the subsurface, it equilibrates with soil gases and may become more acidic because of a higher soil concentration of dissolved CO_2. Acidic groundwater has an increased capacity for dissolving minerals. The higher the CO_2 concentration in soil air, the lower is the pH of groundwater. Acidic groundwater may become buffered, minimizing pH changes, by dissolution of soil minerals, particularly calcium carbonate. Limestone (calcium carbonate, $CaCO_3$) is particularly susceptible to dissolution by low pH waters. Limestone caves are formed when low pH groundwaters move through limestone deposits and dissolve the limestone minerals.

RULES OF THUMB

1. Unpolluted rainwater is acidic, pH 5.7, because of dissolved CO_2 from the atmosphere. Rainwater appears to become more acidic as atmospheric CO_2 levels increase, mainly due to increasing burning of fossil fuels.
2. Acid rain has lower pH values, reaching pH 2.0 or lower, because of dissolved sulfuric, nitric, and hydrochloric acids which result mainly from industrial air emissions.
3. Dissolved carbonate species, $CO_2(aq)$ (equivalent to H_2CO_3), HCO_3^-, and CO_3^{2-}, are present in any natural water system near the surface of the earth. The relative proportions depend on pH.
4. At pH values between 7.0 and 10.0, bicarbonate is the dominant DIC species in water. Between pH 7.8 and 9.2, bicarbonate is close to 100%; carbonate and dissolved CO_2 concentrations are essentially zero.
5. In subsurface soil pore space, oxygen is depleted and CO_2 is increased, compared to the atmosphere. Oxygen typically decreases from 21% in atmospheric air to 15% or less in soil pore space air, and CO_2 typically increases from ~360 ppmv in atmospheric air to between 1,000 and 100,000 ppmv in soil pore space air. Thus, unpolluted groundwaters tend to be more acidic than unpolluted surface waters because of higher dissolved concentrations of CO_2.

3.5   ACIDITY AND ALKALINITY

3.5.1   BACKGROUND

The alkalinity of water is its acid-neutralizing capacity. The acidity of water is its base-neutralizing capacity. Both parameters are related to the buffering capacity of water (the ability to resist changes in pH when an acid or base is added). Water with high alkalinity can neutralize a large quantity of acid without large changes in pH; on the other hand, water with high acidity can neutralize a large quantity of base without large changes in pH.

3.5.2   ACIDITY

Acidity is determined by measuring how much standard base must be added to raise the pH to a specified value. Acidity is the net effect of the presence of several constituents, including dissolved CO_2, dissolved multivalent metal ions, strong mineral acids such as sulfuric, nitric, and hydrochloric acids, and weak organic acids such as acetic acid. Dissolved CO_2 is the main source of acidity in unpolluted waters. Acidity from sources other than dissolved CO_2 is not commonly encountered in unpolluted natural waters and is often an indicator of pollution.

Titrating an acidic water sample with base to pH 8.3 measures phenolphthalein* acidity or total acidity. Total acidity measures the neutralizing effects of essentially all the acid species present, both strong and weak.

Titrating with base to pH 3.7 measures methyl orange* acidity. Methyl orange acidity primarily measures acidity due to the presence of strong mineral acids, such as sulfuric (H_2SO_4), hydrochloric (HCl), and nitric (HNO_3) acids.

3.5.3   ALKALINITY

In natural waters that are not highly polluted, alkalinity is more commonly found than acidity. Alkalinity is often a good indicator of the total DIC (bicarbonate and carbonate anions) present. All unpolluted natural waters can be expected to have some degree of alkalinity. Since all natural waters contain dissolved CO_2, they all will have some alkalinity contributed by carbonate species—unless acidic pollutants have consumed the alkalinity. It is not unusual for alkalinity to range from 0 to 750 mg/L as $CaCO_3$. For surface waters, alkalinity levels less than 30 mg/L are considered low, and levels greater than 250 mg/L are considered high. Average values for rivers are around 100–150 mg/L. Alkalinity in environmental waters is beneficial because it minimizes pH changes, reduces the toxicity of many metals by forming complexes with them, and provides nutrient carbon for aquatic plants.

Alkalinity is determined by measuring how much standard acid must be added to a given amount of water to lower the pH to a specified value. Like acidity, alkalinity is the net effect of the presence of several constituents, but the most important are the bicarbonate (HCO_3^-), carbonate (CO_3^{2-}), phosphate (PO_4^{3-}), and hydroxyl (OH^-) anions. Alkalinity is often taken as an indicator for the concentration of these

---

* Phenolphthalein and methyl orange are pH-indicator dyes that change color at pH 8.3 and 3.7, respectively.

constituents. There are other, usually minor, contributors to alkalinity, such as ammonia, phosphates, borates, silicates, and other basic substances.

Titrating with acid a water sample having pH greater than 8.3 down to pH 8.3 measures phenolphthalein alkalinity. Phenolphthalein alkalinity primarily measures the amount of carbonate ion (CO_3^{2-}) present. Titrating with acid to pH 3.7 measures methyl orange alkalinity or total alkalinity. Total alkalinity measures the neutralizing effects of essentially all the bases present.

Because alkalinity is a property caused by several constituents, some convention must be used for reporting it quantitatively as a concentration. The usual convention is to express alkalinity as the ppm or mg/L of calcium carbonate ($CaCO_3$) that would produce the same alkalinity as measured in the sample. This is done by calculating how much $CaCO_3$ would be neutralized by the same amount of acid as was used in titrating the water sample when measuring either phenolphthalein or methyl orange alkalinity. Whether it is present or not, $CaCO_3$ is used as a proxy for all the base species that are actually present in the water. The alkalinity value is equivalent to the mg/L of $CaCO_3$ that would neutralize the same amount of acid as does the actual water sample and is written as, for example, alkalinity = 1200 mg/L as $CaCO_3$.

3.5.4 IMPORTANCE OF ALKALINITY

Alkalinity is important to fish and other aquatic life because it buffers both natural- and human-induced pH changes. The chemical species that cause alkalinity, such as carbonate, bicarbonate, hydroxyl, and phosphate ions, can form chemical complexes with many toxic heavy metal ions, often reducing their toxicity. Water with high alkalinity generally has a high concentration of DIC (in the forms of HCO_3^- and CO_3^{2-}), which can be converted to biomass by photosynthesis. A minimum alkalinity of 20 mg/L as $CaCO_3$ is recommended for environmental waters and levels between 25 and 400 mg/L are generally beneficial for aquatic life. More productive waterfowl habitats correlate with increased alkalinity above 25 mg/L as $CaCO_3$. A range between 100 and 250 mg/L is considered normal for surface waters.

3.5.5 WATER QUALITY CRITERIA AND STANDARDS FOR ALKALINITY

Naturally occurring levels of alkalinity reaching at least 400 mg/L as $CaCO_3$ are not considered a health hazard. An upper limit of 500 mg/L is considered safe for livestock. EPA guidelines recommend a minimum alkalinity of 20 mg/L as $CaCO_3$, and that natural background alkalinity is not reduced by more than 25% by any discharge. For waters where the natural level is less than 20 mg/L, alkalinity should not be further reduced. Changes from natural alkalinity levels should be kept to a minimum. The volume of sample required for alkalinity analysis is 100 mL.

RULES OF THUMB

1. Alkalinity is the mg/L of $CaCO_3$ that would neutralize the same amount of acid as does the actual water sample.

(Continued)

RULES OF THUMB (Continued)

2. Phenolphthalein alkalinity (titration with acid to pH 8.3) measures the amount of carbonate ion (CO_3^{2-}) present.
3. Total or methyl orange alkalinity (titration with acid to pH 3.7) measures the neutralizing effects of essentially all the bases present.
4. Surface- and groundwaters draining carbonate mineral formations become more alkaline due to dissolved minerals.
5. High alkalinity can partially mitigate the toxic effects of heavy metals to aquatic life.
6. Alkalinity greater than 25 mg/L $CaCO_3$ is beneficial to water quality.
7. Surface waters without carbonate buffering may be more acidic than pH 5.7 (the value established by equilibration of dissolved CO_2 with CO_2 in the atmosphere) because of water reactions with metals and organic substances, biochemical reactions, and acid rain.

3.5.6 CALCULATING ALKALINITY

Although alkalinity is usually determined by titration, the contribution from carbonate species (carbonate alkalinity) is readily calculated if the total carbonate (bicarbonate plus carbonate) concentration is known. If only the bicarbonate or carbonate concentration is known, the pH can be used with Figure 3.2 to estimate the total carbonate concentration, as in Example 3. Carbonate alkalinity is equal to the sum of the concentrations of bicarbonate and carbonate ions, expressed as the equivalent concentration of $CaCO_3$.

EXAMPLE 3

A groundwater sample contains 300 mg/L of bicarbonate at pH 10.0. The carbonate concentration was not reported. Calculate the total carbonate alkalinity as $CaCO_3$.

Answer:

1. Use the measured values of bicarbonate and pH, with Figure 3.2, to determine the carbonate concentration. At pH 10.0, total carbonate is about 73% bicarbonate ion and 27% carbonate ion. Although these percentages are related to moles/L rather than mg/L, the molecular weights of bicarbonate and carbonate ions differ by only about 1.7%; therefore, mg/L can be used in the calculation without significant error.

$$\text{Total carbonate } \left(HCO_3^- + CO_3^{2-} \right) = \frac{300 \text{ mg/L}}{0.73} = 411 \text{ mg/L}$$

$CO_3^{2-} = 0.27 \times 411 = 111$ mg/L, or alternatively, $411 - 300 = 111$ mg/L.

2. Determine the equivalent weights of HCO_3^-, CO_3^{2-}, and $CaCO_3$ (or use Table 1.1, page 18).

$$\text{Eq. wt.} = \frac{\text{Molecular or atomic weight}}{\text{Magnitude of ionic charge or oxidation number}}$$

$$\text{Eq. wt. of HCO}_3^- = \frac{61.0}{1} = 61.0 \text{ g/eq}$$

$$\text{Eq. wt. of CO}_3^{2-} = \frac{60.0}{2} = 30.0 \text{ g/eq}$$

$$\text{Eq. wt. of CaCO}_3 = \frac{100.1}{2} = 50.0 \text{ g/eq}$$

3. Determine the multiplying factors to obtain the equivalent concentrations of $CaCO_3$.

$$\text{Multiplying factor for HCO}_3^- \text{ as CaCO}_3 = \frac{\text{Eq. wt. of CaCO}_3}{\text{Eq. wt. of HCO}_3^-} = \frac{50.0 \text{ g/eq}}{61.0 \text{ g/eq}} = 0.820$$

$$\text{Multiplying factor for CO}_3^{2-} \text{ as CaCO}_3 = \frac{\text{Eq. wt. of CaCO}_3}{\text{Eq. wt. of CO}_3^{2-}} = \frac{50.0 \text{ g/eq}}{30.0 \text{ g/eq}} = 1.667$$

4. Use the multiplying factors and concentrations to calculate the carbonate alkalinity, expressed as mg/L of $CaCO_3$.

$$\text{Carbonate alk. (as CaCO}_3) = 0.820\left(\text{HCO}_3^-, \text{mg/L}\right) + 1.667\left(\text{CO}_3^{2-}, \text{mg/L}\right) \qquad (3.17)$$

$$\text{Carbonate alk.} = 0.820(300 \text{ mg/L}) + 1.667(111 \text{ mg/L}) = 431 \text{ mg/L CaCO}_3$$

Equation 3.17 may be used to calculate carbonate alkalinity whenever pH and the total carbonate concentration are known.

3.5.7 Calculating Changes in Alkalinity, Carbonate, and pH

A detailed calculation of how pH, total carbonate, and total alkalinity are related to one another is moderately complicated because of the three simultaneous carbonate equilibria reactions, Equations 3.13 through 3.15. However, the relations can be conveniently plotted on a total alkalinity/pH/total carbonate graph, also called a Deffeyes diagram, or capacity diagram (see Figures 3.3 and 3.4). Details of the construction of the diagrams may be found in Stumm and Morgan (1996) and Deffeyes (1965).

In a total alkalinity/pH/total carbonate graph, shown in Figure 3.3, a vertical line represents adding strong base or acid without changing the total carbonate. The added base or acid changes the pH and, therefore, shifts the carbonate equilibrium, but does not add or remove any carbonate. The amount of strong base or acid in meq/L equals the vertical distance on the graph. You can see from Figure 3.3 that if the total carbonate is small, the system is poorly buffered; so a little base or acid makes large changes in pH. If total carbonate is large, the system buffering capacity is similarly large and it takes much more base or acid for the same pH change.

A horizontal line represents changing total carbonate, generally by adding or losing CO_2, without changing alkalinity. For alkalinity to remain constant when total carbonate changes, the pH must also change. Changes caused by adding bicarbonate or from simple dilution are indicated in the figure.

Figure 3.4 is a total acidity/pH/total carbonate graph. Note that changes in composition, caused by adding or removing CO_2 and carbonate, are indicated by

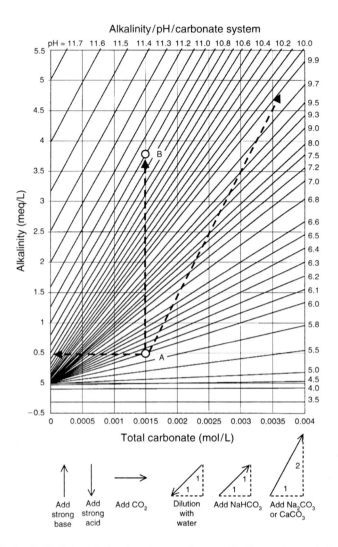

FIGURE 3.3 Total alkalinity/pH/total carbonate diagram (Deffeyes diagram). Relationships among total alkalinity, pH, and total carbonate are shown here. If any two of these quantities are known, the third may be determined from the plot. The composition changes indicated refer to Example 4.

different movement vectors in the acidity and alkalinity graphs. Examples below illustrate the uses of the diagrams.

EXAMPLE 4

Designers of a wastewater treatment facility for a meat rendering plant planned to control ammonia concentrations in the wastewater by raising its pH to 11, in order to convert about 90% of the ammonia to the volatile form. The wastewater would then

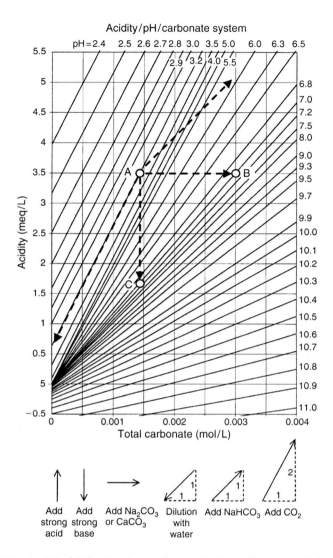

FIGURE 3.4 Total acidity/pH/total carbonate diagram (Deffeyes diagram). Relationships among total acidity, pH, and total carbonate are shown here. If any two of these quantities are known, the third may be determined from the plot. The composition changes indicated refer to Example 5.

be passed through an air-stripping tower to transfer the ammonia to the atmosphere. Average initial conditions for alkalinity and pH in the wastewater were expected to be about 0.5 meq/L and 6.0, respectively.

In the preliminary design plan, four options for increasing the pH were considered:

1. Raise the pH by adding NaOH, a strong base.
2. Raise the pH by adding calcium carbonate, $CaCO_3$, in the form of limestone.
3. Raise the pH by adding sodium bicarbonate, $NaHCO_3$.
4. Raise the pH by removing CO_2, perhaps by aeration.

ADDITION OF NaOH

In Figure 3.3, we find that the intersection of pH 6.0 and alkalinity 0.5 meq/L occurs at total carbonate 0.0015 mol/L (point A). Assuming that no CO_2 is lost to the atmosphere, addition of the strong base NaOH represents a vertical displacement upward from point A. Enough NaOH must be added to intersect with the pH 11.0 contour at point B. In Figure 3.3, the vertical line between points A and B has a length of about 3.3 meq/L. Thus, the quantity of NaOH needed to change the pH from 6.0 to 11.0 is 3.3 meq/L (132 mg/L).

ADDITION OF CaCO₃

Addition of $CaCO_3$ is represented by a line of slope +2 from point A. The carbonate addition line rises by 2 meq/L of alkalinity for each increase of 1 mol/L of total carbonate (because 1 mole of carbonate = 2 equivalents). Note, in Figure 3.3, that the slope of the pH 11.0 contour is nearly 2. The $CaCO_3$ addition vector and the pH 11.0 contour are nearly parallel. Therefore, a very large quantity of $CaCO_3$ would be needed, making this method impractical.

ADDITION OF NaHCO₃

Addition of $NaHCO_3$ is represented by a line of slope +1 from point A. Although this vector is not shown in Figure 3.3, it is evident it cannot cross the pH 11 contour. Therefore, this method will not work.

REMOVING CO₂

Removal of CO_2 is represented by a horizontal displacement to the left. Loss or gain of CO_2 does not affect the alkalinity. Note that if CO_2 is removed, total carbonate is decreased correspondingly. However, pH and $[OH^-]$ also increase correspondingly, resulting in no net change in alkalinity. We see from Figure 3.3 that removal of CO_2 to the point of zero total carbonate cannot achieve pH 11.0. Therefore, this method also will not work.

Of the four potential methods considered for raising the wastewater pH to 11.0, only addition of NaOH is useful.

EXAMPLE 5

A large excavation at an abandoned mine site has filled with water. Because pyrite minerals were exposed in the pit, the water is acidic with pH 3.2. The acidity was measured at 3.5 meq/L. Because the pit overflows into a stream during heavy rains, managers of the site must meet the conditions of a discharge permit, which include a requirement that pH of the overflow water be between 6.0 and 9.0. The site manager decides to treat the water to pH 7.0 to provide a safety margin. Use Figure 3.4 to evaluate the same options for raising the pH as considered in Example 4.

ADDITION OF NaOH

In Figure 3.4, we find that the intersection of pH 3.2 and acidity 3.5 meq/L occurs at about total carbonate 0.0014 mol/L, point A. Assuming that no CO_2 is lost to the atmosphere, addition of the strong base NaOH represents a vertical displacement downward from point A to point C. Enough NaOH must be added to intersect with the pH 7.0 contour. The vertical line between points A and C has a length of about 1.8 meq/L. Thus, the quantity of NaOH needed to change the pH from 3.0 to 7.0 is 1.8 meq/L (72 mg/L).

ADDITION OF CaCO₃

In the acidity diagram, addition of $CaCO_3$ is represented by a horizontal line to the right. In Figure 3.4, the $CaCO_3$ addition line intersects the pH 7.0 contour at point B, where total carbonate is 0.0030 mol/L. Therefore, the quantity of $CaCO_3$ required to reach pH 7.0 is $0.0030 - 0.0014 = 0.0016$ mol/L (160 mg/L).

ADDITION OF NaHCO₃

The addition of $NaHCO_3$ is represented by a line of slope $+1$ (the vector upward to the right from point A in Figure 3.4). Note that the slope of the pH 7.0 contour is just a little greater than $+1$. The $NaHCO_3$ addition vector and the pH 7.0 contour are nearly parallel. Therefore, a very large quantity of $NaHCO_3$ would be needed, making this method impractical.

REMOVING CO₂

In the acidity diagram, the removal of CO_2 is represented by a line downward to the left with slope $+2$. We see from Figure 3.4 that removal of CO_2 to the point of zero total carbonate cannot achieve pH 7.0. Therefore, this method will not work.

Of the four potential methods considered for raising the wastewater pH to 7.0, addition of either NaOH or $CaCO_3$ will work. The choice will be based on other considerations, such as costs or availability.

EXAMPLE 6

Figures 3.3 and 3.4 are useful for finding properties of mixed waters. As a simple example, consider an industry that requires its process water to be at pH 7.0. It uses two different water sources for filling a storage tank for its process water. Let water A be the remaining water in the half-full tank that needs replenishing. Water A has a pH 7.0 and an alkalinity 2.5 meq/L. Water B, used to fill the tank, has pH 8.0 and alkalinity 3.5 meq/L.

a. What is the pH of the mixture after Water B fills the tank? Assume no CO_2 is lost to the atmosphere.
b. What would be a reasonable way to bring the mixture back to pH 7.0?

Answer:

a. From Figure 3.3, for water A, at the intersection of pH 7.0 and alkalinity 2.5 meq/L, total carbonate is 0.0030 mol/L, and for water B, at the intersection of pH 8.0 and alkalinity 3.5, total carbonate is 0.0036. Because the volumes of waters A and B are equal and no CO_2 is assumed to be lost, the mixture will have alkalinity and total carbonate values that are the average of waters A and B, i.e., for the mixture, alkalinity is 3.0 meq/L and total carbonate is 0.0033 meq/L. From Figure 3.3, the mixture's pH is very close to 7.3.

b. Adding 0.2 meq/L of strong acid will lower the pH to 7.0.

Note: If unequal volumes of water are mixed, as is usually the case, the only change needed in the above approach is to use weighted averages of alkalinity and total carbon for the mixture.

3.6 HARDNESS

3.6.1 BACKGROUND

Originally, water hardness was a measure of the ability of water to precipitate soap. It was measured by the amount of soap needed for adequate lathering and served as an indicator of the rate of scale formation in hot water heaters and boilers. Soap is precipitated as a gray "bathtub ring" deposit mainly by reacting with the calcium and magnesium cations (Ca^{2+} and Mg^{2+}) present, although other polyvalent cations may play a minor role.

Hardness has some similarities to alkalinity. Like alkalinity, hardness is a property of water that is not attributable to a single constituent and, therefore, some convention must be adopted to express hardness quantitatively as a concentration. As with alkalinity, hardness is usually expressed as an equivalent concentration of $CaCO_3$. However, hardness is a property of cations (Ca^{2+} and Mg^{2+}), while alkalinity is a property of anions (HCO_3^- and CO_3^{2-}).

3.6.2 CALCULATING HARDNESS

Current practice is to define total hardness as the sum of the calcium and magnesium ion concentrations in mg/L, both expressed as calcium carbonate. Hardness is usually calculated from separate measurements of calcium and magnesium, rather than measured directly by colorimetric titration.

Calcium and magnesium ion concentrations are converted to equivalent concentrations of $CaCO_3$ as follows:

1. Find the equivalent weights of Ca^{2+}, Mg^{2+}, and $CaCO_3$ (also, see Table 1.1, page 18).

$$\text{Eq. wt.} = \frac{\text{Molecular or atomic weight}}{\text{Magnitude of ionic charge or oxidation number}}$$

$$\text{Eq. wt. of } Ca^{2+} = \frac{40.08 \text{ g/mol}}{2 \text{ eq/mol}} = 20.04 \text{ g/eq}$$

$$\text{Eq. wt. of } Mg^{2+} = \frac{24.31 \text{ g/mol}}{2 \text{ eq/mol}} = 12.15 \text{ g/eq}$$

$$\text{Eq. wt. of } CaCO_3 = \frac{100.09 \text{ g/mol}}{2 \text{ eq/mol}} = 50.04 \text{ g/eq}$$

2. Determine the multiplying factors to obtain the equivalent concentration of $CaCO_3$.

$$\text{Multiplying factor for } Ca^{2+} \text{ as } CaCO_3 = \frac{\text{Eq. wt. of } CaCO_3}{\text{Eq. wt. of } Ca^{2+}} = \frac{50.04 \text{ g/eq}}{20.04 \text{ g/eq}} = 2.497$$

$$\text{Multiplying factor of } Mg^{2+} \text{ as } CaCO_3 = \frac{\text{Eq. wt. of } CaCO_3}{\text{Eq. wt. of } Mg^{2+}} = \frac{50.04 \text{ g/eq}}{12.15 \text{ g/eq}} = 4.119$$

3. Calculate the total hardness with Equation 3.18.

$$\text{Total hardness (as } CaCO_3) = 2.497 \, (Ca^{2+}, \text{mg/L}) + 4.118 \, (Mg^{2+}, \text{mg/L}) \quad (3.18)$$

Equation 3.18 may be used to calculate hardness whenever Ca^{2+} and Mg^{2+} concentrations are known.

EXAMPLE 7

Calculate the total hardness as $CaCO_3$ of a water sample in which $Ca^{2+} = 98$ mg/L and $Mg^{2+} = 22$ mg/L.

Answer:

Use Equation 3.18

$$\text{Total hardness} = 2.497(98 \text{ mg/L}) + 4.118(22 \text{ mg/L}) = 335 \text{ mg/L } CaCO_3$$

Both alkalinity and hardness are expressed in terms of an equivalent concentration of calcium carbonate. As noted before, alkalinity results from reactions of the anions, CO_3^{2-} and HCO_3^-, whereas hardness results from reactions of the cations, Ca^{2+} and Mg^{2+}.

It is possible for hardness as $CaCO_3$ to exceed the total alkalinity as $CaCO_3$. When this occurs, the portion of the hardness that is equal to the alkalinity is referred to as "carbonate hardness or temporary hardness," and the amount in excess of alkalinity is referred to as "non-carbonate hardness or permanent hardness."

3.6.3 IMPORTANCE OF HARDNESS

Hardness is sometimes useful as an indicator proportionate to the total dissolved solids (TDS) present, since Ca^{2+}, Mg^{2+}, CO_3^{2-}, and HCO_3^- often represent the largest part of the TDS. No human health effects due to hardness have been proven; however, an inverse relation with cardiovascular disease has been reported. Higher levels of drinking water hardness correlate with lower incidence of cardiovascular disease. High levels of water hardness may limit the growth of fish; on the other hand, low hardness (soft water) may increase fish sensitivity to toxic metals. In general, higher hardness is beneficial by reducing metal toxicity to fish. Aquatic life water quality standards for many metals are calculated by using an equation that includes water hardness as a variable.

The main advantages in limiting the level of hardness (by softening water) are economical: less soap requirements in domestic and industrial cleaning, and less scale formation in pipes and boilers. Water treatment by reverse osmosis (RO) often requires a water-softening pretreatment to prevent scale formation on RO membranes. Increased use of detergents, which do not form precipitates with Ca^{2+} and Mg^{2+}, has lessened the importance of hardness for soap consumption. On the other hand, a drawback to soft water is that it is more "corrosive" or "aggressive" than hard water. In this context, "corrosive" means that soft water more readily dissolves metal from plumbing systems than does hard water. Thus, in plumbing systems where brass, copper, galvanized iron, or lead solders are present, a soft water system will carry higher levels of dissolved copper, zinc, lead, and iron, than will a hard water system.

RULES OF THUMB

1. The higher the hardness, the more tolerant are many surface water metal standards for aquatic life.
2. Hardness above 100 mg/L can cause significant scale deposits to form in boilers.
3. The softer the water, the greater the tendency to dissolve metals from the pipes of water distribution systems.
4. An ideal quality goal for total hardness is about 70–90 mg/L. Municipal treatment sometimes allows up to 150 mg/L of total hardness in order to reduce chemical costs and sludge production from precipitation of calcium and magnesium carbonates.

Water will be "hard" wherever groundwater passes through calcium and magnesium carbonate mineral deposits. Such deposits are very widespread and hard to moderately hard groundwater is more common than soft groundwater. Very hard groundwater occurs frequently. Calcium and magnesium carbonates are the most common carbonate minerals and are the main sources of hard water. A geologic map showing the distribution of carbonate minerals serves also as an approximate map of the distribution of hard groundwater. The most common sources of soft water are where rainwater is used directly, or where surface waters are fed more by precipitation than by groundwater.

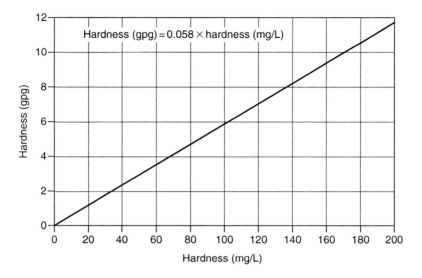

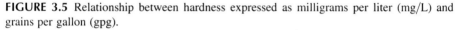

FIGURE 3.5 Relationship between hardness expressed as milligrams per liter (mg/L) and grains per gallon (gpg).

RULES OF THUMB

Degree of Hardness	mg CaCO₃/L	Effects
Soft	<75	May increase toxicity to fish of dissolved metals
		May increase corrosivity of water to metals
		No scale deposits
		Efficient use of soap
Moderately hard	75–120	Not objectionable for most purposes
		Requires somewhat more soap for cleaning
		Above 100 mg/L may deposit significant scale in boilers
Hard	120–200	Considerable scale buildup and staining.
		Generally softened if >200 mg/L
Very hard	>200	Requires softening for household or commercial use

In industrial usage, hardness is sometimes expressed as grains per gallon (gpg). The conversion between gpg and mg/L is shown in Figure 3.5.

3.7 DISSOLVED OXYGEN

3.7.1 BACKGROUND

Sufficient dissolved oxygen (DO) is important for high-quality water. DO is crucial for the survival of fish and most other aquatic life forms. It oxidizes many sources of objectionable tastes and odors. Oxygen becomes dissolved in surface waters by diffusion from the atmosphere and from aquatic–plant photosynthesis.

On average, most oxygen dissolves into water from the atmosphere; only a little net DO is produced by aquatic–plant photosynthesis. Although water plants produce oxygen during the day, they consume oxygen at night as an energy source. When they die and decay, dead plant matter serves as an energy source for microbes, which consume additional oxygen. The net change in DO is small during the life cycle of aquatic plants.

RULES OF THUMB

1. The solubility of oxygen in water decreases as water temperature increases.
2. Saturation concentration of O_2 in water at sea level is 14.7 mg/L (ppm) at 0°C:

$$8.3 \text{ mg/L (ppm) at } 25°C$$

$$7.0 \text{ mg/L (ppm) at } 35°C$$

Dissolved oxygen is consumed by the degradation (oxidation) of organic matter in water. Because the concentration of DO is never very large, oxygen-depleting processes can rapidly reduce it to near zero in the absence of efficient aeration mechanisms. Fish need at least 5–6 ppm DO to grow and thrive. They stop feeding if the level drops to around 3–4 ppm and die if DO falls to 1 ppm. Many fish kills are not caused by the direct toxicity of contaminants but instead by a deficiency of oxygen caused by the biodegradation of organic contaminants (Table 3.2).

Typical state aquatic life standards for DO are

- 7.0 ppm for cold water spawning periods
- 6.0 ppm for class 1 cold water biota
- 5.0 ppm for class 1 warm water biota

TABLE 3.2

Dissolved Oxygen and Water Quality

Water Quality	Dissolved Oxygen (mg/L)
Good	Above 8.0
Slightly polluted	6.5–8.0
Moderately polluted	4.5–6.5
Heavily polluted	4.0–4.5
Severely polluted	Below 4.0

3.8 BIOLOGICAL OXYGEN DEMAND AND CHEMICAL OXYGEN DEMAND

3.8.1 BACKGROUND

Biological oxygen demand (BOD) refers to the amount of oxygen potentially consumed if all the biologically degradable organic matter in a given volume of water were biodegraded. BOD is an indicator of the potential for a water body to become depleted in oxygen and possibly become anaerobic because of biodegradation. BOD measurements do not take into account reoxygenation of water by naturally occurring diffusion from the atmosphere or mechanical aeration. Water with a high BOD and an active microbial population can become depleted in oxygen and may not support aquatic life, unless there is a means for rapidly replenishing DO.

Chemical oxygen demand (COD) refers to the amount of oxygen consumed when all the organic matter in a given volume of water is chemically oxidized to CO_2 and H_2O by a strong chemical oxidant, such as permanganate or dichromate. COD is sometimes used as a measure of general pollution. For example, in an industrial area built on fill dirt, COD in the groundwater might be used as an indicator of organic materials leached from the fill material. Leachate from landfills often has high levels of COD.

BOD is a subset of COD. The COD analysis oxidizes organic matter that is both chemically and biologically oxidizable. If a reliable correlation between COD and BOD can be established at a particular site, the simpler COD test may be used in place of the more complicated BOD analysis.

3.8.2 BOD$_5$

BOD_5 refers to a particular empirical test, accepted as a standard, in which a specified volume of sample water is seeded with bacteria and nutrients (nitrogen and phosphorus) and then incubated for 5 days at 20°C in the dark. BOD_5 is measured as the decrease in DO (in mg/L) after 5 days of incubation. The BOD_5 test originated in England, where any river contaminant not decomposed within 5 days will reach the ocean.

Water surface turbulence helps to dissolve oxygen from the atmosphere by increasing the water surface area. A BOD_5 of 5 mg/L in a slow-moving stream might be enough to produce anaerobic conditions, while a turbulent mountain stream might be able to assimilate a BOD_5 of 50 mg/L without appreciable oxygen depletion (Figure 3.6).

3.8.3 BOD CALCULATION

EXAMPLE 8

When 1 L water sample is collected for analysis, an insect weighing 50.0 mg is accidentally trapped in the bottle. The initial DO is 7.0 mg/L. Assume that 15% of the insect's fresh weight is readily biodegradable and has the approximate unit formula CH_2O^*. Also,

---

* Organic biomass contains carbon, hydrogen, and oxygen atoms approximately in the ratio of 1:2:1, so that CH_2O serves as a convenient "unit molecule" of organic matter.

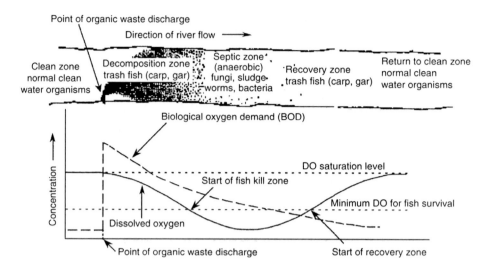

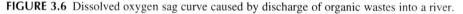

FIGURE 3.6 Dissolved oxygen sag curve caused by discharge of organic wastes into a river.

assume that microbes are present that will metabolize the insect. If the laboratory does not analyze the sample until biodegradation is complete, what DO will they measure?

Answer:

The chemical reaction for oxidation of organic matter is

$$CH_2O + O_2 \rightarrow CO_2 + H_2O \qquad (3.19)$$

Equation 3.19 shows that 1 mole of O_2 oxidizes 1 mole of CH_2O. Therefore, the moles of CH_2O biodegraded will equal the moles of O_2 consumed during biodegradation. First, find the moles of O_2 initially present and the biodegradable moles of organic matter in the insect.

$$\text{Moles } O_2 \text{ initially present as DO} = \frac{7.0 \times 10^{-3} \text{ g/L}}{32 \text{ g/mol}} = 2.19 \times 10^{-4} \text{ mol/L}$$

Initial weight of insect organic matter (CH_2O) present = 50.0 mg
Weight of insect matter that is biodegradable = $0.15 \times (50.0 \text{ mg}) = 7.50$ mg
Molecular weight of $CH_2O = 12 + 2 + 16 = 30$ g/mol

$$\text{Moles biodegradable organic matter} = \frac{7.50 \times 10^{-3} \text{g/L}}{30 \text{ g/mol}} = 2.50 \times 10^{-4} \text{ mol}$$

The moles of biodegradable organic matter exceed the moles of DO in the 1 L sample. Therefore, biodegradation of the insect will consume all of the DO in the sample and there will be some biodegradable organic matter left over.

Moles of organic matter biodegraded = moles of O_2 consumed = 2.19×10^{-4} mol/L
Mass of organic matter biodegraded = $(2.19 \times 10^{-4} \text{ mol/L}) \times (30 \text{ g/mol}) = 6.57 \times 10^{-3}$ g
Mass of biodegradable organic matter remaining = $7.50 - 6.57$ mg = 0.093 mg

Since all of the DO has been consumed the laboratory will find the water anaerobic, with no DO. There will be $(50.0 - 6.57 \text{ mg}) = 43.4$ mg of insect remaining, of which 0.093 mg is still biodegradable. Note that the concentration of BOD degraded (6.6 mg/L) is similar to the concentration of DO consumed (7.0 mg/L), being about 6% less.

RULE OF THUMB

The mg/L of BOD biodegraded in a BOD_5 test $\approx$ mg/L of O_2 consumed by microbes in the sample.

Note: This convenient approximation is valid because, by Equation 3.19, the moles of O_2 consumed in a BOD_5 test equals the moles of BOD consumed, and the MW of O_2 ($MW_{O_2} = 32$ g/mole) is close to that of the "unit molecule" of organic matter (CH_2O, $MW_{CH_2O} = 30$ g/mole). This means that the weight concentration ratio of BOD to DO consumed in a BOD_5 test is 30/32, or 0.9375, very close to 1/1.

3.8.4 COD CALCULATION

EXAMPLE 9

COD levels of 60 mg/L were measured in groundwater. It is suspected that fuel contamination is the main cause. What concentration of hydrocarbons from fuel is necessary to account for all of the COD observed?

Answer:

For simplicity, assume fuel hydrocarbons to have an average unit formula of CH_2. The oxidation reaction is

$$CH_2 + 1.5 \ O_2 \rightarrow CO_2 + H_2O \tag{3.20}$$

For each carbon atom in the fuel, 1.5 oxygen molecules are consumed.

Weight of 1 mole of $CH_2 = 12 + 2 = 14$ g
Weight of 1.5 mole of $O_2 = 1.5 \times 32 = 48$ g
Weight ratio of oxygen to fuel consumed during oxidation is $\dfrac{48 \text{ g}}{14 \text{ g}} = 3.4$
A COD of 60 mg/L requires $\dfrac{60 \text{ mg/L}}{3.4} = 18$ mg/L fuel hydrocarbons.

If dissolved fuel hydrocarbons in the groundwater are around 18 mg/L or greater, the fuel alone could account for all the measured COD.

If dissolved fuel hydrocarbons in the groundwater are much less than 18 mg/L, then fuels could account for only a part of the COD; other organic substances, such as pesticides, fertilizers, solvents, PCBs, etc., must account for the rest.

3.9 NITROGEN: AMMONIA, NITRITE, AND NITRATE

3.9.1 BACKGROUND

Nitrogen compounds of greatest interest to water quality are those that are biologically available as nutrients to plants or exhibit toxicity to humans or aquatic life.

Atmospheric nitrogen (N_2) is the primary source of all nitrogen species, but it is not directly available to plants as a nutrient because the N≡N triple bond is too strong to be broken by photosynthesis. Atmospheric nitrogen must be converted to other nitrogen compounds before it can become available as a plant nutrient.

The conversion of atmospheric nitrogen to other chemical forms is called nitrogen fixation and is accomplished by certain bacteria that are present in water, soil, and root nodules of alfalfa, clover, peas, beans, and other legumes. Atmospheric lightning is another significant source of fixed nitrogen because the high temperatures generated in lightning strikes are sufficient to break N_2 and O_2 bonds, making possible the formation of nitrogen oxides. Nitrogen oxides created within lightning bolts dissolve in rainwater and are absorbed by plant roots, thus entering the nitrogen nutrient subcycles (see Figure 3.7).

The rate at which atmospheric nitrogen can enter the nitrogen cycle by natural processes is too low to support today's intensive agricultural production. The shortage of fixed nitrogen must be made up with fertilizers containing nitrogen fixed by industrial processes, which are dependent on petroleum fuel. Modern large-scale farming has been called a method for converting petroleum into food.

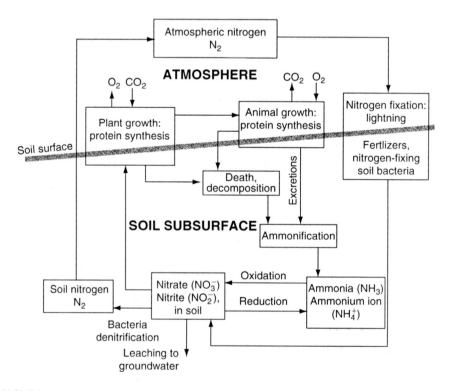

FIGURE 3.7 Nitrogen cycle.

3.9.2 NITROGEN CYCLE

In the nitrogen cycle (Figure 3.7) plants take up ammonia and nitrogen oxides dissolved in soil pore water and convert them into proteins, DNA, and other nitrogen compounds. Animals get their nitrogen by eating plants or other plant-eating animals. Once in terrestrial ecosystems, nitrogen is recycled through repeated biological birth, growth, death, and decay steps. There is a continual and relatively small loss of fixed nitrogen when specialized soil bacteria convert fixed nitrogen back into nitrogen gas (denitrification), which is then released to the atmosphere, from which it can eventually reenter the nutrient subcycles again.

When nitrogen is circulating in the nutrient subcycles, it undergoes a series of reversible oxidation-reduction reactions that convert it from nitrogenous organic molecules, such as proteins, to ammonia (NH_3), nitrite (NO_2^-), and nitrate (NO_3^-). Ammonia is the first product in the oxidative decay of nitrogenous organic compounds. Further oxidation leads to nitrite and then to nitrate. Ammonia is naturally present in most surface- and wastewaters. Under aerobic conditions, ammonia is then oxidized to nitrites and nitrates, consuming dissolved oxygen (Equation 3.21).

$$\text{Organic N} \longrightarrow NH_3 \xrightarrow{\text{O}_2} NO_2^- \xrightarrow{\text{O}_2} NO_3^- \tag{3.21}$$

3.9.3 AMMONIA/AMMONIUM ION

In water, ammonia reacts as a base, raising the pH by generating OH^- ions (Equation 3.22).

$$NH_3 + H_2O \leftrightarrow NH_4^+ + OH^- \tag{3.22}$$

The equilibrium of Equation 3.22 depends on pH and temperature (see Figure 3.8). In a laboratory analysis, total ammonia ($NH_3 + NH_4^+$) is measured and the distribution between unionized ammonia (NH_3) and ionized ammonia (NH_4^+) is calculated from the knowledge of water pH and temperature at the sampling site. Since the unionized form is far more toxic to aquatic life than the ionized form, field measurements of water pH and temperature at the sampling site are very important.

The two forms of ammonia have different mobilities in the environment. Ionized ammonia is strongly adsorbed on mineral surfaces, where it is effectively immobilized. In contrast, unionized ammonia is only weakly adsorbed and is transported readily by water movement. If suspended sediment carrying sorbed NH_4^+ is carried by a stream into a zone with a higher pH, a portion will be converted to unionized NH_3, which can then desorb and become available to aquatic life forms as a toxic pollutant. Unionized ammonia is also volatile and a fraction of it is transported as a gas.

As discussed above, nitrogen passes through several different chemical forms in the nutrient subcycle. To allow quantities of these different forms to be directly compared with one another, analytical results often report their concentrations in terms of their nitrogen content. For example, 10.0 mg/L of unionized ammonia may

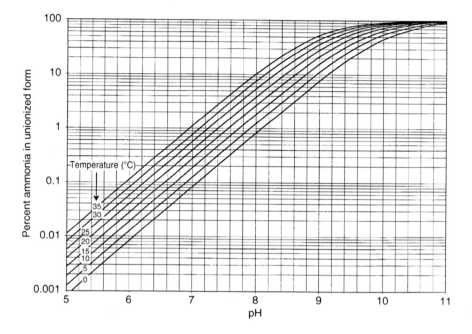

FIGURE 3.8 Percent unionized ammonia (NH_3) as a function of pH and temperature.

be reported as 8.23 mg/L NH_3–N (ammonia nitrogen); 10.0 mg/L of nitrate may be reported as 2.26 mg/L NO_3–N (nitrate nitrogen).

RULES OF THUMB

1. Ammonia toxicity increases with pH and temperature.
2. At 20°C and pH > 9.4, the equilibrium of Equation 3.22 is to the left, favoring NH_3, the toxic form.
3. At 20°C and pH < 9.4, the equilibrium of Equation 3.22 is to the right, favoring NH_4^+, the nontoxic form.
4. Temperature increase shifts the equilibrium to the left, favoring the NH_3 form.
5. NH_3 concentrations >0.5 mg NH_3–N/L cause significant toxicity to fish.
6. The unionized is volatile (air-strippable), and the ionized form is nonvolatile.

Changes in environmental conditions can cause an initially acceptable concentration of total ammonia to become unacceptable and in violation of a stream standard. For example, consider a wastewater treatment plant that discharges its effluent into a detention pond that, in turn, periodically releases its water into a stream. The treatment plant meets its discharge limit for unionized ammonia when its effluent is measured at the end of its discharge pipe. However, the detention pond that receives the effluent supports algal growth. In such a situation, it is common for algae to grow to a level that influences the pond's pH. During daytime photosynthesis, algae may remove enough dissolved CO_2 from the pond to raise the pH and shift the equilibrium of Equation 3.22 to the left far enough that the pond concentration of NH_3 becomes higher than the discharge permit limit. In this case, discharges from the pond could exceed the stream standard for unionized ammonia even though the total ammonia concentration is unchanged.

EXAMPLE 10

Ammonia is removed from an industrial wastewater stream by an air-stripping tower. To meet the effluent discharge limit of 5 ppm ammonia, the influent must be adjusted so that 60% of the total ammonia is in the volatile form. To what pH must the influent be adjusted if the wastewater in the stripping tower is at 10°C? Use Figure 3.8.

Answer:

In Figure 3.8, the 60% unionized ammonia gridline crosses the 10°C curve between pH 9.7 and 9.8. Thus, the influent must be adjusted to pH 9.8 or higher to meet the discharge limit.

3.9.4   WATER QUALITY CRITERIA AND STANDARDS FOR AMMONIA

Typical state standards for unionized ammonia (NH_3) are

- Aquatic life: Cold water biota $= 0.02$ mg/L NH_3–N, chronic; warm water biota $= 0.06$ mg/L NH_3–N, chronic; acute standard calculated from temperature and pH.
- Domestic water supply: 0.05 mg N/L total ($NH_3 + NH_4^+$), for a 30 day average.

3.9.5   NITRITE AND NITRATE

Ammonia and other nitrogenous materials in natural waters tend to be oxidized by aerobic bacteria, first to nitrite and then to nitrate. Therefore, all organic compounds containing nitrogen should be considered as potential nitrate sources. Organic nitrogen compounds enter the environment from wild animal and fish excretions, dead animals, human sewage, and livestock manure. Inorganic nitrates come primarily from manufactured fertilizers containing ammonium nitrate and potassium nitrate, and from nitrate-based explosives and rocket fuels.

In oxygenated waters, nitrite is rapidly oxidized to nitrate, so normally there is little nitrite present in surface waters. Both nitrite and nitrate are important nutrients for plants, but they are toxic to fish and humans at sufficiently high concentrations. Nitrates and nitrites are very soluble, do not adsorb readily to mineral and soil surfaces, and are very mobile in the environment. Consequently, where soil nitrate levels are high, contamination of groundwater by nitrate leaching is a serious problem. Unlike ammonia, nitrites and nitrates do not evaporate and remain in water until they are consumed by plants and microorganisms.

RULES OF THUMB

1. Unpolluted, oxygenated surface waters normally contain only trace amounts of nitrite.
2. Measurable nitrite concentrations in groundwater are more common because of low oxygen concentrations in the soil's subsurface.
3. Nitrate and nitrite leach readily from soils to surface and ground-waters.
4. High concentrations (>1–2 mg/L) of nitrate or nitrite in surface or groundwater generally indicate agricultural contamination from fertilizers and manure seepage.
5. Greater than 10 mg/L of nitrite and nitrate in drinking water is a human health hazard.

Drinking water standards for nitrate are strict because the nitrates can be reduced to nitrites in human saliva and in the intestinal tracts of infants during the first 6 months of life. Nitrite oxidizes iron in blood hemoglobin from ferrous iron (Fe^{2+}) to ferric iron (Fe^{3+}). The resulting compound, called methemoglobin, cannot carry oxygen. The resulting oxygen deficiency is called methemoglobinemia. It is especially dangerous in infants (blue baby syndrome) because of their small total blood volume.

3.9.6 Water Quality Criteria and Standards for Nitrate

Typical state water quality standards for nitrate (NO_3^-) are

- Agriculture MCLs: Nitrate, 100 mg/L NO_3–N; Nitrite, 10 mg/L NO_2–N (1 day average).
- Domestic water supply MCLs: Nitrate, 10 mg/L NO_3–N; Nitrite, 1.0 mg/L NO_2–N (1 day average).

EXAMPLE 11

COD Caused by Sodium Nitrite Disposal

A chemical company applied for a permit to dispose of 250,000 gal of water containing 500 mg/L of nitrite into a municipal sewer system. The manager of the municipal wastewater treatment plant had to determine whether this waste might be detrimental to the operation of his plant. Calculate the increase in COD in the sewer system that would be caused by the chemical company's proposed wastewater release.

Under oxidizing conditions that exist in the treatment plant, nitrite is oxidized to nitrate as follows:

$$2NO_2^- + O_2 \rightarrow 2NO_3^- \tag{3.23}$$

The consumption of oxygen shown in Equation 3.23 makes the use of nitrite compounds to deoxygenate water useful as a rust-inhibiting additive in boilers, heat exchangers, and storage tanks. When nitrite is added to a wastewater stream, it is the same as adding chemical oxygen demand. More oxygen will be needed to maintain aerobic treatment steps at their optimum performance level. In addition, it will produce additional nitrate that may have to be denitrified before it can be discharged.

Calculation:

In a wastewater stream of 250,000 gal containing 500 mg/L of nitrite, the net weight of nitrite is

$$500 \times 10^{-3} \text{ g/L} \times 250,000 \text{ gal} \times 3.79 \text{ L/gal} = 474,000 \text{ g of nitrite}$$

From Equation 3.23, stoichiometric consumption of oxygen is 1 mole (32 g) for each 2 moles (92 g) of nitrite oxidized, resulting in a 32/92 g = 0.35 ratio of O_2 to NO_2^- by weight. Therefore, 474,000 g of nitrite will potentially consume

$$474,000 \text{ g } NO_2^- \times 0.35 = 165,000 \text{ g of DO} = \text{Additional COD load}$$

Whether this represents a significant additional amount of COD depends on the operating specifications of the treatment plant.

In addition, from Equation 3.23, stoichiometric production of nitrate is 2 moles (124 g) for each 2 moles (92 g) of nitrite oxidized, resulting in a 124/92 g = 1.35 ratio of NO_3^- to NO_2^- by weight. Therefore, oxidation of 474,000 g of nitrite will produce

$$474,000 \text{ g } NO_2^- \times 1.35 = 639,000 \text{ g } (1,409 \text{ lb}) \text{ of nitrate.}$$

Depending on the limit for nitrate in the treatment plant's discharge permit, additional capacity for denitrification might be required.

3.9.7 Methods for Removing Nitrogen from Wastewater

After the activated sludge treatment stage, municipal wastewater generally still contains some nitrogen in the forms of organic nitrogen and ammonia. Additional treatment may be required to remove nitrogen from the waste stream.

3.9.7.1 Air-Stripping Ammonia

Air stripping can follow the activated sludge process (see Example 10). pH must be raised with lime to about 10 or higher to convert all ammoniacal nitrogen to the volatile NH_3 form. At higher temperatures, the removal efficiency will be higher. Lime (CaO) is the least expensive way to raise the pH, but it generates $CaCO_3$ sludge. NaOH is more expensive but it does not generate sludge. Scaling, icing, and air pollution are some of the disadvantages of air stripping; whereas an advantage is that raising the pH with lime also precipitates any phosphorus present in the form of calcium phosphate compounds.

3.9.7.2 Nitrification–Denitrification

This is a two-step process:

1. Ammonia and organic nitrogen are first biologically oxidized completely to nitrate under strongly aerobic conditions (nitrification). This is achieved by extensive aeration of the sewage:

$$2NH_4^+ + 3O_2 \xrightarrow{\text{Nitrosomas}} 4H^+ + 2NO_2^- + 2H_2O \tag{3.24}$$

$$2NO_2^- + O_2 \xrightarrow{\text{Nitrobacter}} 2NO_3^- \tag{3.25}$$

2. Nitrate is then biologically converted to gaseous nitrogen under anaerobic conditions (denitrification). This requires a carbon nutrient source. Water that is low in total organic carbon (TOC; shown as CH_2O below) may require the addition of methanol or other carbon source.

$$4NO_3^- + 5CH_2O + 4H^+ \xrightarrow{\text{Denitrifying bacteria}} 2N_2(g) + 5CO_2(g) + 7H_2O \tag{3.26}$$

3.9.7.3 Breakpoint Chlorination

Chlorination can be used to remove dissolved ammonia and ammonium ion from wastewater by the chemical reactions 3.27a and 3.27b. The chemical reaction of ammonia with dissolved chlorine results in denitrification by first converting ammonia to chloramines, Equations 3.27a and 3.27b. With continued addition of Cl_2, nitrogen gas is formed, Equation 3.28. Any chloramine remaining serves as a weak disinfectant and is relatively nontoxic to aquatic life.

$$NH_3 + Cl_2 \rightarrow NH_2Cl + Cl^- + H^+ \tag{3.27a}$$

$$NH_4^+ + Cl_2 \rightarrow NH_2Cl + Cl^- + 2H^+ \tag{3.27b}$$

The ratio by weight of chlorine to NH_3–N is 5:1 and ammonia is converted stoichiometrically to monochloramine (NH_2Cl) at 1:1 molar ratio. $NHCl_2$ (dichloramine) and NCl_3 (nitrogen trichloride or trichloramine) may also be formed, depending on pH and small excesses of chlorine. Further addition of chlorine leads to conversion of chloramines to nitrogen gas. The reaction for conversion of monochloramine is

$$2NH_2Cl + Cl_2 \rightarrow N_2(g) + 4H^+ + 4Cl^- \tag{3.28}$$

The overall reaction for complete nitrification of ammonia by chlorine oxidation is

$$2NH_3 + 3Cl_2 \rightarrow N_2(g) + 6H^+ + 6Cl^- \tag{3.29}$$

Equation 3.29 is theoretically complete at a molar ratio of 3:2 and a weight ratio of 7.6:1 of Cl_2 to NH_3–N. This process is called breakpoint chlorination. The reaction is very fast and both ionized (NH_4^+) and unionized (NH_3) forms of ammonia are removed.

RULES OF THUMB

1. The rate of ammonia removal is most rapid at pH 8.3.
2. The rate decreases at higher and lower pH. Since the reactions lower the pH, additional alkalinity as lime might be needed if $[NH_3] > 15$ mg/L.

> Add alkalinity as $CaCO_3$ in a weight ratio of about 11:1 of $CaCO_3$ to NH_3–N.
> 3. The rate also decreases at temperatures below 30°C.
> 4. The chlorine "breakpoint" (see Figure 3.9) occurs theoretically at a Cl_2:NH_3–N weight ratio of 7.6.
> 5. In actual practice, ratios of 10:1 to 15:1 may be needed if oxidizable substances other than NH_3 are present (such as Fe^{2+}, Mn^{+2}, S^{-2}, and organics).

EXAMPLE 12

CALCULATE THE CHLORINE NEEDED TO REMOVE AMMONIA

A waste treatment plant handles 1,500,000 L/day of sewage that contains an average of 50 mg/L of NH_3–N. How many grams of Cl_2(aq) must be present daily in the wastewater to remove all of the ammonia?

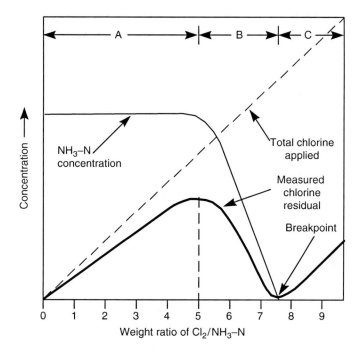

FIGURE 3.9 Breakpoint chlorination curves showing removal of ammonia from wastewater. In Region A, easily oxidizable substances such as Fe^{2+}, H_2S, and organic matter react. Ammonia reacts to form chloramines. Organics react to form chloroorganic compounds. In Region B, adding more chlorine oxidizes chloramines to N_2O and N_2. At the breakpoint, virtually all chloramines and a large part of chloroorganics have been oxidized. In Region C, further addition of chlorine results in a free residual of HOCl and OCl⁻.

Answer:

By Equation 3.29, three moles of chlorine are needed for every two moles of ammonia nitrogen.

$$2NH_3 + 3Cl_2 \rightarrow N_2(g) + 6H^+ + 6Cl^-$$

Molecular weights are $Cl_2 = 71$ and $N = 14$

$$3 \text{ moles of } Cl_2 = 3 \times 71 = 213 \text{ g}$$

$$2 \text{ moles of } N = 2 \times 14 = 28 \text{ g}$$

Thus, the stoichiometric weight ratio is $213/28 = 7.6$ g Cl_2 per gram of N (as ammonia). One mole of NH_3 contains 14 grams of N and 3 grams of H. Thus, 50 mg/L of NH_3 contains $14/17 \times 50$ mg/L $= 41.2$ mg/L of N. In 1,500,000 L there will be

$$1,500,000 \text{ L} \times 41.2 \text{ mg/L} = 61,800,000 \text{ mg N or } 61,800 \text{ g N/day}$$

The theoretical amount of chlorine required is

$$\frac{7.6 \text{ g } Cl_2}{1 \text{ g N}} = 61,800 \text{ g N} = 470 \text{ kg } Cl_2/\text{day or about 1,036 lb/day}$$

Depending on the quantity of other oxidizable substances in the wastewater, the plant operator should be prepared to use up to twice this amount of chlorine.

3.9.7.4   Ammonium Ion Exchange

This is a good alternative to air stripping because an exchange resin, the natural zeolite clinoptilolite, is selective for ammonium ion. NH_4^+ is exchanged for Na^+ or Ca^{2+} on the resin. The zeolite can be regenerated with sodium or calcium salts.

3.9.7.5   Biosynthesis

The removal of biomass, produced in the sewage treatment system by filtering to reduce suspended solids, results in a net loss of nitrogen that has been incorporated in the biomass cell structure.

3.10   SULFIDE AND HYDROGEN SULFIDE

3.10.1   BACKGROUND

Hydrogen sulfide (H_2S) is a colorless gas with the characteristically disagreeable odor of rotten eggs. It is formed when sulfide ion (S^{2-}) reacts reversibly with water to form HS^- and H_2S by the reverse reaction of Equation 3.30, written below with only the sulfide species.

$$S^{2-} \leftrightarrow HS^- \leftrightarrow H_2S$$

S^{2-} and HS^- are soluble, nonvolatile anions with no odor. H_2S is a gas with a strong odor of rotten eggs. Analytically, the three sulfur species S^{2-}, HS^-, and H_2S are collectively called sulfide.

Sulfide is often present naturally in groundwater as the dissolved anion S_2^-, especially in natural hot springs. There, it arises from soluble sulfide minerals and anaerobic bio-reduction of dissolved sulfates. Sulfide is also formed in surface waters from anaerobic decomposition of organic matter containing sulfur. It is a common product of wetlands and eutrophic lakes and ponds. Sulfide reacts with water to form H_2S a colorless, highly toxic gas that smells like rotten eggs. The human nose is very sensitive to the odor of low levels of H_2S. The odor threshold for H_2S dissolved in water is 0.03–0.3 mg/L.

There are two important sources of H_2S in the environment: the anaerobic decomposition of organic matter containing sulfur, and the reduction of mineral sulfates and sulfites to sulfide. Both mechanisms require reducing, or anaerobic, conditions, and are strongly accelerated by the presence of sulfur-reducing bacteria. H_2S is not formed in the presence of an abundant supply of oxygen.

Surface waters become oxygenated (made aerobic) by oxygen diffusion from the atmosphere and by oxygen released from photosynthesizing aquatic plants. In standing water, where oxygen diffusion from the atmosphere is slow and where dead organic matter (leaf and grass litter, insect and animal waste, etc.) can accumulate, biodegradation of the organic matter can consume the DO faster than it is replenished by diffusion and create anaerobic conditions. If sulfate is present, it will be reduced to sulfide and produce H_2S gas, whose odor is an indicator of this process. Another indicator is the formation of blackened soils, sludge, and sediments in locations with standing water, which usually results from the reaction of hydrogen sulfide with dissolved iron to form precipitated ferrous sulfide (FeS), along with other metal sulfides.

3.10.1.1 Formation of H_2S in Detention Ponds, Wetlands, and Sewers

Water conditions promoting the formation of H_2S are sulfate >60 mg/L (or presence of sulfur-containing organic matter such as protein), oxidation-reduction potential <200 mV, and pH <6–7. These conditions frequently occur in standing or slowly moving water, such as in detention ponds, wetlands, sewers, etc. where organic litter can accumulate and where the water or soil contains sulfate. Since surface waters seldom contain more than 8–12 mg/L of dissolved oxygen (more often less), decay of organic matter can quickly reduce dissolved oxygen to anaerobic levels (<1 mg/L). Such waters often develop a bottom layer of black sediments containing iron and other metal sulfides along with organic matter in various stages of decay. In still water, oxygen diffusion into this sediment layer is slow and anaerobic conditions can be maintained with minimal water cover, less than 10 in. in depth. If all water is removed and the soil allowed to dry, diffusion of oxygen into the sediment quickly oxidizes the sulfides to sulfate and H_2S disappears.

Blackening of soils, wastewater, sludge, and sediments in locations with standing water, in addition to the odor of rotten eggs, is an indication that sulfide is present. The black material results from a reaction of H_2S with dissolved iron and other metals to form precipitated FeS, along with other metal sulfides.

H_2S can have two stages of dissociation under reducing conditions in water, depending on the pH:

$$H_2S \leftrightarrow H^+ + HS^- \leftrightarrow 2H^+ + S^{2-} \qquad (3.30)$$

- At pH 5, about 99% of dissolved sulfide is in the form of H_2S, the unionized form.
- At pH 7, dissolved sulfide is 50% HS^- and 50% H_2S.
- At pH 9, about 99% is in the form of HS^-.
- S^{2-} becomes measurable only above pH 12.

H_2S is the most toxic and volatile form; HS^- and S^{2-} are nonvolatile and much less toxic. $H_2S > 2.0$ mg/L constitutes a long-term hazard to fish.

RULES OF THUMB

1. Well water smelling of H_2S is usually a sign of sulfate-reducing bacteria. Look for a water redox potential below -200 mV and a sulfate (SO_4^{2-}) concentration in groundwater >100 mg/L.
2. A typical concentration of H_2S in unpolluted surface water is <0.25 mg/L.
3. $H_2S > 2.0$ mg/L constitutes a chronic hazard to aquatic life.
4. In aerated water, H_2S is bio-oxidized to sulfates and elemental sulfur.
5. Unionized H_2S is volatile and air-strippable. The ionized forms, HS^- and S^{2-}, are nonvolatile.

3.10.1.2 Typical Water Quality Criteria and Standards for H_2S

- Aquatic life (cold and warm water biota): 2.0 mg/L (30 day average).
- Domestic water supply: 0.05 mg/L (30 day average).

3.10.2 CASE STUDY

3.10.2.1 Odors of Biological Origin in Water (Mostly Hydrogen Sulfide and Ammonia)

Odors from anaerobic surface waters, groundwater, and domestic wastewater are usually from inorganic and organic gases generated by biological activity. Anaerobic decomposition of nitrogenous or sulfurous organic matter often produces gases that contain sulfur or nitrogen. Such gases are frequent causes of odors in water. The most common inorganic gases in water are carbon dioxide (CO_2), methane (CH_4), hydrogen (H_2), hydrogen sulfide (H_2S), ammonia (NH_3), carbon disulfide (CS_2), sulfur dioxide (SO_2), oxygen (O_2), and nitrogen (N_2). Of these inorganic gases, those with an odor always contain N or S in combination with H, C, and O, such as H_2S, NH_3, CS_2, and SO_2.

Hydrogen sulfide from the anaerobic reduction of sulfate (SO_4^{2-}) by bacteria usually is the most prevalent odor in natural waters and sewage. Sulfate, formed from the aerobic biodegradation of sulfur-containing proteins, is commonly present in domestic wastewater between 30 and 100 mg/L. Sulfate can arise in natural waters from sulfate minerals and aerobic decomposition of organic material. In addition to H_2S, other disagreeable odorous compounds may be formed by anaerobic decomposition of organics. The particular compounds that are formed depend on the types of bacteria and organic compounds present. Table 3.3 lists a number of common odiferous inorganic and organic compounds with their odor characteristics and odor threshold concentrations when dissolved in water. Sewage carrying industrial wastes may contain other volatile organic chemicals that can contribute additional odors.

3.10.2.2   Environmental Chemistry of Hydrogen Sulfide

Under anaerobic aqueous conditions, in the presence of organic matter or sulfate-reducing bacteria, sulfate is reduced to sulfide ion (S^{2-}) (Equation 3.31).

$$SO_4^{2-} + \text{organic matter/sulfate-reducing bacteria} \rightarrow S^{2-} + H_2O + CO_2 \quad (3.31)$$

TABLE 3.3
Odor Characteristics and Threshold Concentrations in Water

Substance	Formula	Odor Threshold Concentration in Water (mg/L)	Odor Characteristics
Allyl mercaptan	$H_2C = CHCH_2SH$	0.00005	Very disagreeable, garlic-like
Ammonia	NH_3	0.037	Sharp, pungent
Benzyl mercaptan	$C_6H_5CH_2SH$	0.0.00019	Unpleasant
Chlorine	Cl_2	0.010	Pungent, irritating
Chlorophenol	ClC_6H_4OH	0.00018	Medicinal
Crotyl mercaptan	$CH_3CH = CHCH_2SH$	0.000029	Skunk-like
Diphenyl sulfide	$(C_6H_5)_2S$	0.00005	Unpleasant
Ethyl mercaptan	CH_3CH_2SH	0.00019	Decayed cabbage
Diethyl sulfide (ethyl sulfide)	$(CH_3CH_2)_2S$	0.000025	Nauseating, ethereal
Hydrogen sulfide	H_2S	0.0011	Rotten egg
Methyl mercaptan	CH_2SH	0.0011	Decayed cabbage
Dimethyl sulfide (methyl sulfide)	$(CH_3)_2S$	0.0011	Decayed vegetables
Pyridine	C_6H_5N	0.0037	Disagreeable, irritating
Skatole	C_9H_9N	0.0012	Fecal, nauseating
Sulfur dioxide	SO_2	0.009	Pungent, irritating
Thiocresol	$CH_3C_6H_4SH$	0.001	Rancid, skunk-like
Thiophenol	C_6H_5SH	0.000062	Putrid, nauseating

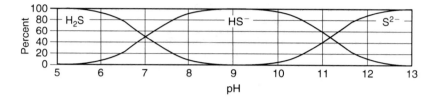

FIGURE 3.10 pH distribution of hydrogen sulfide species in water.

Sulfide ion is a strong base, reversibly reacting rapidly in water to form HS^- and gaseous hydrogen sulfide (Equation 3.32).

$$S^{2-} + 2H_2O \leftrightarrow OH^- + HS^- + H_2O \leftrightarrow H_2S(g) + 2OH^- \qquad (3.32)$$

HS^- and S^{2-} are nonvolatile with no odor. H_2S is gaseous with a strong odor of rotten eggs. The equilibrium distribution between S^{2-}, HS^-, and H_2S depends mainly on the pH and somewhat on the temperature. In Figure 3.10, $T = 30°C$.

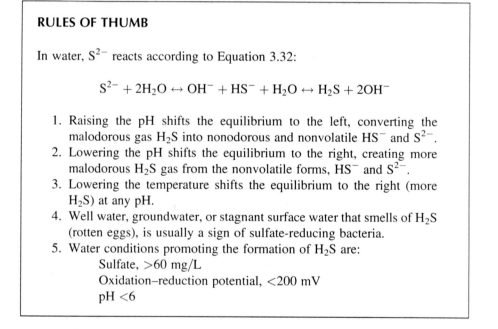

RULES OF THUMB

In water, S^{2-} reacts according to Equation 3.32:

$$S^{2-} + 2H_2O \leftrightarrow OH^- + HS^- + H_2O \leftrightarrow H_2S + 2OH^-$$

1. Raising the pH shifts the equilibrium to the left, converting the malodorous gas H_2S into nonodorous and nonvolatile HS^- and S^{2-}.
2. Lowering the pH shifts the equilibrium to the right, creating more malodorous H_2S gas from the nonvolatile forms, HS^- and S^{2-}.
3. Lowering the temperature shifts the equilibrium to the right (more H_2S) at any pH.
4. Well water, groundwater, or stagnant surface water that smells of H_2S (rotten eggs), is usually a sign of sulfate-reducing bacteria.
5. Water conditions promoting the formation of H_2S are:
   Sulfate, >60 mg/L
   Oxidation–reduction potential, <200 mV
   pH <6

3.10.2.3 Chemical Control of Odors

Depending on the odor-causing compound, chemical control of odors may be accomplished by a combination of pH control, eliminating the causes of reducing conditions, chemical oxidation, and aeration; sorption to activated charcoal; air stripping of volatile species; and chemical conversion (often microbially mediated, as in nitrification).

3.10.2.4   pH control

3.10.2.4.1   Hydrogen sulfide

For odors from H_2S, raising the pH (by adding NaOH or lime) shifts the equilibrium to the left (Equation 3.33):

$$S^{2-} + 2H_2O \leftrightarrow OH^- + HS^- + H_2O \leftrightarrow H_2S + 2OH^- \tag{3.33}$$

This converts gaseous H_2S to the nonodorous ionic forms. However, the pH must be maintained above 9 for complete odor removal. Normally, odor control by removing the sulfur compounds is more practical.

Lowering the pH (by adding acid) shifts the equilibrium to the right, converting the ionic forms to gaseous H_2S. At low pH, the gas can be removed from the water by air stripping (see the following example). This, of course, does not destroy the H_2S; it moves it from the water to the air. Figure 3.11 gives the fraction of hydrogen sulfide that is in the volatile form of H_2S at different pH values and temperatures. Note that H_2S behaves the opposite of NH_3 (Section 3.10.2.4.2). Stripping efficiency is increased with decreasing pH and lower temperatures.

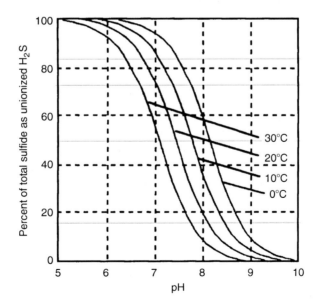

FIGURE 3.11 Fraction of hydrogen sulfide in unionized form (H_2S) as a function of temperature and pH.

3.10.2.4.2 Ammonia

pH control of odors from ammonia is opposite to that for hydrogen sulfide. To use air stripping to remove odor caused by ammonia, the pH must be raised (see Example 10 and Section 3.9.7).

3.10.2.5 Oxidation

Add oxidizing agents such as Cl_2, NaOCl, H_2O_2, O_2, $KMnO_4$, or ClO_2.

They oxidize hydrogen sulfide to odorless sulfate ion, SO_4^{2-}, and ammonia to odorless nitrogen compounds, including elemental nitrogen, N_2.

RULES OF THUMB

1. The usual chlorine dose for odor control is 10–50 mg/L.
2. 8.9 mg of chlorine is required to oxidize 1 mg of hydrogen sulfide, H_2S.

3.10.2.6 Eliminate Reducing Conditions Caused by Decomposing Organic Matter

This often means mechanically cleaning out the organic slime and sludge in a well, sewer, drain, or wetland. Mechanical cleaning will aid the use of oxidizing agents.

It can also be accomplished sometimes by aerating the water. Increasing the DO level will shift conditions from reducing to oxidizing as the oxygen diffuses into the reducing zone of the water. Mixing currents in the water help this process. If the reducing zone is thick and the water stagnant and motionless, aeration control might be very slow.

Drying out wet soil that has an H_2S odor also allows oxygen to diffuse into the organic matter, changing the decomposition processes from anaerobic to aerobic.

3.10.2.7 Sorption to Activated Charcoal

Sorption to powdered or granular activated charcoal is a reliable "last resort" for removing bad tastes and odors. Powdered charcoal can be added as a slurry directly to a waste stream. Gases can be passed through a canister filled with granulated activated charcoal (GAC).

3.11 PHOSPHORUS

3.11.1 BACKGROUND

Phosphorus is a common element in igneous and sedimentary rocks and in sediments but it tends to be a minor element in natural waters because most inorganic phosphorous compounds have low solubility. Dissolved concentrations are generally in the range of 0.01–0.1 mg/L and seldom exceed 0.2 mg/L. The environmental behavior of phosphorus is largely governed by the low solubility of most of its inorganic compounds, its strong adsorption to soil particles, and the fact that it is an essential nutrient for most life forms—animal, plant, and microbial.

Because of its low dissolved concentrations, phosphorus is usually the limiting nutrient in natural waters. The dissolved phosphorous concentration is often low enough to limit algal growth. Because phosphorus is essential to metabolism, it is always present in animal wastes and sewage. Too much phosphorus in wastewater effluent is frequently the main cause of algal blooms and other precursors of eutrophication.

3.11.2 IMPORTANT USES FOR PHOSPHORUS

Phosphorous compounds are used for corrosion control in water supply and industrial cooling water systems. Certain organic phosphorous compounds are used in insecticides. Perhaps the major commercial uses of phosphorous compounds are in fertilizers and in the production of synthetic detergents. Detergent formulations may contain large amounts of polyphosphates as "builders," to sequester metal ions and maintain alkaline conditions. The widespread use of detergents instead of soap makes a major contribution to the available phosphorus in domestic wastewater.

Prior to the use of phosphate detergents, most wastewater inorganic phosphorus was contributed from human wastes; about 1.5 g/day per person is released in urine. As a consequence of detergent use, the concentration of phosphorus in treated municipal wastewaters has increased from 3 to 4 mg/L in predetergent days, to the present values of 10–20 mg/L. Since phosphorus is an essential element for the growth of algae and other aquatic organisms, rapid growth of aquatic plants can be a serious problem when effluents containing excessive phosphorus are discharged to the environment.

3.11.3 PHOSPHOROUS CYCLE

In a manner similar to nitrogen, phosphorus in the environment is cycled between organic and inorganic forms. An important difference is that under certain soil conditions, some nitrogen is lost to the atmosphere by ammonia volatilization and microbial denitrification. There are no analogous gaseous loss mechanisms for phosphorus. While nitrogen in the atmosphere is continually redistributed globally, phosphorus has no such global redistribution mechanism; the closest approaches are by bird migration and international shipping of fertilizers.

Also important are the differences in earthbound mobility of the two nutrients. Both exist in anionic forms (NO_2^-/NO_3^- and $H_2PO_4^-/HPO_4^{2-}$), which are not subject to retention by cation exchange reactions. However, nitrate anions do not form insoluble compounds with metals and, therefore, nitrates leach readily from soil to surface and groundwaters. Phosphate anions are largely immobilized in the soil by the formation of insoluble compounds, chiefly iron, calcium, and aluminum phosphates, and by adsorption to soil particles. Because nitrogen compounds leach from soils more readily than phosphorous compounds, nitrogen is more generally available than phosphorus to water vegetation, a condition that contributes to phosphorous-limited algal growth in most surface waters. The critical level of inorganic phosphorus for forming algal blooms can be as low as 0.01–0.005 mg/L under summer growing conditions but is more frequently around 0.05 mg/L.

Organic compounds containing phosphorus are found in all living matter. Orthophosphate (PO_4^{3-}) is the only form readily used as a nutrient by most plants and organisms. The two major steps of the phosphorous cycle, conversion of organic phosphorus to inorganic phosphorus and back to organic phosphorus, are both bacterially mediated. Conversion of insoluble forms of phosphorus, such as calcium phosphate, $Ca(HPO_4)_2$, into soluble forms, principally PO_4^{3-}, is also carried out by microorganisms. Organic phosphorus in the tissues of dead plants and animals and in animal waste products is converted bacterially to PO_4^{3-}. The PO_4^{3-} thus released to the environment is taken up again into plant and animal tissue.

RULES OF THUMB

1. In surface waters, phosphorous concentrations are influenced by the sediments, which serve as a reservoir for adsorbed and precipitated phosphorus. Sediments are an important part of the phosphorous cycle in streams. Bacteria-mediated exchange between dissolved and sediment-adsorbed forms plays a role in making phosphorus available for algae and therefore contributes to eutrophication.
2. In streams, dissolved phosphorus from all sources, natural and anthropogenic, is generally present in low concentrations, around 0.1 mg/L or less.
3. The natural background of total dissolved phosphorus has been estimated to be about 0.025 mg P/L; that of dissolved phosphates about 0.01 mg P/L.
4. The solubility of phosphates increases at low pH and decreases at high pH.
5. Particulate phosphorus (sorbed on sediments and insoluble phosphorous compounds) is about 95% of the total phosphorus in most cases.
6. In carbonate soils, dissolved phosphorus can react with carbonate to form the mineral precipitate hydroxyapatite (calcium phosphate hydroxide), $Ca_{10}(PO_4)_6(OH)_2$.

3.11.4 MOBILITY IN THE ENVIRONMENT

Phosphorus is an important plant nutrient and is often present in fertilizers to augment the natural concentration in soils. Phosphorus is also a constituent of animal wastes. Runoff from agricultural areas is a major contributor to total phosphorus in surface waters, where it occurs mainly in sediments because of the low solubility of its inorganic compounds and its tendency to adsorb strongly to soil particles.

Dissolved phosphorus is removed from the solution by

* Precipitation
* Strong adsorption to clay minerals and oxides of aluminum and iron
* Adsorption to organic components of soil

Reducing (anaerobic) conditions, as in water-saturated soil, may increase phosphorous mobility because insoluble ferric iron, to which phosphorus is strongly adsorbed, is reduced to soluble ferrous iron, thereby releasing adsorbed phosphorus. In acid soils, aluminum and iron phosphates precipitate, while in basic soils, calcium phosphates precipitate. The immobilization of phosphorus is therefore dependent on soil properties, such as pH; aeration; texture; cation-exchange capacity; the amount of calcium, aluminum, and iron oxides present; and the uptake of phosphorus by plants.

Because of these removal mechanisms for dissolved phosphorus, phosphorous compounds resist leaching, and there is little movement of phosphorus with water drainage through most soils. It is mobilized mainly by sorption to erosion sediments. Phosphorous transport into surface waters is controlled chiefly by preventing soil erosion and controlling sediment transport. In most soils, except for those that are nearly all sand, almost all the phosphorus applied to the surface is retained in the top 1–2 ft. The adsorption capacity for phosphorus has been estimated for several soils to be in the range of 77 to over 900 lb/acre-ft of soil profile. Often, the total phosphorous removal capacity for a soil will exceed the planning life of a typical land application project. If the phosphorus-removing capacity of a soil becomes saturated, it usually can be restored in a few months, during which adsorbed phosphorus is precipitated with metals or removed by crops.

As pH becomes higher, the equilibrium of Equation 3.34 shifts increasingly to the right.

$$H_3PO_4 \leftrightarrow H_2PO_4^- + H^+ \leftrightarrow HPO_4^{2-} + 2H^+ \leftrightarrow PO_4^{3-} + 3H^+ \qquad (3.34)$$

Dissolved phosphate species exhibit the following pH-dependent equilibria (see Figure 3.12):

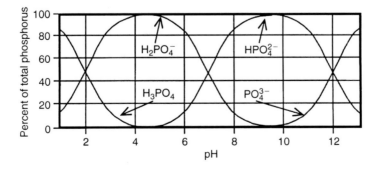

FIGURE 3.12 pH dependence of phosphate species.

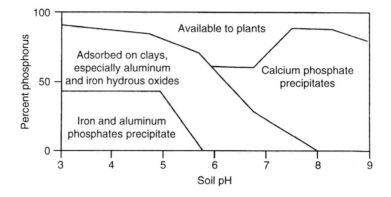

FIGURE 3.13 Forms of immobile phosphorus.

- Below pH 2, H_3PO_4 is the dominant species.
- Between pH 2 and 7, $H_2PO_4^-$ is the dominant species.
- Between pH 7 and 12, HPO_4^{2-} is the dominant species.
- Above pH 12, PO_4^{3-} is the dominant species.

Figure 3.13 shows some general relationships between soil pH and phosphorous reactions:

In the acid pH range, dissolved phosphorus is predominantly $H_2PO_4^-$, and immobile phosphorus is bound with iron and aluminum compounds.

In the basic pH range, dissolved phosphorus is predominantly HPO_4^{2-}, and immobile phosphorus is mainly in the form of calcium phosphate.

Maximum availability of phosphorus for plant uptake (as well as leaching) occurs between pH 6 and 7.

Even when algal growth in lakes is temporarily limited by carbon or nitrogen instead of phosphorus, natural long-term mechanisms act to compensate for these deficiencies. Carbon deficiencies are corrected by CO_2 diffusion from the atmosphere, and nitrogen deficiencies are corrected by changes in biological growth mechanisms. Therefore, even if a sudden increase in phosphorus occurs temporarily causing algal growth to be limited by carbon or nitrogen, eventually these deficiencies are corrected. Then, algal growth becomes proportional to the phosphorous concentration as the system becomes once more phosphorous-limited.

No national criteria have been established for concentrations of phosphorous compounds in water; however, to control eutrophication, the EPA makes the following recommendations:

- Total phosphates should not exceed 50 mg/L (as phosphorus) in a stream at a point where it enters a lake or reservoir.
- Total phosphorus should not exceed 100 mg/L in streams that do not discharge directly into lakes or reservoirs.

RULES OF THUMB

1. The critical level of inorganic phosphorus for algae bloom formation can be as low as 0.01–0.005 mg/L under summer growing conditions but more frequently is around 0.05 mg/L.
2. Lakes are nitrogen-limited if the weight ratio of total nitrogen to total phosphorus (N/P) is <13, nutrient-balanced if $13 < N/P < 21$, and phosphorous-limited if $N/P > 21$. Exact ranges depend on the particular algae species. Most lakes are phosphorous-limited; in other words, additional phosphorus is needed to sustain further algal growth.
3. Different N/P ratios and pH values favor the growth of different kinds of algae.
4. Low N/P ratios favor N-fixing blue–green algae.
5. High N/P ratios, often achieved by controlling phosphorous input by means of additional wastewater treatment, cause a shift from blue–green algae to less objectionable species.
6. Lower pH (or increased CO_2) gives green algae a competitive advantage over blue–green algae.
7. To control eutrophication, EPA recommends that total phosphates should not exceed 50 mg/L (as phosphorus) in a stream where it enters a lake or reservoir, or 100 mg/L in streams that do not discharge directly into lakes or reservoirs.

3.11.5 PHOSPHOROUS COMPOUNDS

Compounds containing phosphorus that are of interest to water quality include

- Orthophosphates (all contain PO_4^{3-})
- Trisodium phosphate, Na_3PO_4
- Disodium phosphate, Na_2HPO_4
- Monosodium phosphate (NaH_2PO_4)
- Diammonium phosphate $(NH_4)_2HPO_4$

Orthophosphates are soluble and are considered the only biologically available form. In the environment, hydrolysis slowly converts polyphosphates to orthophosphates. Analytical methods measure orthophosphates. To measure total phosphate, all forms of phosphate are chemically converted to orthophosphates (hydrated forms).

Polyphosphates (called condensed phosphates, meaning dehydrated) include

- Sodium hexametaphosphate, $Na_3(PO_4)_6$
- Sodium tripolyphosphate, $Na_5P_3O_{10}$
- Tetrasodium pyrophosphate, $Na_4P_2O_7$
- Organic phosphate (biodegradation or oxidation of organic phosphates releases orthophosphates)

Sedimentary phosphorus occurs in the following forms:

- Phosphate minerals: Mainly hydroxyapatite, $Ca_5OH(PO_4)_3$.
- Nonoccluded phosphorus: Phosphate ions (usually orthophosphate) bound to the surface of SiO_2 or $CaCO_3$. Nonoccluded phosphorus is generally more soluble and more available than occluded phosphorus (below).
- Occluded phosphorus: Phosphate ions (usually orthophosphate) contained within the matrix structures of amorphous hydrated oxides of iron, aluminum, and amorphous aluminosilicates. Occluded phosphorus is generally less available than nonoccluded phosphorus.
- Organic phosphorus: Phosphorus incorporated with aquatic biomass, usually algal or bacterial.

3.11.6 REMOVAL OF DISSOLVED PHOSPHATE

Current remedies for phosphate-caused foaming and eutrophication are

- Using lower phosphate formulas in detergents
- Precipitating the phosphate with Fe^{3+}, Al^{3+}, or Ca^{2+}
- Diverting the discharge to a less sensitive location

Municipal wastewater treatment plants in many areas are required to remove phosphorous in their treatment process. While the biological treatment process removes some phosphorus, in most cases precipitation as an insoluble metal phosphate is required to meet discharge regulations. This precipitation step is normally accomplished by the addition of a metallic salt such as ferric sulfate, ferric chloride, or aluminum sulfate in the primary or secondary clarifiers. The usual precipitants for removing phosphate are alum $[Al_2(SO_4)_3]$, lime $[Ca(OH)_2]$, ferric sulfate $[Fe_2(SO_4)_3]$, and ferric chloride ($FeCl_3$). The choice of precipitant depends on the discharge requirements, wastewater pH, and chemical costs.

Pertinent reactions for the precipitation of phosphate with alum, ferric sulfate, ferric chloride, and lime are

$$\text{Alum: } Al_2(SO_4)_3 + 2HPO_4^{2-} \rightarrow 2AlPO_4(s) + 3SO_4^{2-} + 2H^+ \tag{3.35}$$

$$\text{Ferric sulfate: } Fe_2(SO_4)_3 + 2HPO_4^{2-} \rightarrow 2FePO_4(s) + 3SO_4^{2-} + 2H^+ \tag{3.36}$$

$$\text{Ferric chloride: } FeCl_3 + HPO_4^{2-} \rightarrow FePO_4(s) + 3Cl^- + H^+ \tag{3.37}$$

$$\text{Lime: } 3Ca^{2+} + 2OH^- + 2HPO_4^{2-} \rightarrow Ca_3(PO_4)_2(s) + 2H_2O \tag{3.38}$$

$$4Ca^{2+} + 2OH^- + 3HPO_4^{2-} \rightarrow Ca_4H(PO_4)_3(s) + 2H_2O \tag{3.39}$$

$$5Ca^{2+} + 4OH^- + 3HPO_4^{2-} \rightarrow Ca_5(OH)(PO_4)_3(s) + 3H_2O \tag{3.40}$$

Where effluent concentrations of phosphorus up to 1.0 mg/L are acceptable, the use of iron or aluminum salts in a wastewater secondary treatment system is often the process of choice. If very low levels of effluent phosphorus are required, precipitation at high pH by lime in a tertiary unit is necessary. The lowest levels of phosphorus are achieved by adding NaF with lime to form $Ca_5(PO_4)_3F$ (fluorapatite). The operating pH for phosphate removal with lime is usually above 11 because flocculation is best in this range.

If alkalinity is present, aluminum and iron ions are consumed in the formation of metal-hydroxide flocs. This may increase required dosages by up to a factor of 3. Calcium ions react with alkalinity to form calcium carbonate. Thus, the amount of precipitant needed for phosphate precipitation is controlled more by the alkalinity than the stoichiometry of the reaction. In the case of aluminum and iron precipitants, the reaction with alkalinity is not totally wasted because the hydroxide flocs assist in the settling and removal of metal-phosphate precipitates, along with other suspended and colloidal solids in the wastewater.

Biological phosphorus removal can be accomplished by operating an activated sludge process in an anaerobic–aerobic sequence. A number of bacteria respond to this sequence by accumulating large excesses of polyphosphate within their cells in volutin granules. During the anaerobic phase, release of phosphate occurs. In the aerobic phase, the released phosphate and an additional increment is taken up and stored as polyphosphate, giving a net removal, coincident with organic removal and metabolism. Phosphate can be removed from the waste stream as sludge or through use of a second anaerobic step. During the second anaerobic step, the stored phosphate is released in dissolved form. Then, the bacterial cells can be separated and recycled and the released soluble phosphate is removed by precipitation.

3.12   SOLIDS (TOTAL, SUSPENDED, AND DISSOLVED)

3.12.1   Background

The general term "solids" refers to matter that is suspended (insoluble solids) or dissolved (soluble solids) in water. Solids can affect water quality in several ways. Drinking water with high dissolved solids may not taste good and may have a laxative effect. Boiler water with high dissolved solids requires pretreatment to prevent scale formation. Water high in suspended solids may harm aquatic life by causing abrasion damage, clogging fish gills, harming spawning beds, and reducing photosynthesis by blocking sunlight penetration, among other consequences. On the other hand, hard water (caused mainly by dissolved calcium and magnesium compounds) reduces the toxicity of metals to aquatic life.

Total solids (sometimes called residue) are the solids remaining after evaporating the water from an unfiltered sample. It includes two subclasses of solids that are separated by filtering (generally with a filter having a nominal 0.45 μm or smaller pore size):

1. Total suspended solids (TSS, sometimes called filterable solids) in water are organic and mineral particulate matter that do not pass through a 0.45 μm

filter. They may include silt, clay, metal oxides, sulfides, algae, bacteria, and fungi. TSS is generally removed by flocculation and filtering. TSS contributes to turbidity, which limits light penetration for photosynthesis and visibility in recreational waters.

2. Total dissolved solids (TDS; sometimes called nonfilterable solids) are substances that will pass through a 0.45 μm filter. If the water passed through the filter is evaporated, the TDS will remain behind as a solid residue. TDS may include dissolved minerals and salts, humic acids, tannin, and pyrogens. TDS is removed by precipitation, ion exchange, and RO. In natural waters, the major contributors to TDS are carbonate, bicarbonate, chloride, sulfate, phosphate, and nitrate salts. Taste problems in water often arise from the presence of high TDS levels with certain metals present, particularly iron, copper, manganese, and zinc.

The difference between suspended and dissolved solids is a matter of definition based on the filtering procedure. Solids are always measured as the dry weight, and careful attention must be paid to the drying procedure to avoid errors caused by retained moisture or loss of material by volatilization or oxidation.

RULES OF THUMB

1. TSS is detrimental to fish health by decreasing growth, disease resistance, and egg development.
2. Suspended solids should be restricted so they do not reduce the maximum depth of photosynthetic activity by more than 10% from the seasonally established norm.
3. Water with TDS < 1200 mg/L generally has an acceptable taste. Higher TDS can adversely influence the taste of drinking water and may have a laxative effect.
4. In water to be treated for domestic potable supply, TDS < 650 mg/L is a preferred goal.
5. For drinking water, recommended TDS is <500 mg/L; the upper limit is 1000 mg/L.
6. At low concentrations, TSS (in mg/L for soil erosion or μg-chlorophyll/L for algae) is roughly equal to turbidity in NTU (see Table 3.4).

3.12.2 TDS and Salinity

TDS and salinity both indicate dissolved salts. Table 3.5 offers a qualitative comparison between the terms.

TABLE 3.4
Total Suspended Solids Concentration Estimated from Turbidity Measurement

Turbidity (NTU)	2	5	10	20	50
Corresponding TSS due to soil sediment (mg/L)	2.2	6.3	12	24	64
Corresponding TSS due to algae (μg chlorophyll/L)	2.2	4.7	10	36	54

3.12.3 SPECIFIC CONDUCTIVITY AND TDS

Specific conductivity is directly related to TDS and serves as a check on TDS measurements.

RULES OF THUMB

1. Conductivity units are μmhos/cm or μSiemens/cm (μS/cm):

$$1 \ \mu\text{mho/cm} = 1 \ \mu\text{Siemen/cm}$$

2. TDS in mg/L can be estimated from a measurement of specific conductivity.

   a. For seawater (NaCl-based):

$$\text{TDS (mg/L)} \approx (0.5) \times (\text{Sp. Cond. in } \mu\text{S/cm})$$

   b. For surface and ground waters (carbonate or sulfate-based):

$$\text{TDS (mg/L)} \approx (0.55\text{--}0.7) \times (\text{Sp. Cond. in } \mu\text{S/cm})$$

3. If the ratio $\frac{\text{TDS}_{meas}}{\text{Sp. Cond.}}$ is demonstrated to be consistent, the simpler specific conductivity measurement may sometimes be substituted for TDS analysis.

TABLE 3.5
Comparison of TDS and Salinity

TDS	Degree of Salinity
1,000–3,000 mg/L	Slightly saline
3,000–10,000 mg/L	Moderately saline
10,000–35,000 mg/L	Very saline
>35,000 mg/L	Briny

3.12.4   TDS Test for Analytical Reliability

A calculated value for TDS may be used for judging the reliability of a sample analysis if all the important ions have also been measured. The TDS concentration should be equal to the sum of the concentrations of all the ions present plus silica. You can use either of the following equations to calculate TDS from an analysis or to check on the validity of analytical results. All concentrations are in mg/L.

$$TDS = \text{sum of cations} + \text{sum of anions} + \text{silica} \qquad (3.41)$$

or

$$TDS = 0.6(\text{alkalinity}) + Na^+ + K^+ + Ca^{2+} + Mg^{2+} + Cl^- + SO_4^{2-} + SiO_3$$
$$(3.42)$$

In any given analysis, it is unlikely that all the ions have been measured. Frequently, only the major ions (Na^+, K^+, Ca^{2+}, Mg^{2+}, Cl^-, HCO_3^-, SO_4^{2-}) are necessary for the calculations, as other ion concentrations are likely to be insignificant by comparison.

Use the following guidelines for checking accuracy of a TDS analysis:

1. TDS_{meas} should always be equal to or somewhat larger than TDS_{calc} because a significant ion contributor might not have been included in the calculation.
2. An analysis is acceptable if the ratio of measured-to-calculated TDS is in the range

$$1.0 < \frac{\text{measured TDS}}{\text{calculated TDS}} < 1.2$$

3. If $TDS_{meas} < TDS_{calc}$, the sample should be reanalyzed.
4. If $TDS_{meas} > 1.2 \times TDS_{calc}$, the sample should be reanalyzed, perhaps with a more complete set of ions.

3.13   TEMPERATURE

Temperature affects all water uses.

- The solubility of gases such as oxygen and CO_2 decreases as water temperature increases.
- Biodegradation of organic material in water and sediments is accelerated with increased temperatures, increasing the demand on DO.
- Fish and plant metabolism depends on temperature.

Most chemical equilibria are temperature dependent. Important environmental examples are the equilibria between ionized and unionized forms of ammonia, hydrogen cyanide, and hydrogen sulfide.

Temperature regulatory limits are set to maintain a normal pattern of diurnal and seasonal fluctuations, with no changes deleterious to aquatic life. Maximum-induced change is limited to a 3°C increase over a 4 h period, lasting for 12 h maximum.

EXERCISES

1. (a) The measured pH of a seawater sample is 8.30. What is the hydrogen ion (H^+) concentration in mol/L and in mg/L?
   (b) What is the pH of a water sample in which $[H^+] = 1.5 \times 10^{-10}$ M?
2. (a) The $[H^+]$ of water in a stream is 6.1×10^{-8} mol/L. What is the pH?
   (b) The pH of water in a stream is 9.3. What is the hydrogen ion concentration?
3. An engineer requested a water sample analysis that included the parameters: pH, carbonate ion, and bicarbonate ion. Explain why she probably is wasting money.
4. (a) Water sample analysis indicated a total carbonate concentration of 0.003 mol/L and a pH of 6.6. What is the alkalinity in meq/L and in mg/L?
   (b) If 1 L of the sample was diluted with enough pure water to change the total carbonate concentration to 0.0025 mol/L, what would be the new pH and alkalinity? Use Figure 3.3.
5. A water sample contains 150 mg/L of Ca^{2+} and 33 mg/L of Mg^{2+}. Calculate the total hardness of the water.
6. A wastewater treatment plant removed ammonia with a nitrification–denitrification process that also reduced alkalinity. For each gram of NH_3–N removed, the process also removed 7.14 g of alkalinity-$CaCO_3$. If the plant was designed to remove 25 mg/L of NH_3–N and the total throughput was 250,000 gpd, how many pounds of caustic soda (sodium hydroxide, NaOH) must be added each day to restore the alkalinity to its original value before ammonia removal?
7. A wastewater flow contains 30 g/L total ammonia nitrogen and has a 10 g/L discharge limit. An air-stripping tower is to be used. Its temperature varies from 20°C to 30°C and the pH is normally about 9. At what pH must the stripper be operated?
8. A water sample from a lake has a measured alkalinity of 0.8 eq/L.
   In the early morning, a monitoring team measures the lake's pH as part of an acid rain study and finds pH 6.0. The survey team returns after lunch to recheck their data. By this time, algae and other aquatic plants have consumed enough dissolved CO_2 to reduce the lake's total carbonate (C_T) to one-half of its morning value. (a) What was the morning value for C_T? (b) What was the pH after lunch?

9. A groundwater has the following analysis at pH 7.6

Analyte	Concentration (mg/L)
Calcium	75
Magnesium	40
Sodium	10
Bicarbonate	300
Chloride	10
Sulfate	112

Calculate alkalinity, total hardness, carbonate (temporary) hardness, and non-carbonate (permanent) hardness.

REFERENCES

Deffeyes, K.S., 1965, Carbonate equilibria: A graphic and algebraic approach, *Limnology and Oceanography*, 10 (3), 412–426.

Nordstrom, D.K. et al., 2000, Negative pH and extremely acid mine waters from Iron Mountain, California, *Envir. Sci. Techn.*, 34 (2), 254–258.

Stumm, W. and Morgan, J.J., 1996, *Aquatic Chemistry, Chemical Equilibria and Rates in Natural Waters*, 3rd edition, John Wiley & Sons, Inc., New York, 1022 pp.

4 Behavior of Metal Species in the Natural Environment

4.1 METALS IN WATER

4.1.1 BACKGROUND

A casual glance at the periodic table shows that most of the elements (about three-fourths) are metals or metalloids.* It often happens in environmental literature that little or no distinction is made between metals and metalloids, especially for the metalloids arsenic, selenium, and antimony. To discuss their chemical behavior, the elemental metals may be divided into three general classes:

1. Alkali metals: Li, Na, K, Rb, Cs, and Fr (periodic table group 1A).
2. Alkaline metals: Be, Mg, Ca, Sr, Ba, and Ra (periodic table group 2A).
3. Metals not in the alkali or alkaline groups include the transition metals (all the group B periodic table metals), the metals and metalloids in groups 3A through 6A, and metals whose classifications are not based primarily on periodic table groups, the so-called trace or heavy metals.[†]

Heavy metals in surface waters can be from natural or anthropogenic sources. Currently, anthropogenic inputs of metals exceed natural inputs. Living organisms require trace amounts of some heavy metals, including cobalt, copper, iron, manganese, molybdenum, vanadium, strontium, and zinc. Excessive levels of essential metals, however, can be detrimental to the organism. Nonessential heavy metals of

---

* Metalloids are those elements in periodic table groups 3A through 6A that have electrical and chemical properties intermediate between those of metals and nonmetals. They are B, Si, Ge, As, Sb, Te, and Po. For regulatory purposes, it is sometimes useful to group metals and metalloids together, as when they share the same analytical method (e.g., ICP, ion-coupled plasma spectroscopy).

[†] The term "heavy metals" is often encountered in texts and reports, usually meaning metals with atomic numbers equal to or greater than Cu (at. no. 29), especially metals exhibiting toxicity. However, the term heavy metals has no precise definition and its use is inconsistent (Duffus, 2002). Another designation often used is trace metals, generally used for those metals found in the earth's crust with average concentrations less than 1%. Nearly all the metals are included in this class, the exceptions (with average crustal concentrations greater than 1%) being Na, K, Ca, Mg, Fe, and Al.

particular concern because of their toxicity are cadmium, chromium, mercury, lead, arsenic, and antimony.

When metal atoms combine chemically with other metal atoms, the result is a metal substance, either pure elemental metals or alloys. When metal atoms combine with nonmetal atoms, nonmetallic compounds result that range from ionic salts like sodium chloride to volatile, inflammable liquid organometallic compounds like dimethylmercury.

Metals and metal-containing compounds in natural waters may be in dissolved, colloidal, or particulate forms, depending on water quality parameters of pH, redox potential, and the presence of other dissolved species such as sulfide or carbonate which can form compounds with metal ions. This section treats how the solubility of metals in water depends on these different factors.

Dissolved forms* are

- Cations: Ca^{2+}, Fe^{2+}, K^+, Al^{3+}, Ag^+, etc.
- Complexes[†]: $Zn(OH)_4^{2+}$, $Au(CN)_2^-$, $Ca(P_2O_7)^{2-}$, PuEDTA, etc.
- Organometallics: $Hg(CH_3)_2$, $B(C_2H_5)_3$, $Al(C_2H_5)_3$, etc.

Particulate forms are

- Mineral sediments
- Precipitated oxides, hydroxides, sulfides, carbonates, silicates, etc.
- Cations and complexes sorbed to mineral sediments (clays, oxides, hydroxides, sulfides, carbonates, silicates, etc.) and organic matter.

The behavior of metals in natural waters may be described in terms of how they become distributed between dissolved and solid species. Metal species undergo

---

* It is shown in this chapter that cations in water always attract a hydration shell of water molecules because of electrostatic attractions. Although a cation dissolved in water is often written as an elemental ion, e.g., Al^{3+}, it actually is a cluster consisting of the metal ion enclosed within progressively larger surrounding shells of water molecules. The water molecules in the innermost shell are the most strongly attracted to the metal cation and the cluster could be written, for example, as $Al(H_2O)_n^{3+}$, where n is the number of water molecules in the inner hydration shell, from 1 to about 6. One simpler way to indicate a dissolved cation is append the suffix "aq" (for aqueous) to it, as in Al^{3+}(aq). In this text, and many others, whenever a dissolved cation is written without any indication of its hydration shell, e.g., Al^{3+}, the more accurate designation is to be assumed, e.g., Al^{3+}(aq).

[†] A complex is a dissolved chemical species formed by the association of a cation with one or more anions or neutral species (such as water) that contain nonbonding electron pairs. The cation has room for one or more electron pairs in its valence shell and the anions or neutral species it connects with (called *ligands*) have nonbonding electron pairs in their outer shells that can fill the cation electron shell vacancies. Remember that chemical bonds consist of electron pairs positioned between the bonded atoms. A complex differs from a covalent compound in that the ligand brings both electrons of the bonding pair to the bond, while covalent compounds are formed when the connected atoms contribute one electron each. Cation–ligand bonds are called coordinate bonds, and the complexes formed are called coordination compounds.

continuous changes between dissolved, precipitated, and sorbed-to-sediment forms. The rates of adsorption, desorption, and precipitation processes depend on pH, redox potential, water chemistry, and the composition of bottom and suspended sediments. Adsorption of dissolved metal species to sediments removes the metal from the water column and stores it in the sediments, where it is less biologically available. Desorption returns the metal to the water column, where it becomes biologically available again and where water flow may carry the metal to a new location where sorption and precipitation can recur. Metals may be desorbed from sediments if the water undergoes increases in salinity, decreases in redox potential, or decreases in pH. Sorption and desorption processes are discussed further in Chapter 5.

In the water environment, nonradioactive metals are of greatest environmental and health concern when in dissolved forms, where they are more mobile and more biologically available than are particulate forms, although ingestion and inhalation of particulates containing metals also can be a serious health hazard. Radioactive metals are hazardous because of their ionizing emissions as well as their chemical toxicity and may be harmful in both dissolved and particulate forms, even without metal species entering the body.

4.1.2 MOBILITY OF METALS IN THE WATER ENVIRONMENT

Two properties, solubility and the tendency to sorb to soil particles (sorption coefficient) largely govern the mobility of metals in the water environment. Metals can take many forms in environmental soil/water systems and the mobility of each form can depend in a different way on environmental conditions.

The important metal forms are

1. Solid elemental metal precipitates.
   a. These may be particles of colloid size or larger. Colloids remain suspended in water and are mobilized by water movement. Larger particles may settle out and require stronger flows to move them as sediments.
2. Solid metal compounds formed by weathering of minerals and by reactions of dissolved metal cations with water and other dissolved species such as carbonate, fluoride, and sulfate.
   a. These may be colloid size or larger.
3. Dissolved metal cations.
4. Dissolved metal compounds, such as carbonate and hydroxy complexes.
5. Metal species sorbed to solid soil and sediment surfaces.
   a. Sorption processes may be reversible to some degree, resulting in a retardation of dissolved metal movement relative to water flow, or irreversible, resulting in immobilization of metal species, except for erosion mechanisms.
   b. Dissolved or colloidal solids may become sorbed to solid surfaces.

RULES OF THUMB

1. Dissolved forms of metals move with surface water and groundwater flows.
2. Metals in particulate form can be transported with sediments by wind and in moving water.
3. Both dissolved and particulate forms of metals may sorb to organic soil solids, where they can be immobilized or carried along with eroding soils.
4. Metals pose the greatest environmental risks when particulate metals encounter environmental conditions that increase their solubility.

4.1.3   GENERAL BEHAVIOR OF DISSOLVED METALS IN WATER

It can be misleading to think in terms of the solubility of elemental metals. For example, to say that "iron is more soluble under reducing conditions than under oxidizing conditions," does not call attention to the fact that it is not elemental iron that is more soluble; it is the iron compounds that can be formed which may (or may not) be more soluble under reducing conditions.

Reactions of metal cations with water (hydrolysis) are the usual criteria for assessing whether a metal is soluble or insoluble under certain redox and pH conditions. When a metal such as iron is said to be insoluble under oxidizing conditions and soluble under reducing conditions, what actually is meant is that the compounds formed by reaction of the metal cation with water under oxidizing conditions are insoluble; under reducing conditions iron does not hydrolyze in water and can remain as a dissolved cation (with a hydration shell). The same is true for pH conditions; metals tend to be less soluble at high pH because their cations often react in high pH water to form low-solubility hydroxides and oxides. At low pH, where hydroxide concentrations are low (see Chapter 3), they may remain as soluble hydrated cations.

4.1.3.1   Hydrolysis Reactions

The simplest form of a dissolved metal is an elemental cation, such as Fe^{3+} or Zn^{2+}. However, elemental cations cannot exist as such in water solutions. Any charged species in solution will interact with other charged or polar species because of electrical forces. Because water molecules are polar (Chapter 2), metal cations always attract a multilayered hydration shell of water molecules by electrostatic attraction of the negative end of the water molecules (the oxygen end, see Chapter 2) to the positive charge of the cation, as described by Equation 4.1 and illustrated in Figure 4.1.

$$M^{+n} \xrightarrow{\text{H}_2\text{O}} M(H_2O)_x^{+n} \qquad (4.1)$$

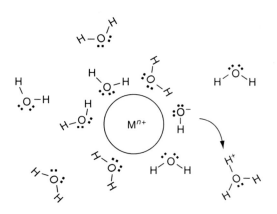

FIGURE 4.1 Water molecules form a hydration shell around dissolved metal cations. Molecules in the hydration shell can lose a proton to bulk water molecules, as indicated by the arrow, leaving a hydroxide group bonded to the metal. In this way, the hydrated metal behaves as an acid. Eventually, the metal may precipitate as a hydroxide compound of low solubility.

where

M is a metal cation

n is the number of positive charges on the cation

x is the maximum number of water molecules in the innermost hydration shell

x is 6 for most cations

Depending on the strength of the electrostatic attraction between the cation and the water molecules, the water molecules closest to the metal cation (in the innermost hydration shell) may bond as a ligand, forming a metal–water complex (see second footnote on page 110). The strength of the electrostatic attraction depends on the magnitude of the cation charge, the cation radius, and, to a lesser extent, the electronegativity of the metal. The strongest bonds are formed with cations having the smallest radius, greatest positive charge, and electronegativity greater than 1.8 (Wulfsburg, 1987).

4.1.3.2 Hydrated Metals as Acids

Hydrated metal ions can behave as acids by releasing protons (H^+) from their water ligands that then become attached to the surrounding free H_2O molecules, forming acidic hydrated protons, H_3O^+ and $H_5O_2^+$ (see Section 4.2.1). The stronger the bond between the water ligand and the metal cation, the more readily a proton is released to surrounding water molecules and the more acidic is the hydrated metal cation. The process can continue stepwise up to n times to make a neutral metal hydroxide.*

---

* For simplicity, hydrated protons will be designated only by the most common form, H_3O^+. Because solvent water molecules are normally not included when balancing a chemical reaction, the stoichiometry of acid-base reactions remains unchanged, regardless of how many water molecules are shown attached to an acidic proton, i.e., whether the hydrated proton is designated by H^+, H_3O^+, $H_5O_2^+$, etc.

$$M(H_2O)_6^{+n} + H_2O \leftrightarrow M(H_2O)_5OH^{+(n-1)} + H_3O^+ \tag{4.2}$$

$$M(H_2O)_5OH^{+(n-1)} + H_2O \leftrightarrow M(H_2O)_4(OH)_2^{+(n-2)} + H_3O^+ \tag{4.3}$$

For example, with Fe^{3+}, it takes three proton transfer steps to form neutral ferric hydroxide:

$$Fe(H_2O)_3^{3+} + H_2O \leftrightarrow Fe(H_2O)_2OH^{2+} + H_3O^+ \tag{4.4}$$

$$Fe(H_2O)_2OH^{2+} + H_2O \leftrightarrow Fe(H_2O)(OH)_2^+ + H_3O^+ \tag{4.5}$$

$$Fe(H_2O)(OH)_2^+ + H_2O \leftrightarrow Fe(OH)_3(s) + H_3O^+ \tag{4.6}$$

Adding Equations 4.4 through 4.6 gives the overall reaction:

$$Fe(H_2O)_3^{3+} + 3H_2O \leftrightarrow Fe(OH)_3(s) + 3H_3O^+ \tag{4.7}$$

With each step, the hydrated metal is progressively deprotonated, forming polyhydroxides and becoming increasingly insoluble. At the same time, the solution becomes increasingly acidic due to the formation of more H_3O^+. Eventually, the metal may precipitate as a low-solubility hydroxide. The degree of acidity induced by metal hydration is the greatest for cations having the greatest electronegativity, which are those of high charge and small size. All metal cations with a charge of $+3$ or more are moderately strong acids. This process is one source of acidic water draining from mines.

RULES OF THUMB

1. Only polyvalent cations (e.g., Fe^{3+}, Zn^{2+}, Mn^{2+}, and Cr^{3+}) have large enough charges to attract water molecules strongly enough to act as acids, by causing the release of H^+ from water molecules in the hydration sphere. Monovalent cations, such as Na^+, do not act as acids at all.

2. The interactions of metal cations with water, Equations 4.2 through 4.7, cause the solubility of metal species in water to be dependent on pH and redox potential.

   a. Low pH (high H_3O^+ concentration and high acidity) increases metal solubility by shifting the equilibria of Equations 4.2 through 4.7 to the left, decreasing the formation of less soluble metal polyhydroxides.

   b. High pH (low H_3O^+ concentration and low acidity) decreases metal solubility by shifting the equilibria of Equations 4.2 through 4.7 to the right, increasing the formation of less soluble metal polyhydroxides.

RULES OF THUMB (Continued)

    c. Low redox potentials (reducing conditions, low-to-zero dissolved oxygen (DO) levels, where electron donors are more common than electron acceptors) increase the solubility of many metals (see Section 4.1.5) by promoting lower oxidation numbers for metal cations (lower positive charge, e.g., Fe^{2+} rather than Fe^{3+}). For cations with lower positive charge, the equilibria of Equations 4.2 through 4.7 are maintained more strongly to the left, resulting in less formation of low-solubility polyhydroxides.

    d. High redox potentials (oxidizing conditions, high DO levels, where electron acceptors are more common than electron donors) decrease the solubility of many metals by promoting higher oxidation numbers for metal cations (higher positive charge, e.g., Fe^{3+} rather than Fe^{2+}). For cations with higher positive charge, the equilibria of Equations 4.2 through 4.7 are maintained more strongly to the right, resulting in greater formation of low-solubility polyhydroxides.

  3. The presence of dissolved species such as sulfide or carbonate, which form low-solubility compounds with metal cations, can largely negate the above generalizations by competing with hydroxide formation.

4.1.4 Influence of pH on the Solubility of Metals

All the reactions (Equations 4.2 through 4.7) are reversible, with H_3O^+ on the right side. This means that the equilibria of these reactions shift to the left if the concentration of H_3O^+ is increased (by adding more acid) and to the right if it is decreased (by adding a base). Thus, the formation of metal hydroxides by hydration of metal cations is sensitive to the solution pH. Considering the overall reaction, Equation 4.7, we see that lowering the pH (increasing the concentration of H_3O^+) shifts the equilibrium of Equation 4.7 to the left, tending to dissolve any solid metal hydroxide that has precipitated. Raising the pH (increasing the concentration of OH^-) consumes H_3O^+ and shifts the equilibrium of Equation 4.7 to the right, precipitating more insoluble metal hydroxide. Thus, one may say that the metal becomes more soluble at lower pH and less soluble at higher pH, even though what actually occurs is that the hydrated metal forms less soluble hydroxide at higher pH.

However, if the pH is raised too high, precipitated metal hydroxides can redissolve (see Figure 4.2). At high pH values, a metal hydroxide may form complexes with OH^- anions to become a negatively charged ion having increased solubility. For example, precipitated $Fe(OH)_3$ can react with OH^- anions as follows:

$$Fe(OH)_3 + OH^- \leftrightarrow Fe(OH)_4^- \tag{4.8}$$

$$Fe(OH)_4^- + OH^- \leftrightarrow Fe(OH)_5^{2-} \tag{4.9}$$

Negatively charged polyhydroxide anions are more soluble because their ionic charge attracts them strongly to polar water molecules. As shown in Figure 4.2,

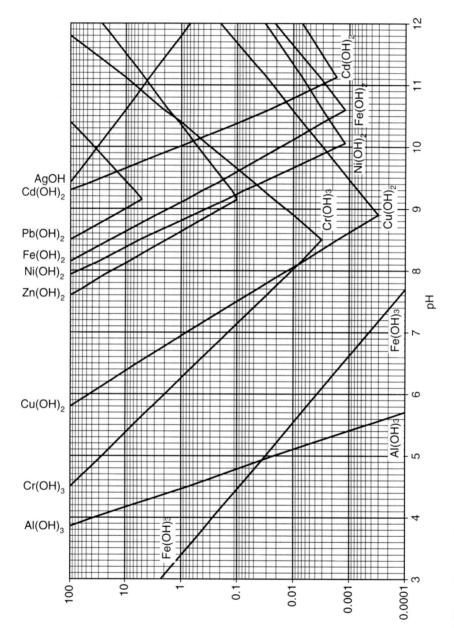

FIGURE 4.2 Theoretical solubilities of some metal hydroxides versus pH.

the high value of pH, where solubility begins to increase again, varies from metal to metal. Alkaline water provides a buffer against pH changes. In alkaline water, the tendency of metals to make water acidic is diminished by reactions like Equations 4.11 through 4.13.

EXAMPLE 1

Effect of Dissolved Metal on Alkalinity

A sample of groundwater contains a high concentration of dissolved iron, about 20 mg/L. At the laboratory, alkalinity is measured to be 150 mg/L for $CaCO_3$. Is this laboratory measurement of alkalinity likely to accurately represent the groundwater alkalinity?

Answer:

Soluble inorganic iron is in the ferrous form, Fe^{2+}. Because of its small charge, loss of protons from the hydration sphere is not a significant process for hydrated ferrous iron Fe^{2+}, denoted in Equation 4.10 by $Fe(H_2O)_6^{2+}$. However, when a groundwater sample is exposed to air, oxygen (an electron acceptor) dissolving from the atmosphere can oxidize Fe^{2+} to the ferric form, Fe^{3+}. This process is often enhanced by aerobic iron bacteria (see reaction step 2 in Figure 4.3). Depending on the pH, hydrated Fe^{3+} can lose protons from its hydration sphere to any bases present, including water molecules and hydroxyl ions (OH^-), forming ferric hydroxide species and making the solution more acidic. The acidic behavior of hydrated Fe^{3+} occurs to a greater extent at higher pH. Equation 4.10 represents the overall oxidation reaction that converts dissolved ferrous iron to precipitated ferric hydroxide:

$$4Fe(H_2O)_6^{2+} + O_2 \leftrightarrow 4Fe(OH)_3(s) + 14H_2O + 8H^+ \tag{4.10}$$

$Fe(OH)_3$ is a yellow to red-brown precipitate often seen on rocks and sediments in surface waters with high iron concentrations.

The molar concentration of H^+ formed by Equation 4.10 can be up to 2 times the Fe^{2+} molar concentration, depending on the final pH. Each H^+ released will neutralize a molecule of base, consuming some alkalinity, by reactions such as

$$H^+ + OH^- \leftrightarrow H_2O \tag{4.11}$$

$$H^+ + HCO_3^- \leftrightarrow H_2CO_3 \tag{4.12}$$

$$H^+ + CO_3^{2-} \leftrightarrow HCO_3^- \tag{4.13}$$

We will assume a worst-case scenario with respect to affecting the alkalinity, where the pH is high enough that the equilibrium of Equation 4.10 goes essentially to completion to the right side as additional oxygen dissolves from the atmosphere. The atomic weights of hydrogen and iron are 1 and 56 g/mol, respectively. If the equilibrium of Equation 4.10 is completely to the right, 1 mole (56 g) of Fe^{2+} will produce 2 moles of H^+ (2 g). At the time of sampling, the concentration of dissolved Fe^{2+} (as $Fe(H_2O)_6^{2+}$) was about 20 mg/L and all is eventually oxidized to Fe^{3+} (as $Fe(OH)_3$). The molar concentration of iron is

$$\frac{0.020 \text{ g/L}}{56 \text{ g/mol}} = 0.00036 \text{ mol/L or } 0.36 \text{ mmol/L}$$

By Equation 4.10, the moles of H^+ produced are two times the moles of iron:

$$\text{Moles of } H^+ = 2 \times 0.36 \text{ mmol/L} = 0.72 \text{ mmol/L}$$

$$\text{Grams of } H^+ = 0.72 \text{ mmol/L} \times 1 \text{ mg/mmol} = 0.72 \text{ mg/L}$$

We must now determine what effect this quantity of H^+ will have on the alkalinity. Alkalinity is measured in terms of a comparable quantity of $CaCO_3$. The molecular weight (MW) of $CaCO_3$ is 100 g/mol, and it dissolves to form the doubly charged ions Ca^{2+} and CO_3^{2-}. Alkalinity is a property of the CO_3^{2-} anion, which consumes acidity by accepting 2 H^+ cations:

$$2H^+ + CO_3^{2-} \leftrightarrow H_2CO_3 \tag{4.14}$$

Therefore, 0.72 mmol/L of H^+ will react with $0.72/2 = 0.36$ mmol/L of CO_3^{2-}, and 0.36 mmol/L of $CaCO_3$ is required as a source of the CO_3^{2-}.* From the definition of alkalinity, the change in alkalinity is equal to the change in concentration of $CaCO_3$, in milligram per liter.

$$0.36 \text{ mmol/L of } CaCO_3 = 0.36 \text{ mmol/L} \times \frac{100 \text{ mg}}{1 \text{ mmol}}$$
$$= 36 \text{ mg/L} = \text{change in alkalinity}$$

$$\text{Groundwater alkalinity at time of sampling}$$
$$= \text{Lab. measured alk.} + \text{alk. lost by Equation 4.10}$$

The maximum possible value for the original alkalinity of the groundwater before exposure to air was

$$150 \text{ mg/L} + 36 \text{ mg/L} = 186 \text{ mg/L as } CaCO_3$$

The laboratory alkalinity measurement was lower than the actual groundwater alkalinity, with a maximum error of about 21%.

In the above example, the pH was assumed high enough to maintain the equilibrium of Equation 4.10 entirely to the right. Under these conditions, essentially all the H^+ added to the solution will react by Equation 4.13 to form H_2CO_3. Thus, there would be no significant net change in aqueous H^+ (as H_3O^+) and little corresponding change in pH. In practice, the changes in concentrations brought about by Equations 4.10 through 4.13 will never cause the equilibria of the reactions to go completely to the right. Thus, the H^+ added will never react completely with the alkalinity and there will always be at least a small net increase in the H_3O^+ concentration, accompanied by a corresponding small decrease in pH.

This example illustrates the pH buffering effect of alkalinity. The addition of H^+ to the solution by hydration of metal ions, as in Equation 4.10, will not change the pH greatly as long as some alkalinity remains, because the added H^+ is taken up by carbonate species in the water. This is also true for the addition of H^+ from other sources, such as mineral and organic acids.

---

* Equation 4.14 can be obtained by adding Equations 4.12 and 4.13.

4.1.5   INFLUENCE OF REDOX POTENTIAL ON THE SOLUBILITY OF METALS

The redox potential (see Chapter 3) influences metal solubility because it can influence the electron structure of metal atoms and, thereby, the metal's ability to react with other substances to form compounds of varying solubilities. This was the case in Example 1, where an increase in DO (an electron acceptor) raised the redox potential. However, the redox potential may also influence other substances, such as sulfate, in a manner more significant than its effect on metal species. For example, the oxidation state of lead is not particularly sensitive to redox changes within common environmental conditions. However, if sulfate is present, a zero or negative value for the redox potential (anaerobic conditions) will cause the reduction of sulfate to sulfide and dissolved lead species will precipitate as insoluble lead sulfide. In the absence of sulfate or sulfide, the lead species may remain relatively soluble.

Although solubility is, perhaps, the most important property governing a contaminant's mobility in the environment, the mobility of a substance does not depend only on its solubility. As discussed in Section 4.1.2, metals can take many forms in environmental soil–water systems and the mobility of each form depends in a different way on environmental conditions.

4.1.5.1   Redox-Sensitive Metals: Cr, Cu, Hg, Fe, Mn

It is useful to classify metals as redox sensitive or insensitive, according to their redox-dependent solubility. Redox-sensitive metals are those that can undergo changes in oxidation state (valence shell electron structure) under common environmental conditions, often resulting in changes in solubility because of the formation of new compounds by reaction with water.

Redox-sensitive metals (Cr, Cu, Hg, Fe, and Mn) can change their oxidation state within the redox conditions common in the environment. Under oxidizing conditions and pH greater than about 5.5, they react with water to form low-solubility* hydroxides and oxides. For example, under oxidizing conditions, concentrations of dissolved iron are limited by precipitation of insoluble $Fe(OH)_3$, dissolved manganese by precipitation of insoluble MnO_2, dissolved chromium by precipitation of insoluble $Cr(OH)_3$, and dissolved copper by precipitation of an insoluble cupric ferrite mineral $CuFe_2O_4$.

Under reducing conditions and pH less than about 7, with no sulfide present, iron is present as the soluble cation Fe^{2+}. Above pH 7, $Fe(OH)_2$, about 10^5 times more soluble than $Fe(OH)_3$, is formed. The other redox-sensitive metals behave similarly under reducing conditions, being present as either the soluble cation or a relatively soluble compound.

4.1.5.2   Redox-Insensitive Metals: Al, Ba, Cd, Pb, Ni, Zn

These metals do not change their oxidation state within the redox conditions common in the environment. In their normal oxidation state, these metals do not react strongly with water to form insoluble oxides and hydroxides.

---

* The solubility table inside the back cover contains definitions for degrees of solubility.

Under oxidizing conditions and the absence of anions with which they can react, they tend to remain as dissolved cations. In the presence of reactants other than water, they can form carbonates, phosphates, sulfates, and oxides/hydroxides whose solubilities depend more on pH than on redox potential.

Although, reducing conditions have little direct effect on these metals as cations, under reducing conditions, with sufficient sulfide present, all can form sulfides of low solubility.

4.1.5.3 Redox-Sensitive Metalloids: As, Se

Arsenic and selenium tend to behave oppositely to the redox-sensitive metals. They are more soluble under oxidizing conditions than under reducing conditions.

Under oxidizing conditions, they form the oxygen anions arsenate (AsO_4^{3-}) and selenite (SeO_3^{2-}). These react with Fe, Mn, and Pb cations to form moderately soluble compounds.

Under reducing conditions, selenium forms insoluble elemental selenium and iron selenide ($FeSe_2$). Arsenic forms sulfides of very low solubility.

4.2   METAL WATER QUALITY STANDARDS

A metal water quality standard may be written for the dissolved, potentially dissolved, total recoverable, or total forms.

- *Dissolved*: Sample is filtered on site immediately after or, preferably, during collection through a 0.45 µm filter and is then acidified to pH 2 for preservation before analysis. Acidification prevents precipitation of any dissolved metal before analysis. This procedure omits from the analysis metals adsorbed on suspended sediments.

  It is essential that the sample be filtered immediately after or during collection. A true measure of dissolved metals in the source water cannot be obtained after transporting an unfiltered sample to a laboratory. An unfiltered sample collected for dissolved metals cannot be acidified for preservation because the lower pH could cause some precipitated metals in the original sample to dissolve. It cannot be transported without acidification because potential gain or loss of dissolved CO_2 or O_2 can result in changes in pH or redox potential that could dissolve or precipitate metals en route, resulting in a measurement not representative of the source water.
- *Potentially dissolved*: Sample is acidified to pH 2, held for 72–90 h, then filtered through a 0.45 µm filter and analyzed. This procedure is intended to simulate the possibility that metals bound in suspended sediments might be transported into more acidic environmental conditions and partially dissolve. It measures the metals dissolved at the time of sampling, in addition to a portion of the metals initially bound to suspended sediments and released during the holding period at low pH.
- *Total recoverable*: Sample is acidified to pH 2 and analyzed without filtering. This procedure measures all metals, dissolved and initially bound to

suspended sediments. Some "unrecoverable" metals may remain as suspended mineral sediments or strongly sorbed to sediments and not be analyzed.

- *Total*: Sample is "digested" in an acidic solution until essentially all the metals present are extracted into soluble forms for analysis.

4.3 CASE STUDY 1

4.3.1 TREATMENT OF TRACE METALS IN URBAN STORMWATER RUNOFF

Stormwater runoff carries solid and dissolved forms of metals as well as other chemical pollutants, including soil sediments and various kinds of debris, such as paper, plastic, garbage, leaf, and plant litter. Because a major constraint on stormwater treatment systems is that they provide passive or near-passive treatment, the biggest challenge for stormwater treatment is the removal of dissolved pollutants by means that require minimal or no operator control or external power sources. The general methods available for water treatment are control of the redox potential and pH, addition of chemical reactants to aid precipitation, and length of detention time in vaults or basins. In a passive or near-passive treatment system.

- The redox potential is regulated by the DO level. Dissolved oxygen is depleted mainly by biodegradation processes that decompose organic matter. Two methods for increasing DO are (passively) by encouraging oxygen diffusion from the atmosphere by inducing turbulence in the feed flow to the treatment basin and (near-passively) by mechanical diffusion via perforated tubing connected to an air pump.
- pH control is usually not appropriate for passive treatment because it normally requires chemical additions. The pH in a stormwater treatment system is usually determined by the prevailing environmental conditions, and normally is in the range of 6–9.
- Addition of chemical reactants other than oxygen is an approach that requires regular maintenance and, therefore, is not an option in passive treatment systems. A passive treatment system will contain all chemical reactants that flow into it from its drainage basin. Since the reactants present can have a strong influence on the successful operation of a stormwater treatment system, it is helpful to determine the typical site-specific chemical composition of water to be treated.
- The retention time of water in the treatment system is important because it determines the amount of time available for chemical reactions to remove dissolved pollutants and for suspended solids to coagulate and settle out. Detention time is determined by the magnitude of water flows that must be handled and the design capacity of the treatment system.

It is evident from the above that a passive/near-passive stormwater treatment system has some control over just two variables: redox potential and retention time. Optimal adjustment of these parameters will depend on site-specific conditions that influence the other parameters of water composition and pH.

Typical stormwater treatment systems, such as stormwater collection vaults, are designed primarily to remove sediment and debris. They serve only incidentally to remove other pollutants such as metals, accomplishing this mainly by having a long enough retention time, with oxidizing conditions, to precipitate a fraction of the dissolved metals as hydroxides and oxides. However, at many locations, incidental metal removal may not be adequate, if nonpoint source contamination by metals is an important secondary (and sometimes primary) cause of water quality impairment.

Ideally, a stormwater vault collects storm runoff, retains the sediments, and immobilizes the pollutants so that they do not enter surface water bodies or groundwater aquifers. Although this goal is never perfectly attained in practice, knowledgeable management of a storm vault's design and operating conditions can go a long way toward assuring that these best management practice (BMP) devices are in fact operating at their best. Unfortunately, the optimal environmental conditions for removing pollutants are often different for different pollutants. For passive treatment, this makes it necessary to find the most efficient compromise for each site. Nevertheless, because most sites have a limited variety of critical pollutants, satisfactory operating conditions are often attainable.

Dissolved pollutants may include metals, nitrogen and phosphorus nutrients, and soluble organic compounds. Although the focus of this case study is the removal of metal compounds from stormwater, it is necessary to understand the behavior of nutrients and organic matter in a stormwater detention basin because the redox potential is strongly affected by biodegradation processes.

- Dissolved organic substances that biodegrade are removed by biodegradation, which converts them to carbon dioxide, water, and a variety of less objectionable organic substances. They are referred to as BOD (biological oxygen demand) because they consume DO in the course of their removal by biodegradation.
- Dissolved metals are removed from stormwater by precipitation as insoluble compounds and adsorption to the surfaces of sediments and organic debris.
- Pesticides and herbicides are removed by biodegradation and adsorption to the surfaces of sediments and organic debris.
- Phosphorus, from fertilizers, cleaning products, and animal wastes, is removed by precipitation as insoluble compounds, adsorption to the surfaces of sediments and organic debris, and utilization as an essential nutrient by biodegrading microbes.
- Nitrogen, from fertilizers and animal wastes, may be present as organic nitrogen, ammonia, nitrites, or nitrates. Organic nitrogen (protein, urea, etc.) is removed by conversion to ammonia. Ammonia is removed by adsorption to solid surfaces and by bacterial conversion to nitrogen gas. If the DO level is high enough, ammonia will be converted to nitrites and nitrates. All nitrite and nitrate compounds have high-solubility and low adsorption potential. Although a small fraction of the nitrates and nitrites present may be utilized as nutrients by biodegrading microbes, the much larger dissolved fraction cannot be removed in a stormwater vault or detention basin.

Good operation of a stormwater treatment BMP means maintaining conditions that

- Trap sediments so they are not washed out with overflow releases.
- Promote the conversion of dissolved metals to solid forms by precipitation and sorption to sediments.
- Retard dissolution of metals already precipitated and desorption of metals sorbed to sediments.
- Encourage biodegradation of dissolved and solid organic substances, such as pesticides and petroleum products.

Unfortunately, no single set of operating conditions can realize all these goals simultaneously. For example, maintaining adequate oxidizing conditions (greater than about 2 ppm of DO) in a stormwater vault encourages biodegradation of organic matter and precipitation of certain metals such as iron, manganese, and copper. On the other hand, some pollutants such as arsenic, selenium, lead, and zinc tend to be present as soluble species under oxidizing conditions. These pollutants are best retained as solids when reducing conditions exist (less than 1 ppm of DO). When sufficient sulfate is present, reducing conditions are best for retaining most metals because the sulfate is reduced to sulfide, and metals precipitate as insoluble metal sulfides. However, if the organic load is high, reducing conditions can cause odor problems because of the generation of ammonia and hydrogen sulfide.

The bottom line is that proper operation of a stormwater vault is site-specific and may require some compromises. In some cases, it might be necessary to operate one or more stormwater vaults in a manner that provides sequential treatment in separate oxidizing and reducing zones.

4.3.2 BEHAVIOR OF COMMON STORMWATER POLLUTANTS UNDER OXIDIZING AND REDUCING CONDITIONS

The behavior of the trace metals most commonly measured in urban runoff (Pb, Cu, Zn, Cd, Cr, and Ni; plus the metalloid As) is strongly affected by the presence of iron and manganese, two generally more abundant metals not often measured in urban runoff.

For most trace metals (except for Cr, Cu, and Hg), the conditions of greatest solubility are low pH and oxidizing positive redox potential. The most reduced form of metals is the insoluble elemental form, but environmental conditions seldom reach low enough redox potentials to precipitate elemental metals. In general, less severe reducing conditions that do not precipitate elemental metals help to retain trace metal cations sorbed to organic sediments, as evidenced by the efficiency of wetlands in this respect. In addition, under reducing conditions, sulfate is reduced to sulfide, one of the most efficient precipitants of dissolved metals. As long as the pH remains between 6.5 and about 9, none of the trace metals commonly monitored in stormwater increases in solubility with decreasing oxidation potential.

However, the more abundant iron and manganese behave differently. Both of these metals pass through a soluble reduced state at environmental pH values when the aquatic oxidation potential goes negative. Therefore, precipitated Fe and Mn may be dissolved from sediments under reducing conditions. Because dissolved trace metals tend to sorb strongly to iron and manganese precipitates, the formation of Fe and Mn precipitates often serves as a sink for dissolved trace metals. If the Fe and Mn solids dissolve under reducing conditions, the sorbed trace metals can become mobilized. Thus, if Fe and Mn are present in high concentrations, reducing conditions could mobilize trace metals. If Fe and Mn concentrations are low, reducing conditions are not likely to mobilize trace metals.

Two conditions could serve to retain trace metals on sediments under reducing conditions in the presence of Fe and Mn solids:

- If the sulfate concentration is high, it will be reduced to sulfide and react to form insoluble Fe and Mn sulfides, which will remain efficient scavengers of dissolved trace metals. In addition, all trace metal sulfides are insoluble.
- If the organic content of the sediments is high, trace metals will sorb preferentially to most organic solids and be retained under reducing conditions, as in a wetland.

However, if the organic content of trapped sediments is high, the combination of oxidizing conditions with high COD or BOD levels may be of concern. These are the conditions that promote oxidation of organic matter with the concurrent generation of dissolved carbon dioxide, resulting in lowering pH to values as low as three or four. Under conditions in a sediment trap, low pH could be the most significant environmental factor for mobilizing metals. Thus, low DO, which will slow biodegradation processes, may be a benefit in the case of high BOD levels in sediment traps. The use of stormwater vaults for passive treatment of runoff will always represent performance compromises, but by adapting to site conditions, these BMPs can make important improvements to water quality.

4.4   CASE STUDY 2

4.4.1   Acid Rock Drainage

The main cause of acid rock drainage* is oxidation of iron pyrite. Iron pyrite, FeS_2, is the most widespread of all sulfide minerals and is found in many ore bodies. The somewhat complicated, but very efficient, process by which acid rock drainage develops is first described below in detail and then summarized to help make the overall process more clear.

During mining operations, particularly coal mining, iron pyrite in the ore becomes exposed to air and water, causing it to be oxidized to sulfuric acid and ferrous ion:

$$FeS_2 + \frac{7}{2}O_2 + H_2O \leftrightarrow Fe^{+2} + 2SO_4^{-2} + 2H^+ \qquad (4.15)$$

---

* Also called by the less general term acid mine drainage. Strictly speaking, acid mine drainage refers to water exiting from underground mine workings, while acid rock drainage includes water that drains through tailings piles and other waste rock depositories.

There is usually enough moisture in mine wastes and mine workings to allow Equation 4.15 to occur, which releases acidity (H^+) and dissolved ferrous iron into the water.

Next, dissolved ferrous iron (Fe^{+2}) is oxidized slowly by DO to ferric iron (Fe^{+3}), consuming some acidity:

$$Fe^{+2} + \frac{1}{4}O_2 + H^+ \leftrightarrow Fe^{+3} + \frac{1}{2}H_2O \qquad (4.16)$$

Above pH 4 and in the absence of iron-oxidizing bacteria, Equation 4.16 is the rate-limiting step in the reaction sequence. However, below pH 4 and in the presence of iron-oxidizing bacteria, the rate of Equation 4.16 is greatly accelerated, a millionfold or more. Ferric iron formed in Equation 4.16 can further oxidize pyrite, as in Equation 4.17, where ferric iron is reduced back to Fe^{+2}, releasing much more acidity.

$$FeS_2(s) + 14Fe^{+3} + 18H_2O \leftrightarrow 15Fe^{+2} + 2SO_4^{-2} + 16H^+ \qquad (4.17)$$

By Equation 4.17, eight times more acidity is generated when ferric iron oxidizes pyrite than when DO serves as the oxidant in Equation 4.15 (16 equivalents compared to 2 equivalents, per mole of FeS_2). In the pH range from 2 to 7, pyrite oxidation by Fe^{+3} (Equation 4.17) is kinetically favored over abiotic oxidation by oxygen (Equation 4.16). In addition, Equation 4.17 returns soluble Fe^{+2} to the reaction cycle where it consumes acidity via Reaction 2.

Overall, four equivalents of acid are formed for each mole of FeS_2 oxidized in the cyclic reaction sequence of Equations 4.16 and 4.17. If bacterially mediated oxidation is occurring in Equation 4.16, the reaction cycle can be accelerated over a millionfold.

Ferric iron also hydrolyzes (reacts with water), releasing more acid to the water and forming insoluble ferric hydroxide, which can coat streambeds with the yellow-orange deposits known as "yellow boy":

$$Fe^{+3} + 3H_2O \leftrightarrow Fe(OH)_3(s) + 3H^+ \qquad (4.18)$$

Precipitated $Fe(OH)_3$ serves as a reservoir for dissolved Fe^{+3}. If the generation of Fe^{+3} by Equation 4.17 is stopped because of lack of oxygen, then Fe^{+3} is supplied by dissolution of solid $Fe(OH)_3$ and is available to react via Equation 4.17.

4.4.1.1 Summary of Acid Formation in Acid Rock Drainage

The steps of acid formation in acid rock drainage are summarized below and illustrated in Figure 4.3.

Step 1: Iron pyrite, dissolved or solid, is oxidized by DO (Equation 4.15) producing Fe^{+3}, SO_4^{-2}, and lowering the pH.

Step 2: Fe^{+2} formed in step 1 is oxidized slowly by DO to Fe^{+3} (Equation 4.16). This is the rate-limiting step in the reaction sequence in the absence of

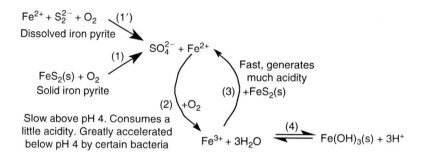

FIGURE 4.3 Reaction scheme for generation of acid mine drainage by pyrite oxidation.

iron-oxidizing bacteria. The abiotic rate decreases with lower pH. However, iron-oxidizing bacteria can greatly accelerate this step when the pH falls below 4.

Step 3: Fe^{+3} from step 2 is reduced rapidly back to Fe^{+2} by pyrite (Equation 4.17) generating much acidity. Ferrous iron, Fe^{+2}, generated in step 3 reenters the reaction cycle via step 2.

Step 4: A portion of ferric iron, Fe^{3+}, reacts with water (Equation 4.18) to form ferric hydroxide precipitate, $Fe(OH)_3(s)$, releasing more acidity. When the Fe^{3+} concentration diminishes, some $Fe(OH)_3$ precipitate can dissolve, acting as a reservoir for replenishing Fe^{3+} and maintaining the acid producing cycle.

As pH is lowered, step 1 becomes less important and the abiotic rate of step 2 decreases.

However, step 2 can be greatly accelerated by certain iron-oxidizing bacteria, Metallogenium, Ferrobacillus, Thiobacillus, and Leptospirillum. These bacteria derive energy from the oxidation of Fe^{+2} to Fe^{+3}. Below pH 4, iron-oxidizing bacteria can catalyze step 2, speeding up the overall reaction rate by a factor as large as one million, lowering the pH to two or less. Furthermore, these bacteria can tolerate high concentrations of dissolved metals (e.g., 40,000 mg/L Zn and Fe; 15,000 mg/L Cu) before experiencing toxic effects. They thrive in acid rock drainage waters as long as a minimal amount of oxygen is present. Once bacterial acceleration of iron pyrite oxidation occurs, it is hard to reverse.

RULE OF THUMB

Oxidation of iron pyrite is the most acidic of all common weathering reactions. The production of acid rock drainage can be a rapid, self-propagating, cyclic process that is accelerated by low pH and the presence of iron-oxidizing bacteria. The process will continue as long as oxygen, pyrite, and water are present.

4.4.1.2 Non-iron Metal Sulfides Do Not Generate Acidity

The abiotic oxidation by dissolved O_2 of non-iron metal sulfides does not generate significant amounts of acidity. The metals are released as dissolved cations, but acidity is not produced, for example:

$$CuS(s) + 2O_2 \leftrightarrow Cu^{+2}(aq) + SO_4^{-2}(aq)$$

$$ZnS(s) + 2O_2 \leftrightarrow Zn^{+2}(aq) + SO_4^{-2}(aq)$$

$$PbS(s) + 2O_2 \leftrightarrow Pb^{+2}(aq) + SO_4^{-2}(aq)$$

Two possible reasons for the lack of acid formation when non-iron sulfides are oxidized are

- The oxidation state of sulfur is different in iron pyrite than in other sulfides, occurring as S_2^{-2} in iron pyrite and as S^{-2} in other sulfides. The respective oxidation reactions of these two sulfur forms indicate that acid is produced only with S_2^{-2}:

    (iron pyrite) $S_2^{-2} + 7/2O_2 + H_2O \leftrightarrow 2\ SO_4^{-2} + 2\ H^+$
    (other sulfides) $S^{-2} + 2O_2 \leftrightarrow SO_4^{-2}$

- Cu^{+2}, Zn^{+2}, Pb^{+2}, etc., do not hydrolyze as extensively as Fe^{+3} does, so non-iron sulfides do not react significantly by reactions equivalent to Equation 4.17:

$$Fe^{+3} + 3H_2O \leftrightarrow Fe(OH)_3(s) + 3H^+$$

Ferric iron, Fe^{+3}, can oxidize other metal sulfides, such as ZnS (sphalerite), CuS (covellite), PbS (galena), and chalcopyrite ($CuFeS_2$), in a similar fashion to its oxidation of FeS_2, releasing metal cations into the water, but not generating acidity.

4.4.1.3 Acid-Base Potential of Soil

Acid-base potential (ABP) is a measure of how effectively the alkalinity (neutralization potential) in a soil or rock sample can neutralize the acid-producing potential resulting from the presence of pyrite in the sample. The ABP is equal to the equivalents of calcium carbonate ($CaCO_3$) in excess of the amount needed to neutralize the acid that could potentially be produced from oxidation of pyritic sulfur.

4.4.1.4 Determining the Acid-Base Potential

The presence of iron pyrite, oxygen, and water will not necessarily result in acid formation if sufficient alkalinity to buffer the water is also present. A semiempirical quantity called ABP can be used to estimate the likelihood that a soil or rock sample will generate acidity when exposed to air and water. The ABP might be used, for example, to evaluate a soil cap used to cover and impound waste materials. If precipitation passing through the soil cap were made more acidic because of iron pyrite in the soil, it might mobilize metals or other contaminants in the wastes and create a new hazard.

The ABP is calculated by

$$ABP = (\text{alkalinity}) - (31.25)\ (\text{wt\% pyritic sulfur}) \qquad (4.19)$$

where the ABP units are tons of acid-neutralizing substances (as $CaCO_3$-equivalents) per 1000 tons of solid material.

Any rock or earth material with a negative ABP is likely to have acidic leachate. If the ABP is −5 or more negative, the earth material has an acid-neutralizing deficiency of 5.0 tons $CaCO_3/1000$ tons material, and may be considered a potentially hazardous waste.

RULES OF THUMB

1. If the ABP is positive, leachate from the sample is likely to be basic.
2. If the ABP is negative, leachate is likely to be acidic.
3. If the ABP is −5, or more negative, the earth material is likely to be acidic enough to be defined as a potentially hazardous waste.

4.5   CASE STUDY 3

4.5.1   IDENTIFYING METAL LOSS AND GAIN MECHANISMS IN A STREAM

This study is modeled after parts of the report by Balistrieri et al. (1995).

Background: Measurements of pH and metal concentrations in a stream receiving acid rock drainage indicated that the stream was acidic and had elevated levels of chloride and metals just below the confluence with the mine drainage. The metal and chloride concentrations decreased again farther downstream. The downstream decrease of metal concentrations could be due to

- Dilution by higher quality water entering the stream or
- Chemical reactions such as sorption or precipitation.

Problem: How can you evaluate which mechanisms, dilution, evaporation, new water inputs, or chemical reactions, cause changes in stream metal concentrations?

Solution: Collect samples from several selected sites on the stream, including one site above the confluence with the mine drainage, and analyze them for the metals of interest and, also, a conservative constituent that is comparatively unreactive. In this case, we will use Cu and Zn as the metals and chloride anion (Cl^-) as the conservative constituent. Conservative constituents, such as chloride, do not adsorb or

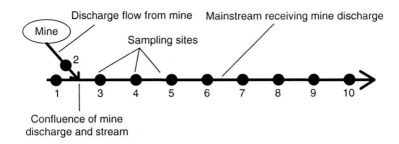

FIGURE 4.4 Line diagram of stream with mine discharge and sampling locations.

precipitate significantly; assuming no new sources, their concentration will change only because of dilution and water evaporation.

Take two chloride solutions with different starting concentrations of chloride and mix them in various ratios. When the chloride concentration of any of the mixtures is plotted against the mixing ratio (e.g., 1:1, 2:1, 3:1, etc.), the concentration of chloride in the mixture should lie on the straight line that connects the highest and lowest chloride concentrations. This is true for any conservative constituent. In this example, flow downstream of the confluence of stream and the mine drainage serves to mix their chloride concentrations and each location downstream represents a particular mixing ratio, where the chloride can only change by evaporation of stream water or by new water inputs.

Figure 4.4 is a line diagram of a stream receiving mine drainage. Sampling sites 1 through 10 are indicated schematically. It is known that the mine discharge (sampling point 2) carries higher concentrations of dissolved copper, zinc and chloride than are in the stream upstream at sampling point 1. For each sampling site, plot the dissolved metal concentration versus the chloride concentration and connect the points for sites 1 and 2, the sites having the lowest and highest chloride and metal concentrations, with a straight line, as in Figures 4.5 and 4.6. Chloride and metal concentrations are lowest at site 1, above the confluence of the stream and the mine drainage, and highest at site 2, in the mine drainage just before it flows into the stream. The straight line between these site points represents a wide range of possible dilution factors arising from the mine drainage mixing into the stream flow and from tributary drainages into the stream. This line represents the mixing curve for the conservative analyte chloride.

Site 1 is in the stream above the confluence with the mine drainage.

Site 2 is in the mine drainage just before it enters the stream.

Site 3 is in the stream just below the confluence.

Sites 4–10 are sequentially positioned farther downstream from the confluence.

If a metal concentration at a site downstream of the confluence lies on the chloride mixing line (within 25% of the predicted value), the metal concentration at that site results from the same causes as the site concentration of chloride, i.e., dilution or evaporation. If the point lies below the line, loss mechanisms other than

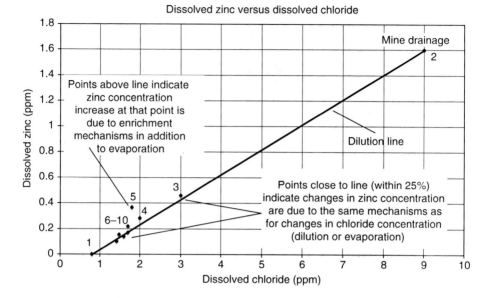

FIGURE 4.5 Plot of dissolved zinc versus dissolved chloride ratios. Numbered points indicate sampling sites in Figure 4.4.

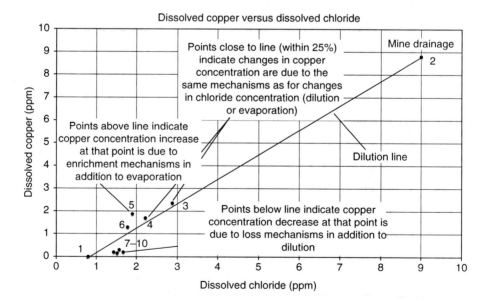

FIGURE 4.6 Plot of dissolved copper versus dissolved chloride ratios. Numbered points indicate sampling sites in Figure 4.4.

dilution have contributed to the metal decrease at that site. If the point lies above the line, enrichment mechanisms other than evaporation have caused a metal increase. This will be true even if additional dilution or evaporation occurs downstream from the confluence.

For Zn, Figure 4.5 shows that all points except for site 5 lie within 25% of the line-predicted value. At site 5, Zn concentrations are about 80% higher than predicted for a conservative substance and probably is caused by inputs of Zn from bank erosion and shallow groundwater inflows carrying dissolved zinc between site 5 and the confluence of the receiving stream and the mine drainage.

For Cu, Figure 4.6, sites 3 and 4 below the confluence show a decrease that is mainly due to dilution. As with Zn, Cu concentrations at site 5 are about 80% higher than what was predicted from simple mixing, again suggesting a source of metals between sites 4 and 5. The Cu increase is still apparent at site 6, although additional dilution has occurred. At sites 7–10, Cu concentrations are about 50%–73% lower than predicted by simple mixing. This suggests that chemical removal processes such as precipitation and adsorption to sediments are effective in this region of the stream.

Sediment measurements, not indicated in Figures 4.4 through 4.6, showed that the stream sediments are enriched significantly in Cu from the mine drainage confluence all the way downstream to site 10. On the other hand, the sediments are not much enriched in Zn just below the confluence, but Zn sediment enrichment increases farther downstream below site 5. The small chemical removal of Zn just below the confluence and enrichment of Zn in sediments farther downstream suggest that solid phase transport as stream sediment may be an important transport mechanism for Zn.

EXERCISES

1. Is a solution of ferric sulfate ($Fe_2(SO)_3$) in water expected to be acidic, basic, or neutral? Explain your answer.
2. Figure 4.2 shows that the solubilities in water of metal hydroxides decrease as pH increases up to a point. The solubility plots go through a minimum and then increase if pH continues to increase.
   a. Explain why this occurs.
   b. Would you expect a plot of metal sulfides to show similar behavior? Explain your answer.
3. A groundwater sample contains dissolved Fe^{2+}. What difficulties might be encountered in using this sample for determining the in situ pH of the groundwater?
4. In Figure 4.2, The least soluble hydroxides, below the pH where the minimum solubility is reached, are $Fe(OH)_3$, $Al(OH)$, and $Cr(OH)_3$. These are all trivalent, whereas all the divalent and univalent hydroxides are more soluble. Discuss why this behavior is expected. The small discrepancies with $Cr(OH)_3$ and $Cu(OH)_2$ may be due to differences in cation size or to experimental uncertainty.

5. An industrial discharge permit has a limit for chromium, copper, and cadmium of 0.01 mg/L. Assuming that the untreated discharge will always exceed this limit, what difficulties are encountered if you try to comply by using only pH adjustment? Use Figure 4.2, assuming that the only complexing anion present is OH^-. How might you meet the limit for all three metals?

REFERENCES

Balistrieri, L.S., Gough, L.P., Severson, R.C., and Archuleta, A., 1995, The biogeochemistry of wetlands in the San Luis Valley, Colorado: The effects of acid drainage from natural and mine sources, in Posey, H.H., Pendleton, J.A., and Van Zyl, D., eds., Summitville Forum Proceedings, Colorado Geological Survey, Special Publication 38, pp. 219–226.

Duffus, J.H., 2002, "Heavy metals" a meaningless term? *Pure Appl. Chem.*, 74 (5), 793–807.

Wulfsburg, G., 1987, *Principles of Descriptive Inorganic Chemistry*, Brooks/Cole–Wadsworth, Belmont, CA.

5 Soil, Groundwater, and Subsurface Contamination

5.1 NATURE OF SOILS

Water is always a potential conveyor of contaminants, whereas soil can be either an obstacle to contaminant movement or a contaminant transporter. The stationary soil matrix slows the passage of groundwater and provides solid surfaces to which contaminants can sorb, delaying or stopping their movement. On the other hand, soil can also move, carried by wind, water flow, and construction equipment. Moving soil, like moving water, transports the contaminants it carries. Predicting and controlling pollutant behavior in the environment requires understanding how soil, water, and contaminants interact. That is the subject of this chapter.

5.1.1 SOIL FORMATION

Soil is the weathered and fragmented outer layer of the earth's solid surface, initially formed from the original rocks and then amended by growth and decay of plants and organisms. The initial step from rock to soil is destructive "weathering." Weathering is the disintegration and decomposition of rocks by natural physical and chemical processes.

5.1.1.1 Physical Weathering

Physical weathering causes fragmentation of rocks, increasing the exposed surface area and, thereby, the potential for further, more rapid, weathering. Common causes of physical weathering are

- Expansion and contraction caused by heating and cooling.
- Stress forces caused by mineral crystal growth and the expansion and contraction of water when it freezes and melts in cracks and pores.
- Penetration of tree and plant roots.
- Scouring and grinding by abrasive particles carried by wind, water, and moving ice.
- Unloading forces that arise when rock-confining pressures are lessened by geologic uplift, erosion, or changes in fluid pressures. Unloading can cause cracks at thousands of feet below the surface.

5.1.1.2   Chemical Weathering

Chemical weathering of rocks causes changes in their mineral composition. Common causes of chemical weathering are

- Hydrolysis and hydration reactions (water reacting with mineral structures)
- Oxidation (usually by oxygen in the atmosphere and in water) and reduction (usually by microbes)
- Dissolution and dissociation of minerals
- Immobilization by precipitation, e.g., the formation of solid oxides, hydroxides, carbonates, and sulfides
- Loss of mineral components by leaching and volatilization
- Chemical exchange processes, such as cation exchange

Physical and chemical weathering processes often produce loose materials that can be deposited elsewhere after being transported by wind (aeolian deposits), running water (alluvial deposits), or glaciers (glacial deposits).

The next steps, after the rocks have been fractured and broken down, are the formation of secondary minerals (e.g., clays, mineral precipitates, etc.) and changes caused by plants and microorganisms.

5.1.1.3   Secondary Mineral Formation

Secondary minerals are formed within the soil itself by chemical reactions of the primary (original) minerals. The reactions forming secondary minerals are always in the direction of greater chemical stability under local environmental conditions. These reactions are facilitated by the presence of water, which dissolves and mobilizes different components of the original rocks, allowing them to react to form new compounds.

5.1.1.4   Roles of Plants and Soil Organisms

Plants and soil organisms play many complex roles. Roots form extensive networks permeating soil. They can exert pressures that compress aggregates in one location and separate them in another. Water uptake by roots causes differential dehydration in soil, initiating soil shrinkage and opening of many small cracks.

The plant root zone in the soil is called the rhizosphere. It is the soil region where plants, microbes, and other soil organisms interact. Soil organisms include thousands of species of bacteria, fungi, actinomycetes, worms, slugs, insects, mites, etc. The number of organisms in the rhizosphere can be 100 times larger than in non-rhizosphere soil zones. Root secretions and dead roots promote microbial activity that produces humic cements. Root secretions contain various sugars and aliphatic, aromatic, and amino acids, as well as mucigel, a gelatinous substance that lubricates root penetration. These substances and dead root material are nutrients for rhizosphere microorganisms. The root structure itself provides surface area for microbial colonization.

5.2 SOIL PROFILES

A vertical profile through soil, Figure 5.1, tells much about how the soil was formed. It usually consists of a succession of more or less distinct layers, or strata. The layers can form from aeolian or alluvial deposition of material, or from in situ weathering processes.

5.2.1 SOIL HORIZONS

When the layers develop in situ by the weathering processes described above, they form a sequence called horizons. The horizons are designated by the U.S. Department of Agriculture by the capital letters: O, A, E, B, C, and R, in order of farthest distance from the surface (see Figure 5.1).

O-horizon: Organic

- The top horizon: starts at the soil surface
- Formed from surface litter
- Dominated by fresh or partly decomposed organic matter

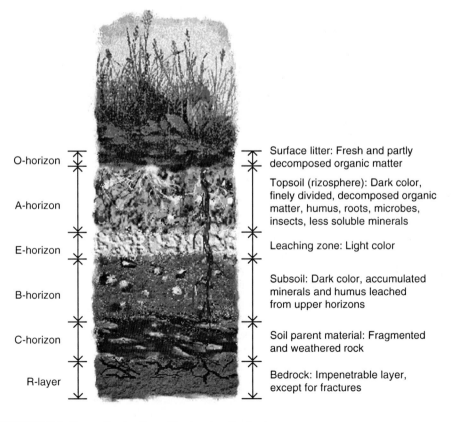

O-horizon	Surface litter: Fresh and partly decomposed organic matter
A-horizon	Topsoil (rizosphere): Dark color, finely divided, decomposed organic matter, humus, roots, microbes, insects, less soluble minerals
E-horizon	Leaching zone: Light color
B-horizon	Subsoil: Dark color, accumulated minerals and humus leached from upper horizons
C-horizon	Soil parent material: Fragmented and weathered rock
R-layer	Bedrock: Impenetrable layer, except for fractures

FIGURE 5.1 Generalized soil profile showing the horizon sequence.

A-horizon: Topsoil

- The zone of greatest biological activity (rhizosphere)
- Contains an accumulation of finely divided, decomposed, organic matter, which imparts a dark color
- Clays, carbonates, and most metal cations are leached out of the A-horizon by downward percolating water; less soluble minerals (such as quartz) of sand or silt size become concentrated in the A-horizon

E-horizon: Leaching zone (sometimes called the A-2 horizon)

- Light-colored region below the rhizosphere where clays and metal cations are leached out and organic matter is sparse

B-horizon: Subsoil

- Dark-colored zone where downward migrating materials from the A-horizon accumulate

C-horizon: Soil parent material

- Fragmented and weathered rock, either from bedrock or base material that has deposited from water or wind

R-layer: Bedrock

- Below all the horizons; consists of consolidated bedrock
- Impenetrable, except for fractures

5.2.2 SUCCESSIVE STEPS IN THE TYPICAL DEVELOPMENT OF A SOIL AND ITS PROFILE (PEDOGENESIS)

1. Physical disintegration (weathering) of exposed rock formations forms the soil parent material, the C-horizon.
2. The gradual accumulation of organic residues near the surface begins to form the A-horizon, which might acquire a granular structure, stabilized to some degree by organic matter cementation. This process is retarded in desert regions where organic growth and decay are slow.
3. Continued chemical weathering (oxidation, hydrolysis, etc.), dissolution, and precipitation begin to form clays.
4. Clays, soluble salts, chelated metals, etc. migrate downward through the A-horizon, carried by permeating water, to accumulate in the B-horizon.
5. The C-horizon, now below the O-, A-, and B-horizons, continues to undergo physical and chemical weathering, slowly transforming into B- and A-horizons, deepening the entire horizon structure.
6. A quasi-stable condition is approached in which the opposing processes of soil formation and soil erosion are more or less balanced.

5.3 ORGANIC MATTER IN SOIL

Soil organic matter influences the weathering of minerals, provides food for soil organisms, and provides sites to which ions are attracted for ion exchange. Only two types of organisms can synthesize organic matter from nonorganic materials. These are certain bacteria called autotrophs and chlorophyll-containing plants. Organic matter is developed in soil from the metabolism, wastes, and decay products of plants and soil organisms. For example, soil fungi metabolism produces excellent complexing agents such as oxalate ion and citric and other chelating organic acids. These promote the dissolution of minerals and increase nutrient availability. Some soil bacteria release the strong organic chelating agent 2-ketogluconic acid. This reacts with insoluble metal phosphates to solubilize the metal ions and release soluble phosphate, a plant nutrient.

As another example, oxalate ion is a metabolite of certain soil organisms. In calcium soils, oxalate forms calcium oxalate, $Ca(C_2O_4)$, which then reacts with precipitated metals (particularly Fe or Al) to complex and mobilize them. The reaction with precipitated aluminum is

$$3H^+ + Al(OH)_3(s) + 2Ca(C_2O_4) \rightarrow Al(C_2O_4)_2^-(aq) + 2Ca^{2+}(aq) + 3H_2O \quad (5.1)$$

Because hydrogen ions are consumed, this reaction raises the pH of acidic soil. It also weathers minerals by dissolving some metals and provides Ca^{2+} as a plant and biota nutrient. Similar processes with silicate minerals release K^+ and other nutrient cations.

The amount of organic matter in soil has a strong influence on soil properties and on the behavior of soil contaminants. For example, plants compete with soil for water. In sandy soils, pore space is large and particle surface area is small. Water is not strongly adsorbed to sands and is easily available to plants. However, in sandy soils the water drains off quickly. On the other hand, water binds strongly to organic matter in soil. Soils with high organic content hold more water; but the water is less available to plants.

RULE OF THUMB

Organic matter is typically less than 5% in most soils, but is critical for plant productivity. Peat soils can be 95% organic matter. Mineral soils can be less than 1% organic matter.

5.3.1 HUMIC SUBSTANCES

The most important organic substance in soil is humus, a collection of variously sized polymeric molecules consisting of soluble fractions (humic and fulvic acids) and an insoluble fraction (humin). Humus is the near-final residue of plant biodegradation and consists largely of protein and lignin. Humus is what remains after the more easily degradable components of plant biomass have degraded, leaving only the parts most resistant to further degradation. Humic materials are not well-defined chemically and have variable composition. Percent by weight for the most abundant elements are C: 45%–55%, O: 30%–45%, H: 3%–6%, N: 1%–5%, and S: 0%–1%.

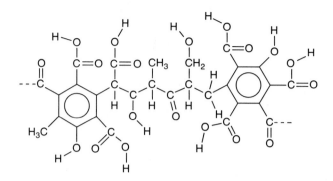

FIGURE 5.2 Characteristic structural portion of an unionized humic or fulvic acid.

The exact chemical structure depends on the source plant materials and the history of biodegradation.

Humic and fulvic acids are soluble organic acid macromolecules containing many –COOH and –OH functional groups that ionize in water, releasing H^+ ions and providing negative charge centers on the macromolecule to which cations are strongly attracted (see Figure 5.2). Humic materials are the most important class of natural soil complexing agents and are found where vegetation has decayed.

5.3.2 SOME PROPERTIES OF HUMIC MATERIALS

5.3.2.1 Binding to Dissolved Species

Humic materials are effective at removing metals from water by sorption to negative charge sites, mainly at the structural oxygen atoms. Polyvalent metal cations are sorbed especially strongly. The cation-exchange capacity of humic materials can be as high as 500 meq/100 mL. Humic materials may sorb metals like uranium in concentrations 10,000 times greater than adjacent water. Humic materials also bind organic pollutants, especially low-solubility compounds like DDT and atrazine. Much of the utility of wetlands for water treatment arises from their high concentrations of humic materials. Figure 5.3 shows several ways by which metal cations bind to humic and fulvic acids.

5.3.2.2 Light Absorption

Humic materials absorb sunlight in the blue region (transmitting yellow) and can transfer the solar energy to sorbed molecules, initiating reactions. This energy transfer process can be effective in degrading pesticides and other organic compounds.

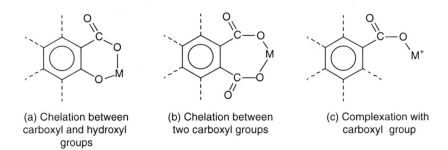

(a) Chelation between
carboxyl and hydroxyl
groups

(b) Chelation between
two carboxyl groups

(c) Complexation with
carboxyl group

FIGURE 5.3 Some types of binding metal ions (M^{2+}) to humic or fulvic acids.

5.4   SOIL ZONES

In discussions of groundwater movement, the soil subsurface is commonly divided into three zones, based on their air and water content (see Figure 5.4). From the ground surface down to an aquifer water table, soils contain mostly air in the pore spaces, with some adsorbed and capillary-held water. This region is called the water-unsaturated zone or vadose* zone. From the top of the water table to bedrock, soils contain mostly water in the pore spaces. This region is called the saturated zone.

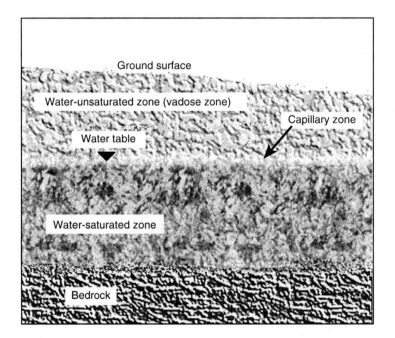

FIGURE 5.4 Soil zones in the subsurface region.

---

* From the Latin *vadosus*, meaning shallow.

Between the vadose and saturated zones, there is a transition region called the capillary zone, where water is drawn upward from the water table by capillary forces. The thickness of the capillary zone depends on the soil texture—the smaller the pore size, the greater the capillary rise. In fine gravel (2–5 mm grain size), the capillary zone will be of the order of 2.5 cm thick. In fine silt (0.02–0.05 mm grain size), the capillary zone can be 200 cm or greater.

The saturated zone lies above the solid bedrock, which is impermeable except for fractures and cracks. The region of the subsurface overlying the bedrock is generally unconsolidated porous, granular mineral material.

5.4.1 AIR IN SOIL

Air in soil has a different composition from atmospheric air because of biodegradation of organic matter by soil organisms. Biodegradation occurs in many small steps, but the net overall reaction is shown in Equation 5.2, where organic matter in soil is represented by the approximate generic unit formula $\{CH_2O\}$. An actual molecule of soil organic matter would have a formula that is approximately some whole number multiple of the $\{CH_2O\}$ unit.

$$\{CH_2O\} + O_2 \rightarrow CO_2 + H_2O \tag{5.2}$$

Equation 5.2 shows that, for each $\{CH_2O\}$ unit contained within a larger organic molecule, one CO_2 molecule and one H_2O molecule are produced by biodegradation.

Oxygen from soil pore space air is consumed and CO_2 released by microbial metabolism. Much of the soil air is semitrapped in pores and cannot readily equilibrate with the atmosphere. As a result, the O_2 content in soil pore space air is decreased from its atmospheric value of 21% to about 15% and CO_2 content is increased from its atmospheric value of about 0.03% to about 3%. This, in turn, increases the dissolved CO_2 concentration in groundwater, making it more acidic. Acidic groundwater contributes to the weathering of soils, especially calcium carbonate ($CaCO_3$) minerals.

When soil becomes water-saturated, as in the saturated zone, many changes occur:

1. Oxygen becomes used up by respiration of microorganisms.
2. Anaerobic processes lower the oxidation potential of water so that reducing conditions (electron gain) prevail, whereas oxidizing conditions (electron loss) dominate in the unsaturated zone.
3. Certain metals, particularly iron and manganese, become mobilized by chemical reduction reactions that change them from insoluble to soluble forms:

$$Fe(OH)_3(s) + 3H^+ + e^- \rightarrow Fe^{+2}(aq) + 3H_2O \tag{5.3}$$

$$Fe_2O_3(s) + 6H^+ + 2e^- \rightarrow 2Fe^{+2}(aq) + 3H_2O \tag{5.4}$$

$$MnO_2(s) + 5H^+ + 2e^- \rightarrow Mn^{+2}(aq) + 2H_2O \tag{5.5}$$

Groundwater, moving under gravity, can transport dissolved Fe^{+2} and Mn^{+2} into zones where oxidizing conditions prevail, e.g., by surfacing to a spring, stream, or lake. There, Equations 5.3 through 5.5 are reversed and the metals redeposit as solid precipitates, mainly $Fe(OH)_3$ and MnO_2. Precipitation of $Fe(OH)_3$ often causes "red water" and red or yellow deposits on rocks and soil. MnO_2 deposits are black. These deposits can clog underdrains in fields and water treatment filters.

5.5 CONTAMINANTS BECOME DISTRIBUTED IN WATER, SOIL, AND AIR

In the environment, contaminants always contact water, air, and soil. No matter where it originated, a contaminant moves across the interfaces between water, soil, and air phases to become distributed, to different degrees, into every phase it contacts. Partitioning of a pollutant from one phase into other phases serves to deplete the concentration in the original phase and increase it in the other phases. The movement of contaminants through soil is a process of continuous redistribution among the different phases it encounters. It is a process controlled by gravity, capillarity, sorption to surfaces, miscibility with water, and volatility.

5.5.1 VOLATILIZATION

The main partitioning process from liquids and solids to air is volatilization, which moves a contaminant across the liquid–air or solid–air interface into the atmosphere or into air in soil pore spaces. Volatilization is an important partitioning mechanism for compounds with high vapor pressures. For liquid mixtures such as gasoline, the most volatile components are lost first, causing the composition and properties of the remaining liquid mixture to change over time. For example, the most volatile components of gasoline are generally the smallest molecules in the mixture. The remaining larger molecules have stronger London attractive forces. Hence, as gasoline weathers and loses the smaller molecules by volatilization, its vapor pressure decreases and its viscosity and density increase.

5.5.2 SORPTION

The main partitioning process from liquids and air to solids is sorption, which moves a contaminant across the liquid–solid or air–solid interface to organic or mineral solid surfaces. Sorption from the water phase is most important for compounds of low solubility. Once a contaminant is sorbed to a surface, it undergoes chemical and biological transformations at different rates and by different pathways than if it were dissolved.

EXAMPLE 1

ESTIMATING SOME RELATIVE PHYSICAL PROPERTIES

Suppose you need to compare the relative vapor pressures, water solubilities, and Henry's law constants of the compounds tabulated below, but are only able to find melting point data. Estimate their relative values of these parameters based on their melting points and structures.

- Vapor pressure (P_v) is a measure of the tendency for molecules to partition from a pure substance into the atmosphere.
- Solubility (S_w) is a measure of the tendency of molecules to partition from a pure substance into water.
- Henry's law constant (K_H) is a measure of a compound's tendency to partition between water and air. It may be considered to be the vapor pressure of a substance dissolved in water.

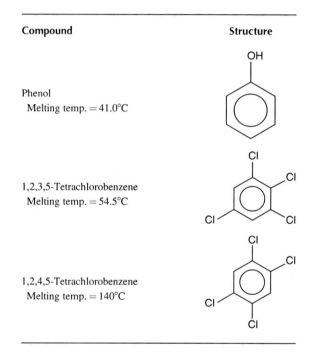

Compound	Structure
Phenol Melting temp. $= 41.0°C$	
1,2,3,5-Tetrachlorobenzene Melting temp. $= 54.5°C$	
1,2,4,5-Tetrachlorobenzene Melting temp. $= 140°C$	

Answer:

Vapor pressure varies inversely with intermolecular attraction. Substances with high vapor pressure have weak intermolecular attractive forces. Therefore, vapor pressure will tend to vary inversely with melting point, because a high melting point indicates strong intermolecular attractive forces.*

Solubility varies with polarity and the ability to form hydrogen bonds to water molecules. The more polar the molecule and the more hydrogen bonding to water, the more soluble it will be, because of stronger attraction to water molecules. It also varies with

---

* The correspondence between melting point and vapor pressure is only approximate, because the melting point may also depend on the crystal lattice energy, a function of the molecular geometry of the solid. However, it can serve as a first approximation where more accurate data are not available.

molecular weight. Higher molecular weight tends to decrease a compound's solubility because London attractive forces are stronger, attracting the compound molecules to one another more strongly.

Henry's law constant depends on two different properties of a substance. It varies inversely with solubility and directly with vapor pressure. In general, solubility will dominate in determining K_H.

Vapor pressure: (Lowest to highest vapor pressure will be, to a first approximation, in the order of highest to lowest melting point.)

$$1,2,4,5\text{-tetrachlorobenzene} < 1,2,3,5\text{-tetrachlorobenzene} < \text{phenol}$$

Solubility: (Lowest to highest solubility will be in the order of lowest to highest polarity, highest to lowest molecular weight, and fewest to most hydrogen bonds to water.)

Least soluble is 1,2,4,5-tetrachlorobenzene (nonpolar because of symmetry; high molecular weight, no hydrogen bonds).

More soluble is 1,2,3,5-tetrachlorobenzene (less symmetrical, therefore more polar; same molecular weight as 1,2,4,5-tetrachlorobenzene, no hydrogen bonds).

Most soluble is phenol (most polar; the only compound that can form hydrogen bonds to water; lightest molecular weight of all).

Henry's law constant: Lowest to highest K_H should vary according to highest to lowest solubility and lowest to highest vapor pressure. Phenol, with the highest polarity, greatest solubility and lowest melting point, will have the lowest K_H. Because the solubilities of 1,2,3,5-tetrachlorobenzene and 1,2,4,5-tetrachlorobenzene are so similar, their K_{HS} are expected to be similar also. However, their vapor pressures differ by a factor of more than 10. The much higher vapor pressure of 1,2,3,5-tetrachlorobenzene might give it a slightly higher K_H than 1,2,4,5-tetrachlorobenzene.

$$\text{Phenol} < 1,2,4,5\text{-tetrachlorobenzene} < 1,2,3,5\text{-tetrachlorobenzene}$$

The measured values in Table 5.1 confirm these relative vapor pressure, solubility, and K_H estimates.

TABLE 5.1
Measured Values for Melting Point, Vapor Pressure, Solubility, and Henry's Law Constant

Compound	MW (g)	T_m (°C)	P_v (atm)	S_w (mg/L)	K_H (atm m³/mol)
Phenol	94.1	41.0	2.6×10^{-4}	8,2000	4.0×10^{-7}
1,2,3,5-Tetrachlorobenzene	215.9	54.5	1.4×10^{-5}	3.5	5.8×10^{-3}
1,2,4,5-Tetrachlorobenzene	215.9	140.0	5.3×10^{-6}	2.2	2.6×10^{-3}

Note: T_m = melting point; P_v = vapor pressure; S_w = solubility in water; K_H = Henry's law constant.

EXAMPLE 2

Rank the four compounds below in order of increasing Henry's law constant (tendency to partition from water into air).

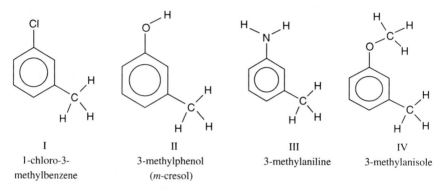

I	II	III	IV
1-chloro-3-	3-methylphenol	3-methylaniline	3-methylanisole
methylbenzene	(*m*-cresol)		

Answer:

All the compounds are similar in molecular weight and their structures differ only in the top functional group. The molecule having the group with the weakest attractive force to water will most readily partition from water into air. Therefore, we want to rank them by their relative attractions to water, i.e., solubilities.

> I. Has no oxygen, nitrogen, or fluorine for H-bonding to water and is the least polar. It will volatilize the most readily.
>
> IV. Has an oxygen, but no hydrogen is attached to it for H-bonding to water. It will be next in volatility from water to air.
>
> II. and III. Both can H-bond. II can form one H-bond while III can form two H-bonds, so, II is third in volatility and III is the least volatile in water.

$$\text{Order of Henry's law constants: } I > IV > II > III$$

Their measured Henry's law constants (dimensionless) are

$$I, K_H = 1.38; \quad IV, K_H = 1.17; \quad II, K_H = 0.0034; \quad III, K_H = 0.0015$$

A larger Henry's law constant means greater tendency to volatilize from water (see Section 5.6.1).

5.6 PARTITION COEFFICIENTS

The tendency for a pollutant to move from one phase to another is often quantified by the use of a partition coefficient, also called a distribution coefficient. Partition coefficients are chemical specific. They can be measured directly or, in some cases, estimated from other properties of the chemical. The simplest form of a partition coefficient is the ratio of the pollutant concentration in phase 1 to its concentration in phase 2:

$$K_{1,2} = \frac{\text{concentration in phase 1}}{\text{concentration in phase 2}} = \frac{C_1}{C_2} \tag{5.6}$$

This expression assumes that a linear relation exists between the concentrations of a substance in different phases, and is often satisfactory for low to moderate concentrations. The phase of the denominator, C_2, is referred to as the reference phase. Water is commonly used as the reference phase to maintain some consistency in published values. Using water as the reference phase, a linear relation gives Equations 5.7 through 5.9 for partitioning between water and air, water and free product (original physical form of the pollutant, such as a layer of oil floating on a river or above the groundwater table; sometimes called the bulk pollutant), and water and soil. In these equations, the pollutant may be a pure substance, like benzene, or one component from a mixture, like benzene from free product liquid gasoline.

$$C_a = K_H C_w \qquad (5.7)$$

K_H is the air–water partition coefficient, also known as Henry's law constant. C_a and C_w are the pollutant concentrations in air and water, respectively.

$$C_p = K_p C_w \qquad (5.8)$$

K_p is the free product–water partition coefficient. C_p and C_w are the pollutant concentrations in the free product and water, respectively.

$$C_s = K_d C_w \qquad (5.9)$$

K_d is the soil–water partition coefficient. C_s and C_w are the pollutant concentrations sorbed on soil and dissolved in water, respectively.

Each value of K depends on properties of the particular pollutant and the temperature. K_d also depends on the type of soil.

5.6.1 AIR–WATER PARTITION COEFFICIENT (HENRY'S LAW)

Henry's law, $C_a = K_H C_w$, describes how a substance distributes itself at equilibrium between water and air. The units of Henry's law constant, $K_H = C_a/C_w$, depends on what units are used to express concentrations in air and water.* For the case of oxygen gas, O_2, at 20°C

- When air and water concentrations both have the same units,

$$K_H(O_2, 20°C) = 26 \, (\text{unitless}) \qquad (5.10)$$

- For water concentration in mol/L or mol/m³, and air in atmospheres,

$$K_H(O_2, 20°C) = 635 \, L \cdot atm/mol = 0.635 \, m^3 \cdot atm/mol \qquad (5.11)$$

- For water concentration in mg/L and air in atmospheres,

$$K_H(O_2, 20°C) = 0.0198 \, L \cdot atm/mg \qquad (5.12)$$

---

* Henry's law constants are tabulated in many references, such as *Handbook of Chemistry and Physics*, CRC Press; Howard (1991); Lyman et al. (1990); Mackay and Shiu (1981); Sander (1999). There are also computer programs that calculate Henry's law constant from other chemical properties.

EXAMPLE 3

If soil pore water is measured to contain 3.2 mg/L of oxygen at 20°C, what is the concentration, in mg/L and in atmospheres, of oxygen in the air of the soil pore space?

Answer:

$$\text{For } O_2 \text{ at } 20°C, \quad K_H = 26 = \frac{C_a}{3.2\,\text{mg/L}}; \quad C_a = (26)(3.2\,\text{mg/L}) = 83.2\,\text{mg/L}$$

Using different units,

$$K_H = 0.0198\,\text{L·atm/mg} = \frac{C_a}{3.2\,\text{mg/L}}$$
$$C_a = (0.0198\,\text{L·atm/mg})(3.2\,\text{mg/L}) = 0.063\,\text{atm}$$

Since the normal atmospheric partial pressure of oxygen at sea level is about 0.2 atm, this result, showing that the soil oxygen concentration is lower than atmospheric, indicates the presence in the soil of microbial activity that has consumed oxygen.

EXAMPLE 4

BOD AND HENRY'S LAW

A certain sewage treatment plant located on a river typically removes 100,000 lb (4.54×10^7 g) of biodegradable organic waste each day. If there were a plant upset and it became necessary to release one day's waste into the receiving river, how many liters of river water could potentially be contaminated to the extent of totally depleting the water of all oxygen?

Answer:

An approximate chemical equation we have used before (Section 5.4.1) as being suitable for biodegradation of organic matter is

$$\{CH_2O\} + O_2 \rightarrow CO_2 + H_2O \tag{5.2}$$

Assume the river water is initially saturated with oxygen from the air at 20°C and that, after the spill, no additional oxygen dissolves from the atmosphere, a worst case scenario.

Necessary Data:

Atm. pressure at the treatment plant = 0.82 atm
Vapor pressure of water at 20°C = 0.023 atm
Percent O_2 in dry air = 21%
From Equation 5.11, $K_H(O_2) = 635$ L·atm/mol

Calculation:

Organic matter is biodegraded, consuming oxygen by Equation 5.2. This chemical equation shows that one mole of O_2 is consumed for each mole of CH_2O biodegraded. The molecular weight of CH_2O is 30 g/mol. Therefore,

$$\text{Moles of } CH_2O \text{ in sewage} = \frac{4.54 \times 10^7\,\text{g}}{30\,\text{g/mol}} = 1.5 \times 10^6\,\text{mol} = \text{moles of } O_2 \text{ consumed}$$

$$\text{Atmospheric pressure} = P_{\text{total}} = 0.82 \text{ atm} = P_{O_2} + P_{N_2} + P_{H_2O}$$

$$P_{\text{dry air}} = \text{atmospheric pressure} - \text{partial pressure of water vapor} = P_{\text{total}} - P_{H_2O}$$

$$\text{Therefore, } P_{O_2} = (0.21)(P_{\text{total}} - P_{H_2O}) = (0.21)(0.82 \text{ atm} - 0.023 \text{ atm}) = 0.17 \text{ atm}.$$

Use Henry's law to find the concentration of dissolved O_2 in the river.

$$K_H(O_2) = 635 \text{ L} \cdot \text{atm/mol} = \frac{C_a}{C_w} = \frac{0.17 \text{ atm}}{C_w}$$

$$C_w = [O_2(aq)] = \frac{0.17 \text{ atm}}{635 \text{ atm} \cdot \text{L/mol}} = 2.7 \ 10^{-4} \text{ mol/L}$$

$$\text{or } [O_2(aq)] = (2.7 \times 10^{-4} \text{ mol/L}) \times (32 \text{ g/mol}) = 8.6 \text{ mg/L}$$

In saturated water at 20°C and 0.82 atm total pressure, $[O_2, aq] = 2.7 \times 10^{-4}$ mol/L

$$\text{Liters of river water depleted of } O_2 = \frac{1.5 \times 10^6 \text{ mol } O_2 \text{ consumed}}{2.7 \times 10^{-4} \text{ mol } O_2/\text{L in river}} = 5.6 \times 10^9 \text{ L}$$

This is a worst-case scenario. Less oxygen than calculated would be lost from the river because of continuous replenishing of oxygen to the river by partitioning from the atmosphere. The rate at which this occurs would depend on several factors such as the surface to volume ratio of the river, water turbulence and cascades, wind velocity, and temperature.

Note that both vapor pressure (proportional to C_a) and solubility (proportional to C_w) of a pure solid or liquid generally increase with temperature, but vapor pressure always increases faster. Therefore, the value of K_H increases with temperature, indicating that, for a gas partitioning between air and water, the atmospheric portion increases and the dissolved portion decreases when the temperature rises. This is consistent with the observation that the water solubility of gases decreases with increasing temperature.

RULE OF THUMB

Estimating K_H:
If a tabulated value for K_H cannot be found, it may be estimated roughly by dividing the vapor pressure of a compound by its aqueous solubility. For some compounds, tabulated values of vapor pressure and solubility may be easier to find than K_H values.

$$\begin{aligned} K_H &= \frac{C_a \text{ (partial pressure in atmospheres)}}{C_w \text{ (mol/L)}} \\ &\approx \frac{\text{vapor pressure (atm)}}{\text{aqueous solubility (mol/L)}} \end{aligned} \qquad (5.13)$$

In this case, the units of K_H are atm $\cdot$ L mol^{-1}

EXAMPLE 5

Estimate Henry's law constants for chlorobenzene and bromomethane using vapor pressure and aqueous solubility.

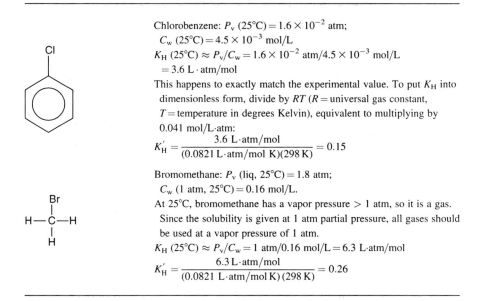

Chlorobenzene: P_v (25°C) $= 1.6 \times 10^{-2}$ atm;
$\quad C_w$ (25°C) $= 4.5 \times 10^{-3}$ mol/L
K_H (25°C) $\approx P_v/C_w = 1.6 \times 10^{-2}$ atm$/4.5 \times 10^{-3}$ mol/L
$\quad = 3.6$ L · atm/mol

This happens to exactly match the experimental value. To put K_H into dimensionless form, divide by RT ($R =$ universal gas constant, $T =$ temperature in degrees Kelvin), equivalent to multiplying by 0.041 mol/L·atm:

$$K'_H = \frac{3.6 \text{ L·atm/mol}}{(0.0821 \text{ L·atm/mol K})(298 \text{ K})} = 0.15$$

Bromomethane: P_v (liq, 25°C) $= 1.8$ atm;
$\quad C_w$ (1 atm, 25°C) $= 0.16$ mol/L.

At 25°C, bromomethane has a vapor pressure > 1 atm, so it is a gas. Since the solubility is given at 1 atm partial pressure, all gases should be used at a vapor pressure of 1 atm.

K_H (25°C) $\approx P_v/C_w = 1$ atm$/0.16$ mol/L $= 6.3$ L·atm/mol

$$K'_H = \frac{6.3 \text{ L·atm/mol}}{(0.0821 \text{ L·atm/mol K})(298 \text{ K})} = 0.26$$

EXAMPLE 6

Estimate relative Henry's law constants for the compounds in Example 1 using vapor pressure and solubility data from Table 5.1. Compare your answer with the measured K_H values in Table 5.1.

Answer:

From Equation 5.13, Henry's law constant varies directly with vapor pressure and inversely with aqueous solubility.

$$K_H \approx \frac{\text{vapor pressure (atm)}}{\text{aqueous solubility (mol/L)}}$$

Use vapor pressure and solubility data from Table 5.1 to estimate approximate Henry's law values.

$$K_H \text{ (phenol)} \approx \frac{2.6 \times 10^{-4} \text{ atm}}{8.2 \times 10^4 \text{ mg/L}} \times \frac{94.1 \text{ g/mol}}{10^{-3} \text{g/mg}} = 3.0 \times 10^{-4} \text{ L·atm/mol}$$

$$K_H \text{ (1,2,3,5-tetrachlorobenzene)} \approx \frac{1.4 \times 10^{-5} \text{ atm}}{3.5 \text{ mg/L}} \times \frac{215.9 \text{ g/mol}}{10^{-3} \text{ g/mg}} = 0.86 \text{ L·atm/mol}$$

$$K_H\,(1,2,4\text{-tetrachlorobenzene}) \approx \frac{5.3 \times 10^{-6}\,\text{atm}}{2.2\,\text{mg/L}} \times \frac{215.9\,\text{g/mol}}{10^{-3}\,\text{g/mg}} = 0.52\,\text{L·atm/mol}$$

$$K_H\,(1,2,3,5\text{-tetrachlorobenzene}) > K_H\,(1,2,4,5\text{-tetrachlorobenzene}) \gg K_H\,(\text{phenol})$$

Table values:*

$$K_H\,(\text{phenol}) = 4.0 \times 10^{-4}\,\text{L·atm/mol}$$

$$K_H\,(1,2,3,5\text{-tetrachlorobenzene}) = 5.8\,\text{L·atm/mol}$$

$$K_H\,(1,2,4,5\text{-tetrachlorobenzene}) = 2.6\,\text{L·atm/mol}$$

The calculated value for phenol is very close to the measured value. The calculated values for the chlorobenzenes are in the correct order but are about one-fifth too small. It cannot be determined why the calculated values for 1,2,3,5-tetrachlorobenzene and 1,2,3,5-tetrachlorobenzene differ from the measured values without examining the original experimental data. It is likely due to experimental error based on the difficulty of accurately measuring such low pressures. This could also explain why the Henry's law constant of the polar 1,2,3,5-tetrachlorobenzene appears to be higher than that of the nonpolar 1,2,4,5-tetrachlorobenzene. The absolute difference in the constants is small and could be accounted for by the limits of error in the vapor pressure measurements, where an uncertainty factor of 3 or 4 is not unusual in tabulated values of small vapor pressures.

It might seem counterintuitive that phenol, the compound with the highest vapor pressure in the pure state, has the lowest Henry's law constant. This example shows the importance of the environment immediately surrounding the molecule. Phenol is much more strongly attracted to water molecules than to other phenol molecules. Consequently, it enters the vapor state more readily from pure liquid phenol than from a water solution.

5.6.2 SOIL–WATER PARTITION COEFFICIENT

Partitioning from water into soil surfaces can be limited by the available soil surface area. The rate of transfer will slow as the soil surface becomes saturated. For this reason, the partitioning of a compound between water and soil may deviate from linearity. This is particularly true for the partitioning of organic compounds. To account for nonlinearity, the corresponding partition coefficient is often written in a modified form called the Freundlich isotherm. The modification consists in introducing an empirically determined exponent to the C_w term.

$$C_s = K_d C_w^n \tag{5.14}$$

where
  C_s is the concentration of sorbed organic compound in solid phase (mg/kg)
  C_w is the concentration of dissolved organic compound in water phase (mg/L)
  $K_d = C_s/C_w^n$ is the partition coefficient for sorption
  n = empirically determined exponential factor

---

* Reference: *Canadian Water Quality Guidelines*, Task Force on Water Quality Guidelines of the Canadian Council of Resource and Environment Ministers, March 1987.

When $C_w \ll C_s$ (the common case for organic compounds of low water solubility), then n is close to unity for most organics. However, n typically is temperature dependent.

Equation 5.14 can be written as an equation for a straight line by taking the logarithm of both sides, as in

$$\log C_s = \log K_d + n \log C_w \qquad (5.15)$$

It is important not to extrapolate the Freundlich isotherm too far beyond the range of experimental data.

EXAMPLE 7

USE OF THE FREUNDLICH ISOTHERM

A power company planned to discharge their power plant cooling water into a small lake. Before purchasing the lake, they tested the water and found the pesticide 2,4-D (2,4-dichlorophenoxyacetic acid) at 0.8 ppt (0.8 parts per trillion, or 0.8×10^{-9} g/L), just a little below the permitted limit of 1 ppt. The company calculated that their operation would raise the water temperature in the mixing zone near their discharge from 5°C to about 25°C. Should they anticipate a problem?

Answer:

2,4-D has low solubility and is denser than water. It will sink in the lake and become sorbed on the bottom sediments. The potential problem is whether the expected increase in temperature will cause the 2,4-D limit to be exceeded because of additional 2,4-D partitioning into the water from the bottom sediments. It is a case for the Freundlich isotherm, because the empirical constant n is a function of temperature and will cause a change in K_d when the temperature changes. A Web search found a study (Means and Wijayratne, 1982) that measured Freundlich isotherm values for 2,4-D at 5°C and 25°C (Table 5.2).

$$C_s = K_d C_w^n \qquad (5.14)$$

At 5°C: $C_s = (6.53)(0.8 \times 10^{-9}\,\text{g/L})^{0.76} = (6.53)(1.22 \times 10^{-7}) = 797 \times 10^{-9}\,\text{g/kg}$

The calculation indicates that there are 797 ppt of 2,4-D sorbed on the sediments at 5°C. Note that the concentration of 2,4-D sorbed to sediments is about 1000 times larger than the dissolved concentration. Even if a temperature rise to 25°C causes a large percentage increase in the dissolved portion, the percentage loss from the sediment fraction will be 1000 times smaller. Assume that the sediment concentration at 25°C is essentially the same as at 5°C. This allows an approximate calculation of C_w at 25°C.

TABLE 5.2
Values for Freundlich Isotherm Parameters of 2,4-D

Temperature (°C)	n	K_d	$\log K_d$
5	0.76	6.53	0.815
25	0.83	5.20	0.716

$$\text{At } 25°C: 797 \times 10^{-9} \text{ g/kg} = (5.20)(C_w)^{0.83}$$

$$C_w(25°C) = \left(\frac{797 \times 10^{-9}}{5.20}\right)^{\frac{1}{0.83}} = 6.2 \times 10^{-9} \text{ g/L} = 6.2 \text{ ppt}$$

The expected temperature rise will cause 2,4-D to desorb from the bottom sediments and raise the water concentration well over the permitted limit of 1 ppt.

5.6.3 DETERMINING K_D EXPERIMENTALLY

The Freundlich Equation 5.14 can be used to determine K_d experimentally, as follows:

1. Prepare samples having several different concentrations of dissolved contaminant in equilibrium with soil from the site of interest.
2. Measure the contaminant concentrations in water, C_w, in each sample.
3. Measure the corresponding contaminant concentrations sorbed to soil, C_s.
4. Plot log C_s versus log C_w to get a straight line with slope $= n$ and intercept $= \log K_d$.

EXAMPLE 8

Prepared water samples containing different concentrations of benzene were equilibrated with clean soil from a site under study. Equilibrium concentrations of dissolved and sorbed benzene are shown in Table 5.3. Find K_d for benzene in this soil.

TABLE 5.3
Benzene Partitioning Data for Soil and Water in Equilibrium

Dissolved Benzene C_w (mg/L)	Sorbed Benzene C_s (mg/kg)
6.59	2.2
10.00	3.1
33.28	8.1
34.57	9.6
68.31	15.00
88.89	26.00
183.74	44.00
340.54	89.00
452.30	119.00
674.79	130.00
819.56	188.00
955.95	247.00

Answer:

    1. Determine the base-10 logarithms of all the concentration values.

Logarithms of C_w and C_s

log C_w	log C_s
0.819	0.342
1.000	0.491
1.522	0.908
1.539	0.982
1.834	1.176
1.949	1.415
2.264	1.643
2.532	1.949
2.655	2.076
2.829	2.114
2.914	2.274
2.980	2.393

    2. Plot log C_w versus log C_s and fit a straight line through the points. The formula of the line is log C_s = log K_d + n log C_w. Therefore, the slope of the line is equal to n and the y-axis intercept is equal to log K_d. The resulting plot is shown in Figure 5.5.

The equation of the least-squares fitted line is log C_s = 0.941 log C_w − 0.468.

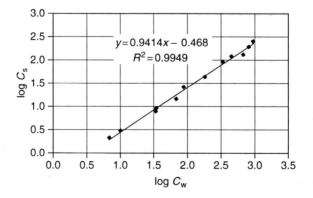

FIGURE 5.5 Freundlich isotherm for benzene partitioning between water and soil.

Therefore,

$$n = 0.941$$

$$\log K_d = -0.468$$

$$K_d = 0.340 \, \text{L/kg}$$

5.6.4 ROLE OF SOIL ORGANIC MATTER

Values for K_d are extremely site- and chemical specific because the extent of sorption depends on several physical and chemical properties of both the soil and the sorbed chemical. For dissolved neutral organic molecules, such as fuel hydrocarbons, sorption to soils is controlled mostly by sorption to the organic portion of the soil. Therefore, the value of the soil–water partition coefficient, K_d, depends on the amount of organic matter in the soil. Expressing the soil–water partition coefficient in terms of soil organic carbon (K_{oc}), rather than total soil mass (K_d), can eliminate a large part, but not all, of the site-specific variability in K_d.

The amount of organic matter in soil is usually expressed as either the weight fraction of organic carbon, f_{oc}, or the weight fraction of organic matter, f_{om}. The amount of organic matter in typical mineral soils is generally between about 1% and 10%, but is typically less than 5%. In wetlands and peat-soils, it can approach 100%. Since soil organic matter is approximately 58% carbon, f_{oc} typically ranges between 0.006 and 0.06. Some characteristic values of f_{oc} for a range of different soil types are given in Table 5.4.

There is a critical lower value for the fraction of organic carbon in the soil, f_{oc}^* (dependent on the dissolved organic compound), below which sorption to inorganic matter becomes dominant. Typical values for f_{oc}^* are between about 10^{-3} and 10^{-4} (0.1%–0.01%).

TABLE 5.4
**Typical Values of Fraction of Organic Carbon
in Different Soils**

Type of Soil	Typical f_{oc} (wt. fraction)
Coarse soil	0.04
Silty loam	0.05
Silty clayey loam	0.03
Clayey silty loam	0.005
Clayey loam	0.004
Sand	0.0005
Glaciofluvial	0.0001

Note: These typical values were collected from many sources. They should only be used as crude estimates when site-specific measurements cannot be obtained. The range of measured values for a single soil type can span a factor of 100.

If we define $K_{oc} = \frac{C_{oc}}{C_w}$, where C_{oc} is the concentration of contaminant sorbed to organic carbon and C_w is the concentration of contaminant dissolved in water, then the soil–water partition coefficient becomes

$$K_d = K_{oc} f_{oc} \qquad (5.16)$$

This relation is useful as long as f_{oc} is greater than about 0.001.

RULES OF THUMB

1. It is often found that typical soil organic matter is about 58% carbon. Using this value to convert between the fraction of organic carbon and the fraction of organic matter leads to the relation $f_{oc} \approx 0.58 f_{om}$.
2. Soil organic carbon content can vary by a factor of 100 for similar soils. Although there have been tabulations of f_{oc} values according to soil type (e.g., Maidment, 1993), it is much better to use values measured at the site for use in Equation 5.16.

EXAMPLE 9

Find K_{oc} for benzene in the soil of Example 8. The soil contains 1.1% organic matter.

Answer:

The soil organic matter was measured to be 1.1%, or $f_{om} = 0.011$. This may be converted to fraction of organic carbon (f_{oc}) by the approximate rule of thumb

$$f_{oc} = 0.58 f_{om}$$

Therefore, $f_{oc} = 0.58 \times 0.011 = 0.0064$. Since $K_d = K_{oc} f_{oc}$, we have

$$K_{oc}(\text{benzene}) = 0.340/0.0064 = 53\,\text{L/kg}$$

5.6.5 OCTANOL–WATER PARTITION COEFFICIENT, K_{ow}

For laboratory experiments, the liquid compound octanol, an eight carbon organic alcohol, $CH_3(CH_2)_7OH$, is a good surrogate for the organic carbon fraction of soils. The partitioning behavior of organic compounds between octanol and water is similar to that between the organic carbon fraction of soil and water. The basic steps for measuring the octanol–water partition coefficient, K_{ow}, are as follows:

1. Combine octanol and water in a bottle. Octanol forms a separate phase floating on top of the water.
2. Add the organic contaminant (e.g., carbon tetrachloride, CCl_4), shake the mixture, and let the phases separate.
3. Measure the contaminant concentrations in the octanol phase and in the water phase.

Then

$$K_{ow} = \frac{C_{octanol}}{C_{water}} \tag{5.17}$$

An empirical equation that relates K_{ow} and organic carbon to K_d is

$$K_d = f_{oc} b K_{ow}^a \tag{5.18a}$$

or

$$\log K_d = a \log K_{ow} + \log f_{oc} + \log b \tag{5.18b}$$

where the empirically determined constants a and b depend on the organic compound.

Equation 5.19, derived from Equations 5.16 and 5.18a, is a relation that allows a calculation of K_{oc} in terms of K_{ow}.

$$K_{oc} = b K_{ow}^a \tag{5.19}$$

The usefulness of these relations is evident in the rules of thumb for K_{ow}. Knowing the value of K_{ow} for a compound allows one to qualitatively estimate many of its environmental properties.

RULES OF THUMB FOR K_{ow}

(High $K_{ow} => 1000$)

(Low $K_{ow} =< 500$)

The higher K_{ow} is for a compound
- The higher is the sorption to soil
- The higher is the bioaccumulation
- The lower is the biodegradation rate
- The lower is the water solubility
- The lower is the mobility

The lower K_{ow} is for a compound
- The lower is the sorption to soil
- The lower is the bioaccumulation
- The higher is the biodegradation rate
- The higher is the solubility
- The higher is the mobility

EXAMPLE 10

Consider the pesticide DDT in terms of the rules of thumb for K_{ow}. For DDT, $K_{ow} = 3.4 \times 10^6$, which is well into the high value range for K_{ow}. From the rules of thumb, one can predict that DDT has low water solubility (its measured solubility is 25 $\mu g/L$), is slowly biodegraded, is persistent in the environment (low mobility and slow biodegradation), is strongly adsorbed to soil, and is strongly bioaccumulated.

On the other hand, phenol has $K_{ow} = 30.2$, a low value. Phenol has high water solubility (its measured solubility is 8.3×10^4 mg/L), biodegrades rapidly, is not persistent in the environment, is weakly sorbed to soil, is highly mobile, and is weakly bioaccumulated.

5.6.6   ESTIMATING K_D USING MEASURED SOLUBILITY OR K_{ow}

Literature values for K_d measurements vary considerably because they are so site specific. Considerable effort has been expended in finding more consistent approaches to soil sorption. EPA has published a comprehensive evaluation of the soil–water partition coefficient, K_d (USEPA, 1999).

The following empirical observations have led to methods for estimating K_d from more easily measured parameters:

*   Water solubility is inversely related to K_d; the lower the solubility, the greater is K_d.
*   Since molecular polarity correlates with solubility, molecules whose structure indicates low polarity (hence, low solubility) may be expected to have a high K_d.
*   For compounds of low solubility, such as fuel hydrocarbons, sorption is controlled primarily by interactions with the organic portion of the solid sorbent.
*   The surface area of the solid is important. The larger the surface area, the larger will be K_d.
*   A simple way to estimate the tendency for a compound to partition between water and organic solids in the soil is to measure K_{ow}, the partition coefficient for the compound between water and octanol. The larger is K_{ow}, the larger is K_d.

Because K_d is site specific, it is generally preferable to calculate it from the more easily obtained quantities K_{ow}, K_{oc}, f_{oc}, and solubility. K_{ow} and solubility (S) are easily measured in a laboratory and K_{oc} may be normalized to f_{oc}, percent organic carbon in soil, which is an easily measured site parameter. There are linear relationships between $\log K_{oc}$, $\log K_{ow}$, and S (Lyman et al., 1990) that vary according to the class of compounds tested (e.g., chlorinated compounds, aromatic hydrocarbons, ionizable organic acids, pesticides, etc.). These relations can be used to calculate K_{oc} from K_{ow} or S in the absence of measured K_{oc} data. The EPA has reviewed the soil–water partitioning literature and selected or calculated the most reliable values in their judgment. In these calculations, EPA used the following relations.

For nonionizable, semivolatile organic compounds

$$\log K_{oc} = 0.983 \log K_{ow} + 0.00028 \qquad (5.20)$$

For nonionizable, volatile organic compounds

$$\log K_{oc} = 0.7919 \log K_{ow} + 0.0784 \qquad (5.21)$$

In addition, EPA suggests the use of the following equation for estimating K_{oc} from solubility (S) or bioconcentration factors (BCF):

$$\log K_{oc} = 0.681 \log \text{BCF} + 1.963 \tag{5.22}$$

Values for K_{oc}, K_{ow}, S, and BCF for many environmentally important chemicals have been collected in look-up tables for use in EPA's soil screening guidance procedures (USEPA, 1996). Table 5.5 is adapted from these EPA tables.

EXAMPLE 11

Benzene Spill:

A benzene leak soaked into a patch of soil. To determine how much benzene was in the soil, several soil samples were taken in a grid pattern across the area of maximum contamination. From analysis of these samples, the average benzene concentration in the soil was 2422 mg/kg (ppm). The soil contained 2.6% organic matter. From Table 5.5, $\log K_{ow}$(benzene) = 2.13. If rainwater percolate down through the contaminated soil, what concentration of benzene might initially leach from the soil and be found dissolved in the water? Assume the water and soil are in equilibrium with respect to benzene.

Calculation:

The general approach to this type of problem is

1. Find a tabulated value of K_{oc} for the chemical of concern, or calculate it from tabulated values for K_{ow} or S.
2. Obtain a measurement of f_{oc} or f_{om} in soil at the site or estimate it from the soil type.
3. Calculate K_d from the above quantities.

Benzene is a volatile, nonionizable organic compound (refer to Group 2 in Table 5.5).

$$\text{Use } K_d = \frac{C_{soil}}{C_w} = K_{oc}f_{oc} \text{ and Equation 5.20:}$$

$$\log K_{oc} = 0.7919 \log K_{ow} + 0.0784$$

$$\log K_{oc} = 0.7919(2.13) + 0.0784 = 1.7651$$

$$K_{oc} = 58.2$$

Since $f_{oc} \approx 0.58 f_{om} = 0.58(0.026) = 0.015$, we have

$$K_d = \frac{C_{soil}}{C_w} = K_{oc}f_{oc} = 58.2(0.015) = 0.88 \, \text{L/kg}$$

When 1 L of water is in equilibrium with 1 kg of soil, $C_{soil} + C_w = 2422$ ppm.

TABLE 5.5

Chemical Properties Used for Calculating Partition Coefficients and Retardation Factors

CAS No.	Compound	Chemical Group[a]	S Solubility of Pure Compound (mg/L)	HLC Henry's Law Constant (atm·m³/L)	H Henry's Law Constant (unitless)	log K_{ow}	log K_{oc}	K_{oc} Organic Carbon Partition Coefficient (L/kg)	T_{bp} Normal Boiling Point (°C)	T_{mp} Normal Melting Point[b] (°C)
83-32-9	Acenaphthene	1	4.24E+00	1.55E−04	6.36E−03	3.92	3.85	7.08E+03	288	93
120-12-7	Anthracene	1	4.34E−02	6.50E−05	2.67E−03	4.55	4.47	2.95E+04	324	215
71-43-2	Benzene	2	1.75E+03	5.55E−03	2.28E−01	2.13	1.77	5.89E−01	178	5.5
56-55-3	Benzo(a)anthracene	1	9.40E−03	3.35E−06	1.37E−04	5.70	5.60	3.98E−05	376	84
205-99-2	Benzo(b)fluoranthene	1	1.50E−03	1.11E−04	4.55E−03	6.20	6.09	1.23E−06	380	168
207-08-9	Benzo(k)fluoranthene	1	8.00E−04	8.29E−07	3.40E−05	6.20	6.09	1.23E−06	401	217
50-32-8	Benzo(a)pyrene	1	1.62E−03	1.13E−06	4.63E−05	6.11	6.01	1.02E+06	380	176
117-81-7	Bis(2-ethylhexyl)phthalate	1	3.40E−01	1.02E−07	4.81E−06	7.30	7.18	1.51E+07	347	−55
56-23-5	Carbon tetrachloride	2	7.93E+02	3.04E−02	1.25E+00	2.73	2.24	1.74E+02	177	−23
57-74-9	Chlordane	2	5.60E−02	4.86E−05	1.99E−03	6.32	5.08	1.20E+05	329	106
108-90-7	Chlorobenzene	2	4.72E+02	3.70E−03	1.52E−01	2.86	2.34	2.19E+02	207	−45
67-66-3	Chloroform (Trichloromethane)	2	7.92E+03	3.67E−03	1.50E−01	1.92	1.60	3.98E+01	168	−64
218-01-9	Chrysene	1	1.60E−03	9.46E−05	3.88E−03	5.70	5.60	3.98E+05	379	258
50-29-3	DDT	1	2.50E−02	8.10E−06	3.32E−04	6.53	6.42	2.63E+06	278	109
53-70-3	Dibenzo(a,h)anthracene	1	2.49E−03	1.47E−08	6.03E−07	6.69	6.58	3.80E+06	395	269
84-74-2	Di-n-butylphthalate	1	1.12E+01	9.38E−10	3.85E−08	4.61	4.53	3.39E+04	323	−35
95-50-1	1,2-Dichlorobenzene	2	1.56E+02	1.90E−03	7.79E−02	3.43	2.79	6.17E+02	234	−17
106-46-7	1,4-Dichlorobenzene	2	7.38E+01	2.43E−03	9.96E−02	3.42	2.79	6.17E+02	231	53
75-34-3	1,1-Dichloroethane (1,1-DCA)	2	5.06E+03	5.62E−03	2.30E−01	1.79	1.50	3.16E+01	166	−97
107-06-2	1,2-Dichloroethane (1,2-DCA)	2	8.52E+03	9.79E−04	4.01E−02	1.47	1.24	1.74E+01	181	−36

75-35-4	1,1-Dichloroethlyene	2	2.25E+03	2.61E-02	1.07E+00	2.13	1.77	5.89E+01	152	-123
156-59-2	cis-1,2-Dichloroethylene	2	3.50E+03	4.08E-03	1.67E-01	1.86	1.55	3.55E+01	168	-80
156-60-5	trans-1,2-Dichloroethlyene (DCE)	2	6.30E+03	9.38E-03	3.85E-01	2.07	1.72	5.25E+01	161	-50
78-87-5	1,2-Dichloropropane	2	2.80E-03	2.80E-03	1.15E-01	1.97	1.64	4.37E+01	187	-70
60-57-1	Dieldrin	2	1.95E-01	1.51E-05	6.19E-04	5.37	4.33	2.14E+04	323	176
121-14-2	2,4-Dinitrotoluene	1	2.70E+02	9.26E-08	3.80E-06	2.01	1.98	9.55E+01	310	71
115-29-7	Endosulfan	2	5.10E-01	1.12E-05	4.59E-04	4.10	3.33	2.14E+03	357	106
72-20-8	Endrin	2	2.50E-01	7.52E-06	3.08E-04	5.06	4.09	1.23E+04	381	200
100-41-4	Ethylbenzene	2	1.69E+02	7.88E-03	3.23E-01	3.14	2.56	3.63E+02	209	-95
206-44-0	Fluoranthene	1	2.06E-01	1.61E-05	6.60E-04	5.12	5.03	1.07E+05	347	108
86-73-7	Fluorene	1	1.98E+00	6.36E-05	2.61E-03	4.21	4.14	1.38E+04	299	115
76-44-8	Heptachlor	1	1.80E-01	1.09E-03	4.47E-02	6.26	6.15	1.41E+06	318	96
58-89-9	γ-HCH (Lindane)	2	6.80E+00	1.40E-05	5.74E-04	3.73	3.03	1.07E+03	314	113
193-39-5	Indeno(1,2,3-cd)pyrene	1	2.20E-05	1.60E-06	6.56E-05	6.65	6.54	3.47E+06	432	162
74-83-9	Methyl bromide	2	1.52E+04	6.24E-03	2.56E-01	1.19	1.02	1.05E+01	136	-94
75-09-2	Methylene chloride	2	1.30E+04	2.19E-03	8.98E-02	1.25	1.07	1.17E+01	156	-96
91-20-3	Naphthalene	1	3.10E+01	4.83E-04	1.98E-02	3.36	3.30	2.00E+03	255	80
108-95-2	Phenol	1	8.28E+04	3.97E-07	1.63E-05	1.48	1.46	2.88E+01	235	41
129-00-0	Pyrene	1	1.35E-01	1.10E-05	4.51E-04	5.11	5.02	1.05E+05	353	151
100-42-5	Styrene	1	3.10E+02	2.75E-03	1.13E-01	2.94	2.89	7.76E+02	215	-33
79-34-5	1,1,2,2-Tetrachloroethane (PCA)	2	2.97E+03	3.45E-04	1.41E-02	2.39	1.97	9.33E+01	215	-44
127-18-4	Tetrachloroethylene (PCE, PERC)	2	2.00E+02	1.84E-02	7.54E-01	2.67	2.19	1.55E+02	201	-22
108-88-3	Toluene	2	5.26E+02	6.64E-03	2.72E-01	2.75	2.26	1.82E+02	195	-95
8001-35-2	Toxaphene	1	7.40E-01	6.00E-06	2.46E-04	5.50	5.41	2.57E+05	347	65 to 90
120-82-1	1,2,4-Trichlorobenzene	2	3.00E+02	1.42E-03	5.82E-02	4.01	3.25	1.78E+03	252	17
71-55-6	1,1,1-Trichloroethane (1,1,1-TCA)	2	1.33E+03	1.72E-02	7.05E-01	2.48	2.04	1.10E+02	175	-30

(Continued)

TABLE 5.5 (Continued)

Chemical Properties Used for Calculating Partition Coefficients and Retardation Factors

CAS No.	Compound	Chemical Group[a]	S Solubility of Pure Compound (mg/L)	HLC Henry's Law Constant (atm·m³/L)	H Henry's Law Constant (unitless)	$\log K_{ow}$	$\log K_{oc}$	K_{oc} Organic Carbon Partition Coefficient (L/kg)	T_{bp} Normal Boiling Point (°C)	T_{mp} Normal Melting Point[b] (°C)
79-00-5	1,1,2-Trichloroethane (1,1,2-TCA)	2	4.42E+03	9.13E−04	3.74E−02	2.05	1.70	5.01E+01	197	−37
79-01-6	Trichloroethylene (TCE)	2	1.10E+03	1.03E−02	4.22E−01	2.71	2.22	1.66E+02	183	−85
67-66-3	Trichloromethane (chloroform)	2	7.92E+03	3.67E−03	1.50E−01	1.92	1.60	3.98E+01	168	−64
75-01-4	Vinyl chloride	2	2.76E+03	2.70E−02	1.11E+00	1.50	1.27	1.86E+01	126	−154
108-38-3	m-Xylene	2	1.61E+02	7.34E−03	3.01E−01	3.20	2.61	4.07E+02	211	−48
95-47-6	o-Xylene	2	1.78E+02	5.19E−03	2.13E−01	3.13	2.56	3.63E+02	214	−25
106-42-3	p-Xylene	2	1.85E+02	7.66E−03	3.14E−01	3.17	2.59	3.89E+02	211	13

Source: Adapted from USEPA, 1996, Soil Screening Guidance: Technical Background Document, Office of Emergency and Remedial Response, Washington, DC, EPA/540/R95/128.

[a] Group 1: Semivolatile nonionizing organic compounds. Fitted to $\log K_{oc} = 0.983 \log K_{ow} + 0.00028$. Group 2: Volatile organic compounds, chlorobenzenes, and certain chlorinated pesticides. Fitted to $\log K_{oc} = 0.7919 \log K_{ow} + 0.0784$.

[b] Compounds solid at soil temperature are defined as those with a melting point >20°C. Compounds liquid at soil temperature are defined as those with a melting point <20°C.

$$\text{Therefore, } K_d = \frac{2422 - C_w}{C_w} = 0.88$$

$$0.88\,C_w = 2422 - C_w$$

$$C_w = \frac{2422\ \text{mg/kg}}{1.88\ \text{L/kg}} = 1{,}288\ \text{mg/L} \approx 1300\ \text{mg/L to two significant figures.}$$

Note that a substantial portion of the sorbed benzene partitions into the water, indicating that benzene has sufficient solubility to be mobile in the environment. This is predictable from the rules of thumb for K_{ow}, which classify $K_{ow}(\text{benzene}) = 2.13$ (from Table 5.5) as a low value, giving benzene correspondingly high solubility and mobility.

5.7 MOBILITY OF CONTAMINANTS IN THE SUBSURFACE

The sorption of a contaminant from water to a solid is a reversible reaction. Just as the contaminant has some probability of sorbing from water to a surface that it contacts, a sorbed contaminant also has a probability of desorbing from the surface back into the water. For strongly sorbed contaminants, the probability of sorption is much greater than the probability of desorption, but both processes continually take place, although at different rates. The rates of sorption and desorption depend on the strength of the bond holding the sorbed compound to the surface and on the concentrations of dissolved and sorbed contaminant.

$$\text{Rate of sorption} = k_{sorb}C_w \qquad (5.23)$$

$$\text{Rate of desorption} = k_{desorb}C_s \qquad (5.24)$$

where
  C_w is the contaminant concentration in water
  C_s is the contaminant concentration sorbed to soil

k_{sorb} and k_{desorb} are the rate constants, which depend on the binding strength of sorption. For strongly sorbed contaminants, $k_{sorb} \gg k_{desorb}$. For weakly sorbed contaminants, $k_{sorb} \ll k_{desorb}$.

Consider the case of a dissolved contaminant sorbing onto initially clean soil particles. Until equilibrium is reached, C_w is continuously decreasing and C_s is continuously increasing as dissolved contaminant molecules move from the water to soil surfaces. Therefore, by Equations 5.23 and 5.24, the rate of sorption continuously decreases and the rate of desorption (which initially started at zero) continuously increases. Eventually, the two rates must reach the condition of equilibrium, where both rates are equal.

The partition coefficient K_d quantifies the equilibrium condition of sorption, where sorption and desorption occur at the same rate. When both rates are equal

$$k_{sorb}\ C_w = k_{desorb}\ C_s$$

and

$$\frac{k_{sorb}}{k_{desorb}} = \frac{C_s}{C_w} = K_d \tag{5.25}$$

Equation 5.25 shows that if $k_{sorb} \gg k_{desorb}$, which is the case of strong sorption, then at equilibrium $C_s \gg C_w$, and more of the contaminant is sorbed than is dissolved. If $k_{sorb} \ll k_{desorb}$, the reverse is true. Thus, strongly sorbed contaminants accumulate to higher concentrations on the soil than do more weakly sorbed contaminants.

In terms of contaminant mobility, this means that strongly sorbed contaminants remain sorbed to soil surfaces longer, on average, than do weakly sorbed contaminants and, consequently, move more slowly through the subsurface than does the groundwater in which they are dissolved. The movement of contaminants dissolved in groundwater through the subsurface is analogous to the movement of analytes through a chromatograph column. Because each analyte binds to the column wall (called the stationary phase) with a unique binding strength, different analytes move through the column at different velocities and eventually become separated in space along the column.

Contaminants dissolved in groundwater are similarly retarded in their down-gradient movement relative to the flow of groundwater. Soil serves as the chromatographic stationary phase. The extent of retardation is related to their value of K_d. The front of a moving groundwater contaminant plume will contain the fastest moving contaminants with the lowest values of K_d (having the weakest sorption strengths) and, moving back upgradient through the plume, one will encounter progressively slower moving contaminants with progressively larger values of K_d.

RULES OF THUMB

The mobility of dissolved organic compounds in groundwater depends on K_d.
- If $K_d = 0$, the organic compound does not sorb to the soils it passes through and moves at the groundwater velocity.
- If $K_d > 0$, movement of the organic compound is retarded.

EXAMPLE 12

Consider the case where groundwater contaminated with benzene and toluene is moving through subsurface soils containing 1.6% organic carbon. Compare benzene and toluene qualitatively with respect to their mobility in the subsurface. Use data from Table 5.5.

Answer:

From Table 5.5, K_{oc}(benzene) = 58.9 L/kg and K_{oc}(toluene) = 182 L/kg.

Using $K_d = K_{oc} f_{oc}$,

K_d(benzene) $= 58.9 \times 0.016 = 0.942$ and K_d(toluene) $= 182 \times 0.016 = 2.91$
K_d(toluene) is approximately three times larger than K_d(benzene), indicating that it sorbs more strongly to the soil and will move significantly slower than benzene through the subsurface.

5.7.1 RETARDATION FACTOR

A retardation factor for contaminant movement can be calculated using K_d (Freeze and Cherry, 1979; Fetter, 1993).

The retardation factor, R, for a contaminant is defined as

$$R = \frac{\text{average linear velocity of groundwater}}{\text{average linear velocity of contaminant}} \quad (5.26)$$

A retardation factor of 10 means that the contaminant moves at one-tenth of the average velocity of the groundwater. The average linear velocity of the contaminant is defined as its velocity measured at the point where its concentration is equal to $1/2$ of its concentration at the contaminant source.

Assuming a linear partition coefficient, $K_d = \frac{C_{soil}}{C_w}$, and the retardation factor becomes (Fetter, 1993)

$$R = 1 + \frac{\rho K_d}{h} \quad (5.27)$$

or for a Freundlich partition coefficient, $K_d = \frac{C_{soil}}{C_w^n}$

$$R = 1 + \left(\frac{K_d}{n}\right)\left(\frac{\rho}{h}\right)\left(C_w^{1/(n-1)}\right) \quad (5.28)$$

where
$\rho =$ soil bulk density (g/cm^3), typically 1.5–1.9 g/cm^3
$h =$ effective soil porosity,* typically 0.35–0.55
$n =$ empirical Freundlich exponential factor from Equation 5.14

Some retardation factors and qualitative mobility classifications determined from data in Table 5.5 are in Table 5.6. The table was developed using Equations 5.16 and 5.27 with typical values for soil f_{oc}, effective porosity, and bulk density.

5.7.2 EFFECT OF BIODEGRADATION ON EFFECTIVE RETARDATION FACTOR

Equations 5.27 and 5.28 are based only on sorption to organic carbon and do not consider the influence of processes like biodegradation, ion exchange, precipitation, and chemical changes. As shown in Figure 5.6, biodegradation causes additional retardation to plume movement.

---

* See Section 5.7.4.

TABLE 5.6
Calculated Retardation Factors and Mobility Classifications

R	Examples	K_d	Mobility Classification
<3	Methylene chloride, MTBE, 1,2-DCA	<0.03	Very mobile
3–9	Benzene, 1,1-DCA, chloroform	0.03–1.2	Mobile
9–30	Ethylbenzene, toluene, xylenes	1.2–4.3	Intermediate
30–100	Styrene, pyrene, lindane	4.3–15	Low mobility
>100	Naphthalene, dioxin, heptachlor	>15	Immobile

Note: Assumed values are $f_{oc} = 0.01$, $\rho = 2.0$ g/cm^3, $h = 0.3$.

RULES OF THUMB

1. Dissolved contaminants generally move more slowly than groundwater because of sorption, ion exchange, precipitation, and biodegradation, but the retardation factor calculated from Equation 5.28 or 5.29 considers sorption only.
2. The movement rate of biodegradable compounds is overestimated by the sorption retardation factor, especially over long periods of time.
3. Sorption to soil not only slows the growth of a contaminant plume, but also slows the rate of cleanup by pump-and-treat methods and increases the water volume that must be extracted to achieve acceptable residual concentrations.

EXAMPLE 13

CALCULATION OF A SORPTION RETARDATION FACTOR

Calculate the retardation factor for 1,2-dichloroethane (1,2-DCA) in a soil with 2.7% total organic carbon (TOC), dry bulk density of 1.7 g/cm^3, and effective soil porosity of 40%.

Answer:

$$K_d = K_{oc} f_{oc}$$

where $K_{oc} = 17.4$ (from Table 5.5)

$$R = 1 + \frac{rK_d}{h}$$

$$r = \text{dry bulk soil density} = 1.7 \, \text{g/cm}^3$$

$$h = \text{effective porosity of soil} = 0.40$$

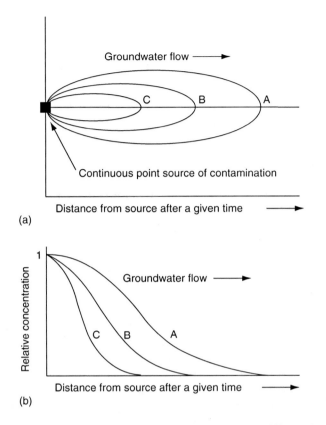

Continuous point source of contamination

Distance from source after a given time

(a)

Distance from source after a given time

(b)

FIGURE 5.6 Positions of contaminant plume front as affected by different flow conditions. (a) Horizontal cross section through plumes from a continuous point source. (b) Decrease in concentration with distance from the source. Curve A represents a nonretarded plume, B represents a plume retarded by sorption, and C represents a plume retarded by sorption and biodegradation.

$$f_{oc} = 0.025$$
$$K_d = 17.4 \times 0.027 = 0.470$$
$$R = 1 + \frac{1.7 \times 0.470}{0.40} = 3.0$$

Divide the groundwater velocity by 3.0 to get the velocity with which 1,2-DCA moves through this particular soil.

EXAMPLE 14

Calculate the retardation factor for tetrachloroethene (PCE*) in a silty clayey loam soil, using data from Table 5.5. Soil density is 2.5 g/cm^3 and effective porosity is 31%.

---

* Other word substitutes for tetrachloroethene are: *tetrachloroethylene* and *PERC*.

Answer:

For tetrachloroethene, $K_{oc} = 155$
Since no measured value for f_{oc} is available, a value from Table 5.4 can be used. For a silty clayey loam soil, $f_{oc} \approx 0.03$.

$$K_d = 155 \times 0.03 = 4.7 \quad \text{and} \quad R = 1 + \frac{2.5 \times 4.7}{0.31} = 38$$

Tetrachloroethene has a K_{oc} nearly 10 times that of 1,2-DCA, which means that it is much less water-soluble. Thus, it is less mobile and has a higher retardation factor.

5.7.3 A Model for Sorption and Retardation

Consider the result of Example 13, where the retardation factor for 1,2-DCA was found to be 3.0 in a particular soil. Since the contaminant is slowed by a factor of 3 relative to the groundwater velocity, it might appear that flushing three pore volumes of water through the impacted soil would completely desorb 1,2-DCA from that part of the subsurface. In other words, the retardation factor might erroneously be interpreted as the number of groundwater pore volumes that must be flushed through the contaminated zone to desorb a contaminant from the impacted soil.

In fact, most low-solubility organic contaminants can never be completely flushed from the soil, because there normally is some fraction of the contaminant that becomes almost irreversibly bound to the organic matter in soil. The part of the contaminant that behaves according to its K_d value governs the movement of the front of the contaminant plume, but there is usually a portion that binds to the soil more strongly and moves more slowly. This part cannot be removed by water flushing in any reasonable time. It can remain in the soil as a slowly diminishing source of groundwater contamination for tens, or even hundreds, of years.

There is an aging process that causes the fraction of contaminant that can be desorbed by flushing to decrease with time. This is why pump-and-treat remediation methods are seldom successful at removing the last part of contamination from soil. When desorption of a contaminant becomes very slow and pump-and-treat methods become ineffective, other approaches such as bioremediation are needed to achieve required cleanup levels.

A conceptual model for sorption has been proposed (Alexander, 1995) that is consistent with time-dependent irreversible sorption and other observations, such as a decrease with time of biodegradation rates.

Model of Sorption

1. A compound with large K_d is initially adsorbed rapidly from water to the external surfaces of soil particles.
2. Measurements of K_d are normally based on the sorption behavior of the initially sorbed contaminant. K_d represents the concentration ratio between sorbed and dissolved contaminant if the elapsed time between sorption and measurement is not too long. The dissolved fraction is

available to microorganisms for biodegradation, and to organisms such as fish and humans that might be susceptible to its toxic effects.

3. As time passes, a portion of the surface-sorbed compound begins to diffuse into micropores in the solid surface, moving away from the surface into the particle interior. Here the compound becomes sequestered within the soil in locations that are remote from the surface and where it is less available for desorption.

Thus, aging appears to be associated with continuous diffusion into more remote sites on the solid particle, where the molecules are retained and rendered less accessible to biological, chemical, and physical changes. After 1–10 years, depending on site-specific soil and pollutant conditions, a large fraction of the sorbed pollutant will not desorb from the soil.

Another limitation of the retardation factor is that the equations unrealistically assume that the soil matrix is homogeneous, which is rarely the case. However, the results are still useful for rough estimates and for estimating relative mobilities of dissolved contaminants.

5.7.4 Soil Properties

The subsurface environment contains inorganic minerals, organic humic materials, air, and water. Also found are plant roots, microorganisms, and burrowing animals, not to mention building foundations, utility service lines, and other man-made structures. All of these disturbances of the natural soil texture can affect the movement of contaminants through the subsurface.

The physical properties of soil that have the greatest effect on the movement of water and contaminants are true porosity, effective porosity, particle size range, and hydraulic conductivity. Representative values for these properties are given in Tables 5.7 through 5.9.

- True porosity is the ratio of the volume of empty space (pore volume) to total volume of soil. It can be expressed as a simple ratio or as a percentage. For a soil sample in which 1/4 of the total volume is empty (or void) space, the porosity may be expressed as 0.25 or 25%.

TABLE 5.7

Representative Values of Effective Porosity for Some Soil Types

Soil Type	Effective Porosity (%)
Well-sorted sand or gravel	25–40
Sand and gravel, mixed	20–35
Medium sand	15–30
Glacial sediments	5–20
Silt	1–20
Clay	1–2

TABLE 5.8

Soil Particle Size Range for Some Soil Types

Soil Type	Particle Size Range (mm)
Clay	<0.002
Silt	0.002–0.04
Very fine sand	0.04–0.10
Fine sand	0.10–0.20
Medium sand	0.20–0.40
Coarse sand	0.40–0.90
Very coarse sand	0.9–2.0
Fine gravel	2.0–10.0
Medium gravel	10.0–20.0
Coarse gravel	20–40
Very coarse gravel	40–80

- Effective porosity is the ratio of the volume of effective void space to the total volume of material, generally expressed as a percentage. Effective porosity accounts for the fact that pore spaces that are not connected, or are too small for the fluid to overcome capillarity, do not contribute to fluid movement. The difference between true porosity and effective porosity is most noticeable with clay, where true porosity is very high, 34%–60% (it can hold a lot of water). Effective porosity, however, is very low, 1%–2%, making clay relatively impermeable.
- Soil particle size range determines how soil particles pack together and, thereby, the average soil pore size. Pore size strongly affects the available soil surface area as well as the capillary attraction between the soil and liquids. The smaller the pore size, the stronger is capillary attraction and the greater is the total soil surface area within a given volume of soil. The larger the surface area, the larger is the volume of liquid immobilized by sorption to soil surfaces. Capillary attractions are an additional force that acts to retard the movement of liquids through soil and can immobilize a fraction of the liquid. Where pore size is small enough, the

TABLE 5.9

Representative Values of Water Hydraulic Conductivity for Some Soils

Soil Type	Hydraulic Conductivity for Water: Typical Range (cm/s)
Clay	10^{-6} to 10^{-9}
Silt	10^{-3} to 10^{-7}
Fine sand	10^{-2} to 10^{-5}
Medium sand	10^{-1} to 10^{-4}
Coarse sand	1 to 10^{-4}
Gravel	10^{2} to 10^{-1}

distance between adjacent soil particles can be small enough that capillary attraction can extend across significant fractions of the pore volume. Thus, silt retards the movement of liquids more than coarser soils like sand and gravel, and will immobilize a larger quantity of liquid than will an equal volume of coarser soil. Clay also is effective at retarding the movement of liquids but, because of its low effective porosity, cannot immobilize large volumes of liquid by sorption and capillarity.
- Hydraulic conductivity indicates the ability of subsurface material to transmit a particular fluid. For example, the hydraulic conductivity of a given soil to transmit water is greater than it would be to transmit a more viscous fluid such as diesel fuel. The hydraulic conductivity of a very porous soil, like loose sand, is greater than for a less porous soil, such as fine silt.

5.8   PARTICULATE TRANSPORT IN GROUNDWATER: COLLOIDS

Contaminants move in groundwater systems as dissolved species in the water, as flowing free-phase liquids, or combined with moving particulates. Particulates that can move through soils with groundwater must be small enough to move through the soil pore spaces. Such particulates are generally less than 2.0 μm in diameter and are called colloids. Colloids are a special class of matter with properties that lie between those of the dissolved state and the solid or immiscible liquid states.

Colloids have a high surface area to mass ratio due to their very small size. Groundwater concentrations of colloidal materials can be as high as 75 mg/L, corresponding to as many as 10^{12} particles/L. This represents a large surface area available for transporting sorbed contaminants.

There are many sources of colloidal material in groundwater. Colloids are formed in soil when fragments of soil, mineral, or contaminant particles become detached from their parent solid because of weathering or by physical abrasion caused by rock movement or drilling and other construction activities. Then they are carried to the groundwater when water from irrigation or precipitation percolates downward through the soil. Colloids form as fine precipitates when dissolved minerals in groundwater undergo pH or redox potential changes. Colloids are often introduced directly into groundwater from landfills, and they can form as emulsions of small droplets from free-phase hydrocarbons or other immiscible liquids.

5.8.1   COLLOID PARTICLE SIZE AND SURFACE AREA

When particle size is reduced to 1–2 μm or smaller, surface forces arising from surface charge or London force attractions begin to exert a significant influence on particle behavior. Consider the effect of reducing the particle size of a given mass of solid. A cube that is 10 mm on a side has a surface area of 6.0×10^{-4} m^2. Cut it in half in each of the three directions perpendicular to its faces to get eight cubes, which are each 5 mm on a side. The surface area is now 12.0×10^{-4} m^2. Continue subdividing until the cubes are 1 μm on an edge. The total surface area is now 6.0 m^2—an increase of 10,000 times over the original cube.

Montmorillinite, a clay mineral, in the dispersed state may break down into small plate-like particles only one unit cell in thickness, about 10^{-9} m. Its specific surface area is about 800 m^2/g. A monolayer of 10 g of this material would cover a football field.

5.8.2 PARTICLE TRANSPORT PROPERTIES

Contaminants of low solubility can move with groundwater as colloids or attached by sorption or occlusion with colloids, resulting in unexpected mobility of low-solubility material. When contaminants are sorbed to colloids, their transport behavior is determined by the properties of the colloid, not the sorbed contaminant. In the low-velocity flow of groundwater, particles larger than 2 μm tend to settle by gravity. Particles smaller than 0.1 μm tend to sorb readily to larger soil particles, becoming retarded or immobilized. Thus, particles in the range 0.1–2.0 μm are the most mobile in groundwater.

Adsorptive Processes
Colloid adsorptive behavior is influenced by

1. Forces acting on colloidal particles
   a. Electrostatic attraction and repulsion
   b. London attractions
   c. Brownian motion
2. Properties of the groundwater
   a. Ionic strength (related to total dissolved solids [TDS] and conductivity)
   b. Ionic composition
   c. Flow velocity
3. Properties of the colloids
   a. Size
   b. Chemical nature
   c. Concentration
4. Properties of the soil matrix
   a. Geologic composition
   b. Particle size distribution
   c. Soil surface area

RULES OF THUMB

1. Usually the most important factor governing colloid behavior is groundwater chemistry and the least important is flow velocity.
2. Generally, colloids are more mobile when the dissolved ion concentration is low and less mobile when the dissolved ion concentration is high.

5.8.3 ELECTRICAL CHARGES ON COLLOIDS AND SOIL SURFACES

The explanation for how water chemistry affects colloid behavior lies in the behavior of charged particles in water solutions. Soil surfaces in an aquifer generally have a net negative charge due to the dominance of silicates in the minerals, which have exposed electronegative oxygen atoms. Colloidal particles also are usually charged, with a sign dependent on the nature of the colloid and the water pH.

- Metal oxide colloids tend to be positively charged.
- Sulfur and the so-called noble metals* tend to be negatively charged.
- Organic macromolecules, such as humic materials, proteins, or resins used in water treatment for flocculation, acquire a charge that depends on the pH. At low pH, hydrogen ions bind to the molecules and make the colloid positive. At high pH, the binding of hydroxyl ions makes the charge negative.

5.8.3.1   Electrical Double Layer

In natural waters, the charge on colloids and on soil surfaces attracts dissolved ions into a configuration known as an electric double layer. Consider the double layer that forms adjacent to a negatively charged surface. The first layer forms when dissolved cations (positive ions) are attracted to the oppositely charged surfaces. Then anions (negative ions) are attracted to the positive ion region of the first layer, to form a second, more diffuse, oppositely charged layer surrounding the first. The net result is an inner layer of cations surrounded by an outer layer of anions. Positively charged surfaces acquire an inner layer of anions and an outer layer of cations. The inner layer effectively neutralizes the surface charge and the outer layer effectively neutralizes the inner layer. Subsequent layers can form in principle, but they are too diffuse to have any effect.

5.8.3.2   Adsorption and Coagulation

If two colloid particles come close enough together, London attractive forces will pull them together into a larger particle and start coagulation. The same thing happens if a colloidal particle comes close enough to a soil surface. London attractive forces will pull them together and the colloid becomes sorbed to the soil surface.

Two colloidal particles of the same material have the same sign of charge in the outer part of their double layer and, therefore, repel one another. For two colloidal particles to coagulate, they must collide with enough energy to force past their repulsive double layers and approach close enough for London attractive forces to be effective. The same is true for adsorption to soil particle surfaces.

Brownian motion provides the collision energy for overcoming the electrical repulsion of the double layer. In high-energy collisions, particle momentum overcomes charge repulsion and allows particles to approach close enough to enter the zone of London attraction, where adsorption and coagulation can occur.

---

* Metals that are resistant to oxidation and corrosion under common environmental conditions are sometimes referred to as noble metals. In addition to gold, silver, and platinum, they may include copper, mercury, aluminum, palladium, rhodium, iridium, tantalum, and osmium.

- At high ionic strength (high dissolved ion concentrations; high TDS), the double layer is thin because the ion atmosphere is dense and there is more charge per unit volume. The higher ion charge density can neutralize the net particle charge with a thinner layer.
- At low ionic strength, a thicker ionic charge layer is required to neutralize the charge on colloidal particles.

Thus, colloidal particles can approach surfaces and other colloids more closely at high TDS than at low TDS, allowing London attraction, which becomes stronger at shorter distances, to be more effective. There are several familiar illustrations of this principle:

- When river water carrying colloidal clay reaches the ocean, the salt water (high TDS) induces coagulation. This is a major cause of silting in estuaries.
- Applying a styptic pencil stops bleeding from small cuts. The styptic pencil contains aluminum salts, often alum (aluminum sulfate). Dissolving high concentrations of aluminum salt into the wound raises the TDS concentration in the external blood and initiates coagulation of colloidal proteins in blood.
- Colloidal material can be "salted out" by dissolving salts into the solution.
- The high TDS of typical landfill leachate provides optimal conditions for immobilizing entrained colloids by sorption and coagulation.

RULE OF THUMB

Coagulation and sorption of colloids are more efficient under high TDS conditions.

5.9   CASE STUDY: CLEARING MUDDY PONDS

Ponds and small lakes sometimes become muddy (or turbid).* Some common causes are

- Stormwater runoff from fields with disturbed soil or sparse vegetation, such as construction sites or fields with active agricultural equipment.

---

* The case study deals with water turbidity caused by soil particles and not turbidity caused by algae or other organic suspensions. Water colored by suspended soils is normally a chocolate brown or red-brown color, while water colored by algae is green.

- Shoreline erosion caused by strong winds and wave action, and by shoreline activity of people and animals.
- Animal activity in the water, cattle or wildlife coming to drink or cool off.
- Heavy use by waterfowl digging in shallow water and eating shoreline vegetation.
- High populations of certain fish that spawn and feed in the bottom sediments.

In addition to being unsightly, muddy water limits sunlight penetration, which, in turn, reduces the amount of aquatic life supported by the pond's food chain; it even can clog the gills of fish, suffocating them. A pond muddied by a singular event like a particularly strong storm or a temporary construction activity will often clear after several days without disturbance. Occasionally however, a muddy pond will fail to clear. When this happens, the pond water must be treated to restore its clarity.

In this case study, a period of severe rainstorm activity caused flooding of a construction site and washed a large quantity of sediment into a pond about 3 acres in surface area. The average depth of the pond was about 8 ft. A quart jar of turbid water collected from the pond was allowed to stand undisturbed for 5 days. After this time, some sediment had settled to the bottom of the jar, but the water was still quite murky.

Suspended sediment that has not settled out after this period is most likely to be colloidal clay particles carrying a negative surface charge that keeps them suspended, as described in Section 5.8. A common treatment for turbid ponds is to induce coagulation by increasing the TDS, especially with cations having high positive charge ($+2$ and $+3$). In addition to thinning the double layer, multivalent cations form bridges between the negatively charged clay particles, greatly assisting the coagulation process and speeding water clarification.

In principle, any salt may be added to facilitate coagulation. For environmental safety however, only compounds not harmful to the ecosystem in the required concentrations can be used. The most commonly used chemicals for clearing muddy ponds are alum (aluminum sulfate, $Al_2(SO_4)_3$) and gypsum (calcium sulfate, $CaSO_4$). Alum is more effective than gypsum (requires smaller concentrations) because the aluminum cation (Al^{3+}) has a greater positive charge than the calcium cation (Ca^{2+}) and forms more extensive bridges between negatively charged clay particles. However, aluminum cations act as acids (see Section 4.1.3.2) and will lower the pond pH. For this reason, buffered alum, with added hydrated lime (calcium hydroxide, $Ca(OH)_2$), should be used for application to ponds with low alkalinity (less than 100 mg/L), or hydrated lime can be added before applying alum (about 20 lb of hydrated lime per acre-foot).

The pond in this case study had an alkalinity of 320 mg/L as $CaCO_3$, high enough that buffered alum was not necessary. The dimensions of the ponds resulted in a volume of about 24 acre-ft. Rule of Thumb number 2 below indicated that 1200 lb of alum would be sufficient. A conservative application of 600 lb was used first, which cleared the pond to a measured underwater visibility of 2.0 ft after 2 days. Further treatment was deemed unnecessary.

RULES OF THUMB

1. For healthy aquatic life, underwater visibility should be at least 1–1.5 ft most of the time.
2. About 50 lb of alum per acre-foot of water will clear most turbid ponds in about 1 week. Alum should be dissolved in water and sprayed over the pond surface on a calm day. For minimum risk to aquatic life, it is prudent to first apply 1/3 to 1/2 of the calculated dose and wait 2 days to see if more alum is needed to increase underwater visibility to 1–1.5 ft.
3. If alkalinity > 50 mg/L and pH > 7, regular alum may be used. Otherwise, use buffered alum or first add about 20 lb of hydrated lime (calcium hydroxide) per acre-foot of water. However, large amounts of alum can lower the alkalinity and pH rapidly. If more than 300–400 lb of alum are required, the alkalinity and pH should be monitored frequently during application.
4. Gypsum can also be used to clear ponds. It will not acidify the water like alum, but it requires a dose about 10 times heavier than alum. About 500 lb of gypsum will have the same effect as 50 lb of alum.
5. To calculate the volume of a pond in acre-feet, estimate its surface area in square feet, multiply by its average depth in feet, and divide this volume by 43,560 ft^3/acre-ft.

5.9.1 Pilot Jar Tests

Sometimes, instead of the trial-and-error approach described above, it is better to use a more scientific approach to determine the correct amount of alum or gypsum (or other coagulant) for clearing a turbid pond. For example, if the pond has a pH beyond the range of 6–8, or the sediment is not common clay or silt but instead contains a large fraction of coal mine dust or, perhaps, arises from animal feedlot runoff, the doses that work for clay and silt might not be appropriate. In such cases, a better approach is to perform a set of jar tests where you test the effects of a series of progressively more concentrated doses of coagulant on samples of the water to be treated. In this way you can conveniently measure the effects of different coagulants at different doses, to find the most effective and economical way to treat the pond.

When working in the field, accurate scales and volumetric glassware are not always available for making up known concentrations of coagulant solutions. Fortunately, precise measurements are seldom critical since modest overdoses of coagulant are generally not harmful and often serve to speed the clearing process. Therefore, the recipes for the pilot tests given below use tablespoons and 1-gallon plastic or glass jars (1-gallon plastic water bottles are fine). If alum is to be used, measure the alkalinity and pH of the pond.

5.9.1.1 Jar Test Procedure with Alum Coagulant

1. Collect six 1-gallon jars of turbid pond water and one 1-gallon jar of bottled drinking water.
2. Number the pond water bottles from 1 to 6.
3. Using a standard measuring spoon, mix 1 tbl.sp. of alum into the 1-gallon jar of drinking water, stirring until the alum forms a slurry.
4. Add 1 tbl.sp. of the alum slurry to jar 1, 2 tbl.sp. to jar 2, 3 tbl.sp. to jar 3, etc. for all six jars.
5. Allow the treated jars to stand undisturbed for at least 12 h.
6. Identify the turbid water jar that required the least number of teaspoons of alum slurry to clear it.
7. Use the table below to determine the proper dose of alum for treating the pond. For example, if the pH >7, alkalinity >100 mg/L, and jars 3, 4, 5, and 6 all cleared up after standing for 12 h, jar 3 required the least amount of alum for clearing. Apply 13 lb/acre-ft of hydrated lime to the pond, followed by 90 lb/acre-ft of alum.

Number of Tablespoons of Alum Slurry Added to Clear Sample in 12 h	Amount of Alum to Apply to Pond (lb/acre-ft)	Amount of Hydrated Lime to Add before Adding Alum If pH >7 and Alkalinity >100 mg/L	Amount of Hydrated Lime to Add before Adding Alum If pH <7 and alkalinity <100 mg/L
1	30	0	13
2	60	0	26
3	90	13	39
4	120	17	52
5	150	21	65
6	180	25	78

5.9.1.2 Jar Test Procedure with Gypsum Coagulant

1. Collect 12 1-gallon jars of turbid pond water and one 1-gallon jar of bottled drinking water.
2. Number the pond water bottles from 1 to 6.
3. Using a standard measuring spoon, mix 2 tbl.sp. of gypsum into the 1-gallon jar of drinking water, stirring until the gypsum forms a slurry.
4. Add 1 tbl.sp. of the gypsum slurry to jar 1, 2 tbl.sp. to jar 2, 3 tbl.sp. to jar 3, etc. for all 12 jars.
5. Allow the treated jars to stand undisturbed for at least 12 h.
6. Identify the turbid water jar that required the least number of teaspoons of gypsum slurry to clear it.
7. Use the table below to determine the proper dose of gypsum for treating the pond. For example, if all jars 8 and above all cleared up after

standing for 12 h, jar 8 required the least amount of gypsum for clearing. Apply 640 lb/acre-ft of gypsum to the pond.

Number of Tablespoons of Gypsum Slurry Added to Clear Sample in 12 h	Amount of Gypsum to Apply to Pond (lb/acre-ft)
1	80
2	160
3	240
4	320
5	400
6	480
7	560
8	640
9	720
10	800
11	880
12	960

EXERCISES

1. What are soil horizons?
2. The pores of soil contain a significant amount of air. How does the composition of soil air differ from that of atmospheric air and why is it so?
3. Many soils have ion-exchange properties. What does this mean?
4. Discuss how fluctuations in the groundwater level can influence the distribution of LNAPL hydrocarbon contamination in the subsurface.
5. A chemical spill has contaminated a lake with carbon tetrachloride (CCl_4). For carbon tetrachloride, Table 5.5 shows that $\log K_{ow} = 2.73$ (thus, $K_{ow} = 5.4 \times 10^2$). The maximum concentration of carbon tetrachloride that a certain fish can accumulate with a low probability of health risks is reported to be 10 ppm. Assuming that the fatty tissue of these fish behaves similarly to octanol with respect to carbon tetrachloride partitioning, at what value should the water quality standard for carbon tetrachloride be set in this lake to protect this fish?

REFERENCES

Alexander, M., 1995, How toxic are toxic chemicals in soil? *Environ. Sci. Technol.*, 29 (11), 2713–2717.

Canadian Water Quality Guidelines, 1987, Canadian Council of Resource and Environmental Ministers (CCREM), Inland Waters Directorate, Environmental Canada, Ottawa, Ontario.

Fetter, C.W., 1993, *Contaminant Hydrogeology*, Macmillan, New York.

Freeze, R.A. and Cherry, J.A., 1979, *Groundwater*, Prentice-Hall, Inc. New Jersey.

Handbook of Chemistry and Physics, 71st Edition, 1990, or later, David R. Lide (Ed.), CRC Press, Boca Raton.

Howard, P.H. (Ed.), 1991, *Handbook of Environmental Fate and Exposure Data for Organic Chemicals*, Vols. I–III, Lewis Publishers, Chelsea, MI.

Lyman, W.J., Reehl, W.F., and Rosenblatt, D.H., 1990, *Handbook of Chemical Property Estimation Methods,* 2nd printing, American Chemical Society, Washington, D.C.

Mackay, D. and Shiu, W.Y., 1981, A critical review of Henry's law constants for chemicals of environmental interest, *J. Phys. Chem. Ref. Data*, 10 (4): 1175–1199.

Maidment, D.R. (Ed.), 1993, *Handbook of Hydrology,* McGraw-Hill, Inc. New York.

Means, J.C. and Wijayratne, R., 1982, Role of Natural Colloids in the Transport of Hydrophobic Pollutants, *Science*, 215, 968–970.

Sander, R., 1999, *Compilation of Henry's Law Constants for Inorganic and Organic Species of Potential Interest in Environmental Chemistry*, http://www.mpch-mainz.mpg. de/~sander/res/henry.html. This Web site is regularly updated with current values and contains a calculator for converting among different units.

USEPA, 1996, *Soil Screening Guidance: Technical Background Document*, Office of Emergency and Remedial Response, Washington, D.C., EPA/540/R95/128.

USEPA, 1999, *Understanding Variation in Partition Coefficient, K_d, Values,* Vol. 1 (EPA 402-R-99-004A) and Vol. 2 (EPA 402-R-99-004B), August.

6 General Properties of Nonaqueous Phase Liquids and the Behavior of Light Nonaqueous Phase Liquids in the Subsurface

6.1 TYPES AND PROPERTIES OF NONAQUEOUS PHASE LIQUIDS

According to the U.S. Environmental Protection Agency (EPA) (USEPA, 2004), as of September 30, 2003, more than 439,000 releases from leaking underground storage tanks (LUSTs) have been reported nationwide. Cleanups have been initiated at more than 403,000 of these sites, and more than 303,000 sites have been cleaned up. The backlog of sites still to be cleaned up is more than 136,000. The majority of these LUSTs contained petroleum hydrocarbons, which are a mixture consisting mainly of compounds having very low solubilities in water. Petroleum hydrocarbons and their components are commonly called nonaqueous phase liquids (NAPLs). This chapter and Chapters 7 and 8 are an introduction to understanding the environmental behavior of these important pollutants.

NAPLs are nonpolar or low-polarity liquids that are minimally soluble in water. Gasoline and diesel fuels, oils, chlorinated solvents, and pesticides are examples. In contact with water, nonpolar liquids do not dissolve appreciably in water, but remain as separate, immiscible liquid phases called NAPLs. NAPL substances are comprised of molecules that are nonpolar, or weakly polar and, therefore, have only weak attraction to polar water molecules but relatively strong London force attraction to other NAPL molecules (see Chapter 2, Section 2.9). Therefore, NAPLs are not very soluble in water and, if mixed into water, separate into a distinct liquid phase with a well-defined boundary between the NAPL and the water that is usually visually observable. The boundary interface is a visible physical dividing surface between the bulk phases of the two liquids.

Many NAPLs are mixtures of different hydrocarbon compounds with varying low solubilities. The phase boundary does not prevent compounds within the NAPL from dissolving into the water. Over time, the more soluble compounds in a NAPL mixture will preferentially move across the interface and dissolve into the surrounding water phase. This changes the composition and physical properties of NAPL mixtures.

NAPL contaminants include a wide range of industrial compounds such as solvents, heating oil, gasoline, chlorinated hydrocarbons, coal tars, and creosote. NAPL compounds are typically stored in underground and aboveground tanks and transported by pipelines, trains, and tank trucks, all of which are susceptible to leaks and spills.

Due to their low solubility, NAPLs in contact with environmental waters dissolve slowly, acting as a long-term source of water contamination. Because of their typically high toxicity, small amounts of NAPL can contaminate very large volumes of soil and groundwater. Remediation of NAPL contaminated sites is made difficult by the low solubility and uneven distribution of NAPLs in the subsurface, and the generally low flow rates of groundwater. The accidental release of NAPL liquids into soils and aquifers is a widespread and serious environmental problem.

NAPLs are further subdivided into light nonaqueous phase liquids (LNAPL) and dense nonaqueous phase liquids (DNAPL).* LNAPLs are those liquid hydrocarbon compounds or mixtures that are less dense than water, such as gasoline and diesel fuels and their individual components. DNAPLs are liquid hydrocarbon compounds or mixtures more dense than water, such as creosote, polychlorinated biphenyls (PCBs), coal tars, and most chlorinated solvents (chloroform, methylene dichloride, tetrachloroethene, etc.).

The distinction between LNAPLs and DNAPLs is important because of their different behavior in the subsurface. When a NAPL spill occurs in the vadose zone, LNAPL will travel downward through soils only to the water table, where they remain "floating" on the water table surface. DNAPLs can sink through the water-saturated zone to impermeable soil structures such as bedrock, where they collect in bottom pools. Different remediation methods are required for LNAPLs and DNAPLs. This chapter discusses LNAPLs and Chapter 7 discusses DNAPLs.

6.2 GENERAL CHARACTERISTICS OF PETROLEUM LIQUIDS, THE MOST COMMON LNAPL

Petroleum liquids are complex mixtures of hundreds of different hydrocarbons, with minor amounts of nitrogen, oxygen, sulfur, and some metals. Nearly all petroleum compounds are nonpolar and not very soluble in water. The behavior of these compounds in a groundwater environment depends on the physical and chemical nature of the particular hydrocarbon blend as well as the particular soil environment. For example, the partition coefficients and migration potential of each individual compound in a mixture depend on the overall composition of the petroleum mixture in which it is found, on the properties of the pure compound, and on the characteristics of the surrounding soil. Furthermore, the composition and properties of petroleum contaminants change with time as the petroleum ages and weathers.

Many nonfuel organic pollutants, such as chlorinated hydrocarbons and pesticides, are more soluble in petroleum than in water. Therefore, if an oil spill occurs where previous organic contamination already exists, the older pollutants tend to concentrate from soil surfaces and pore space water into the fresh oil phase. An oil

---

* See Chapter 7 for a discussion of DNAPLs.

spill into an already contaminated soil can mobilize other pollutants that have been immobilized there by sorption and capillarity. As freshly spilled oil moves downward through the soil, immobilized pollutants sorbed to the soil can dissolve into the moving liquid oil and be carried along with it. Analysis of spilled petroleum products will often detect other organic compounds that were previously sorbed to the soil.

RULE OF THUMB

Because many organic pollutants are soluble in NAPLs, analysis of spilled petroleum products will often detect other organic compounds, such as pesticides, that were previously sorbed to the soil but were not originally present in the NAPL spill being investigated.

6.2.1 TYPES OF PETROLEUM PRODUCTS

The first step in refining crude oil into petroleum products is usually fractional distillation, a process that separates the crude oil components according to their boiling points. The resulting products are groups of mixtures, or fractions, each of which has boiling points within a specified range. All but the lightest fractions (lowest boiling temperature ranges) can contain hundreds of different hydrocarbon compounds. The fractions are often classified into the general groups described in Table 6.1. In addition, several pure petrochemicals may be produced, such as butane, hexane, benzene, toluene,

---

TABLE 6.1
Principal Petroleum Fractions from Fractional Distillation

Boiling Range (°C)	Dominant Composition Range	Fraction	Uses
−160 to +30	C1–C4	Gases	LPG, methane, gaseous fuels, feedstock for plastics
30–60	C5–C7	Petroleum ether	Solvents, gasoline additives
90–130	C6–C9	Ligroin, naphtha	Solvents
40–200	C4–C12	Gasoline	Motor fuel
60–200	C7–C12	Mineral spirits	Solvents
150–300	C10–C16	Kerosene	Jet fuel, diesel fuel, lighter fuel oils
300–350	C16–C18	Fuel oil	Diesel oil, heating oil, cracking stock
>350	C18–C24	Lubricating stock	Lubricating oil, mineral oil, cracking stock
Solid residue	C25–C40	Paraffin wax	Candles, toiletries, wax paper
Solid residue	>C40	Residuum	Roofing tar, road asphalt, waterproofing

Note: The notation used here gives the range of carbon atoms in the fraction compounds. For example, C5–C7 means the petroleum fraction that contains mostly hydrocarbon compounds containing between 5 and 7 carbon atoms. This table indicates that as the number of carbon atoms in a hydrocarbon molecule increases, so do its boiling temperature and its viscosity. Volatility decreases as the number of carbon atoms in a compound increases.

and xylene, for use as solvents, for production of plastics and fibers, and for reblending into fuel mixtures. Refined petroleum products are further modified by catalytic cracking, blending, and reformulation processes to enhance desirable properties.

RULE OF THUMB

The larger the hydrocarbon compound and the more carbon atoms it contains, the higher are its boiling point and viscosity, and the lower is its volatility.

6.2.2 GASOLINE

Gasoline is among the lightest liquid fractions of petroleum and consists mainly of aliphatic and aromatic hydrocarbons in the carbon number range C4–C12.
Aliphatic hydrocarbons consist of

- *Alkanes*: Saturated hydrocarbons in which all carbons are connected by single bonds. They may have linear, branched, or cyclic carbon-chain structures, such as pentane, octane, decane, isobutane, and cyclohexane.
- *Alkenes*: Unsaturated hydrocarbons in which there are one or more double bonds between carbon atoms. They also may have linear, branched, or cyclic carbon-chain structures, such as ethylene (ethene), 1-pentene, and 1,3-cyclohexadiene.
- *Alkynes*: Unsaturated hydrocarbons in which there are one or more triple bonds between carbon atoms. They also may have linear, branched, or cyclic carbon-chain structures, such as acetylene (ethyne), propyne, and 1-butyne.

Aromatic hydrocarbons (also called arenes) are hydrocarbons based on the benzene ring as a structural unit. They include monocyclic hydrocarbons such as benzene, toluene, ethylbenzene, and xylene (the BTEX group, see Figure 6.1) and polycyclic hydrocarbons such as naphthalene and anthracene.

RULES OF THUMB

Gasoline mixtures are volatile, include somewhat soluble components, and are mobile in the groundwater environment.
    Gasolines contain a much higher percentage of the BTEX group of aromatic hydrocarbons (benzene, toluene, ethylbenzene, and the *ortho-* and *para-*xylene isomers) than do other fuels, such as diesel. They contain lower concentrations of heavier aromatics like naphthalene and anthracene than do diesel and heating fuels. Therefore, the presence of BTEX in appropriate concentration ratios* is often a useful indicator of gasoline contamination.

---

* The BTEX concentration ratios that are characteristic of fresh and weathered gasoline are discussed in Section 6.7.3.

RULES OF THUMB (Continued)

Oxygenated compounds such as alcohols (methanol and ethanol) and ethers (methyl-*tert*-butyl ether [MTBE]) are often added as octane boosters and oxygenators. Until recently, MTBE was the most commonly used of these. Gasoline formulations between 1980 and 2003 typically contained around 15% MTBE by volume. EPA began to phase out MTBE around 2003, replacing it with ethanol, because of concerns about the health effects of MTBE groundwater contamination.

6.2.3 MIDDLE DISTILLATES

Middle distillates cover a broad range of hydrocarbons in the range of C6 to about C25. They include diesel fuel, kerosene, jet fuels, and lighter fuel oils. Typical middle distillate products are blends of up to 500 different compounds. They tend to be denser, more viscous, less volatile, less water soluble, and less mobile than gasoline. They contain low percentages of the lighter weight aromatic BTEX group, which may not be detectable in older releases of middle distillates due to degradation or transport.

6.2.4 HEAVIER FUEL OILS AND LUBRICATING OILS

Heavier fuel oils and lubricating oils are composed of heavier molecular weight (MW) compounds than the middle distillates, encompassing the approximate range of C15–C40. They are more viscous, less soluble in water, and less mobile in the subsurface than the middle distillates.

Figure 6.2 relates the carbon number of a petroleum compound to its properties, uses, and the usual instrumental methods used for its analysis.

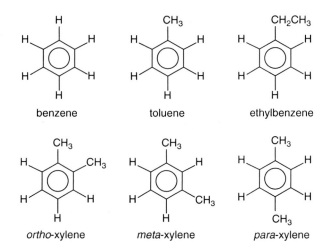

FIGURE 6.1 The BTEX group of aromatic hydrocarbons.

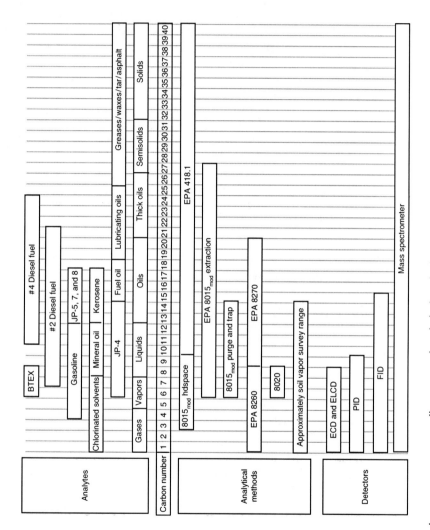

FIGURE 6.2 Hydrocarbon ranges, corresponding uses, and analytical methods.

6.3   BEHAVIOR OF PETROLEUM HYDROCARBONS IN THE SUBSURFACE

6.3.1   SOIL ZONES AND PORE SPACE

As illustrated in Figure 6.3, subsurface soil may be divided into a water-unsaturated zone (also called the vadose zone) extending from the soil surface down to just above the water table, and a water-saturated zone extending from the water table down to bedrock. Capillary forces extend the saturated zone slightly above the water table, establishing a capillary zone of transition between the unsaturated and saturated zones. Capillary zone can vary from a fraction of an inch in coarse-grained sediments to several feet in fine-grained sediments such as clay.

Each zone contains soil particles with pore spaces between them. In permeable soils, most of the pore spaces are continuous, allowing movement of water and liquid contaminants through them. In the absence of contaminants, pore spaces in the unsaturated zone contain air with some water adsorbed to the soil particles. Pore spaces in the saturated zone contain mainly water.

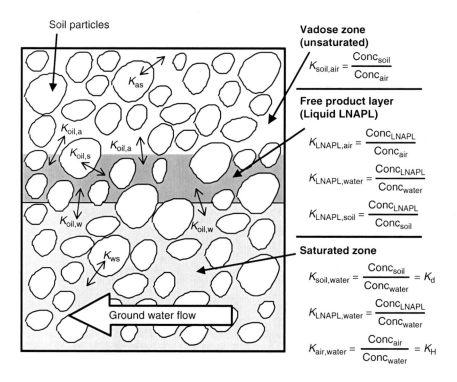

FIGURE 6.3 Soil zones and partitioning behavior of a free product pollutant. All the Ks in the equations are partition coefficients. They quantitatively describe how the pollutant distributes itself among water, soil, air, and free product.

When contaminants enter the subsurface region as spilled liquid petroleum (free product)

- Volatile compounds vaporize from the free product mixture to the atmosphere and to the air in the soil pore spaces.
- The more soluble compounds in the free product begin to dissolve into water contained on soil particle surfaces, into water percolating down from the ground surface, and into groundwater in the saturated zone.
- A small fraction of the free product is taken up by microbiota.
- The remaining free product adsorbs to soil particles and, where free product is abundant, fills the pore spaces.

6.3.2 PARTITIONING OF LIGHT NONAQUEOUS PHASE LIQUIDS IN THE SUBSURFACE

Because of their low water solubilities, most of the compounds classified as petroleum hydrocarbons are generally considered as NAPLs). If mixed into water, NAPLs separate into a distinct liquid phase with a well-defined boundary between the NAPL and the water. However, all petroleum NAPLs contain some compounds of sufficient solubility to represent a pollution hazard if brought into contact with water. In addition, volatile compounds can escape into the atmosphere, causing air pollution and the insoluble, nonvolatile compounds that remain trapped in the soil can render the soil unfit for many uses (Freeze and Cherry, 1979).

Before a petroleum release occurs, the voids of vadose zone earth materials are filled with air and water. After a release, some voids contain immobile petroleum held by capillary forces and sorbed to soil surfaces. There may also be liquid petroleum moving downward through the pore interstices under gravity. If LNAPL reaches the water table, its buoyancy prevents further downward movement and it spreads out horizontally over the water table to form a layer of free product "floating" above the saturated zone. The individual components of the petroleum become partitioned into air, water, and solid phases that are in contact with the free product.

6.3.3 PROCESSES OF SUBSURFACE MIGRATION

After part of the spilled petroleum has partitioned from the free product into other phases, hydrocarbons are present in solid, liquid, dissolved, and vapor phases.

1. Solid phase hydrocarbons are sorbed on soil surfaces or diffused into micropores and mineral grain lattices. They are immobile and degrade very slowly.
2. Liquid phase hydrocarbons exist in the subsurface as
   a. Immobile residual liquids held by capillary forces and as a thin layer sorbed to sediments in the unsaturated and capillary zones.
   b. Free mobile liquids in the unsaturated zone above the capillary zone.
   c. Immobile residual liquids trapped below the water table in the saturated zone.

3. Dissolved phase hydrocarbons are found in
   a. Water infiltrating downward through the unsaturated zone and in residual films of water sorbed to sediments in the unsaturated zone.
   b. Pore water and residual films of water sorbed to sediments in the capillary zone and elsewhere in the HC plume.
   c. Groundwater in the saturated zone.
4. Vapor phase hydrocarbons are found
   a. Mostly in void spaces of the unsaturated zone not occupied by water or liquid hydrocarbons, where they are mobile.
   b. As small bubbles trapped in the HC plume and in the water-bearing zone below the plume, where they are immobile.
   c. Dissolved in the groundwater of the saturated zone, where they move with the groundwater.

6.3.4 PETROLEUM MOBILITY THROUGH SOILS

The concept of NAPL partitioning into other phases in the subsurface is nothing new to oil field workers. Liquid petroleum fields are found in rock formations of 10%–30% porosity. Up to half the pore space contains water. Primary recovery of oil, which relies on pumping out the portion of oil that is mobile and will accumulate in a well, collects only 15%–30% of the oil in the formation. Secondary recovery techniques force water under pressure into the oil-bearing rocks to drive out more oil. Primary and secondary techniques together typically extract somewhat less than 50% of the oil from a formation. Tertiary recovery techniques use pressurized carbon dioxide to lower oil viscosity along with detergents to solubilize the oil. Even with using tertiary techniques, producers expect 40% of the oil to remain immobile and unrecoverable.

Similar difficulties are encountered when trying to "cleanup" soils contaminated by a petroleum spill. There is always a fraction of the oil that is strongly sorbed on soil particles or trapped in soil pore spaces, which is not easily removed by water flushing or air sparging. In the subsurface environment, a significant portion of oil contamination must be regarded as "permanent," with a lifetime of well over 25 years, unless deliberate efforts are made to degrade it or physically remove it by excavation (Bredehoeft, 1992; MacDonald and Kavanaugh, 1994). Immobilized petroleum contaminants in the subsurface act as a long-term source of groundwater contamination, as the more soluble components continue to diffuse to the oil–water interface and dissolve into the groundwater.

6.3.5 BEHAVIOR OF LNAPL IN SOILS AND GROUNDWATER

Light nonaqueous phase liquid movement in the subsurface is a continual process of partitioning different components of a mixture among different phases that are present in the subsurface matrix. Spilled LNAPL at or near the soil surface penetrates and thoroughly saturates the soil because there is little trapped water or air to block its movement. Under the influence of gravity, the LNAPL sinks vertically downward, leaving behind in the soil a trail of residual LNAPL trapped by sorption and capillary

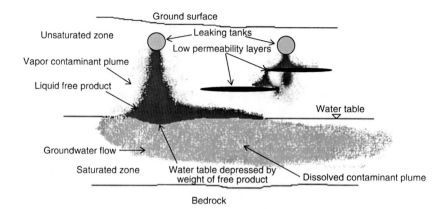

FIGURE 6.4 Light nonaqueous phase liquids (LNAPL) fuel leaking from underground storage tanks migrates downward under gravity. Enough fuel free product has leaked from the left tank to reach the saturated zone and spread out above the water table, moving in the direction of groundwater flow. The smaller spill from the right tank is insufficient to reach the water table and has become immobilized within the unsaturated zone by sorption and capillary forces. The more soluble components of the free product are present in the dissolved plume, which extends below the free product plume into the saturated zone and moves downgradient with groundwater flow. There also is a vapor plume in the unsaturated zone consisting of the most volatile components. The vapor plume migrates away from the liquid free product in the vadose zone in all directions independent of gravity. It may enter underground cavities such as sewers and basements, and may escape through the ground surface into the atmosphere.

forces. Capillary forces cause the LNAPL to spread horizontally as well as vertically downward, creating an inverted funnel-shaped zone of soil contamination (Figure 6.4).

As LNAPL moves downward through soil, a significant portion becomes immobilized by sorption to soil particle surfaces and capillary entrapment in soil pore space. This continually reduces the amount of mobile contaminant. When mobile LNAPL encounters an impermeable soil structure, bedrock, or the water table, its downward movement is halted and it spreads laterally under its internal pressure or down the slope of the retarding surface under gravity. If the LNAPL layer is not replenished by a continuing leak, it eventually is completely depleted by entrapment in the soil and becomes essentially immobilized. However, even when immobilized, the trapped free product continues to lose mass into the water and vapor phases and by biodegradation.

When all the spilled oil has entered the subsurface, the LNAPL "front" continues downward, leaving behind an ever-widening "inverted funnel" of contaminated soil containing residual immobile LNAPL sorbed to soil surfaces and trapped in pore spaces. The term "immobile" is used loosely and really means that, although some mobility may still occur, it will be very slow compared to the remediation time frame of interest.

A spill may or may not reach the water table. If the groundwater table is far enough below the surface or if the amount of spilled LNAPL is small enough, the mobile LNAPL can be completely depleted by entrapment in the soil before it reaches the groundwater. If the spill is large enough or the groundwater table is shallow, mobile LNAPL, commonly called free product, will contact groundwater. The weight of the free product depresses the water table locally below the free product column (see Figure 6.4).

Free product will continue to spread laterally as a layer over the water table, leaving a trail of residual LNAPL entrapped in the soil, until it has spread out to a saturation level so low that it all becomes immobile. Lateral spreading of the free product is influenced by a viscous "frictional" interaction at the water–LNAPL interface, which tends to move the free product preferentially in the direction of groundwater movement, along the hydraulic gradient. The relative downgradient velocities of water and free product depend on their relative viscosities and densities, as well as the soil conductivities for the different liquids.

When the water table rises and falls, the "floating" free product is moved vertically, buoyed above the changing water level. A rising water table can entrap some LNAPL by sorption and capillary forces within the newly saturated zone below the water table. A falling water table allows floating LNAPL to move downward into soils formerly saturated with water and exposes LNAPL formerly below the water table, allowing some of it to resume partitioning into air and, perhaps, to continue its slow migration downward under gravity. This up and down movement "smears" LNAPL into a region thicker than the free product thickness. Still more residual LNAPL becomes immobilized in this "smear zone." The result is that the smear zone of entrapped LNAPL extends above and below the average level of the water table.

6.3.6 SUMMARY: BEHAVIOR OF SPILLED LNAPL

The following list summarizes the behavior of spilled LNAPL:

1. Spilled LNAPL moves downward through the unsaturated zone under gravity.
2. A large fraction sorbs to the subsoil surfaces as trapped residual free product.
3. Some horizontal spreading occurs in the unsaturated zone because of attractive forces to mineral surfaces, and capillary attractions.
4. Free product tends to accumulate and spread horizontally above layers of low permeability (low hydraulic conductivity).
5. At the water-bearing region of the capillary zone, the free liquid phase floats on the water and begins to move laterally. The weight of floating LNAPL displaces water underneath it, depressing the water table in the region below the LNAPL.
6. If the spill is small enough, all the LNAPL may become trapped as residual liquid in the vadose zone, so that no LNAPL reaches the water table. However, any portion of the residual LNAPL that dissolves in downward percolating water from precipitation can be carried to the water table and contaminate it.
7. The vapor phase spreads widely in the unsaturated zone and can escape to the atmosphere and accumulate in cellars, sewers, and other underground air spaces.

6.3.7 WEATHERING OF SUBSURFACE CONTAMINANTS

With time, the composition of immobilized oil changes in the following ways:

- Less viscous components move downgradient through the soils.
- Components that are more viscous remain trapped as residual LNAPL in the vadose zone.
- Volatile components are lost into the atmosphere.
- Soluble components are lost into the groundwater.
- Biodegradable components are degraded by bacterial activity.

However, the total mass of immobilized oil decreases slowly because the loss processes are usually slow, unless they are artificially enhanced as part of a remediation program. The natural rate of depletion becomes progressively slower with time, as the remaining contaminants are increasingly enriched in those components that resist the loss mechanisms. The remaining oil becomes more firmly fixed in the subsurface soil, continually releasing its more soluble components in slowly decreasing concentrations to the groundwater.

RULES OF THUMB

1. Less than 1% of the total mass of a gasoline spill will dissolve into water in the vadose and saturated zones.
2. Since more than 99% of a fuel spill remains as adsorbed or free product LNAPL, it is impossible to clean up groundwater fuel contamination simply by "pump-and-treat" without eliminating the source residual and free product remaining in the soil.

6.3.8 PETROLEUM MOBILITY AND SOLUBILITY

The environmental impact of a contaminant release is determined mainly by mobility and water solubility of the different components of the contaminant. The most important soil and liquid parameters determining the mobility of LNAPL free product are

- Average soil pore size.
- Percent of soil pore space (soil porosity).
- Density and viscosity of the moving LNAPL. Density is mass per unit volume. Most petroleum products have a density less than the density of water, which is 1 g/mL. Viscosity measures resistance of fluid to flow. Gasoline is less viscous than water and can flow through pores and fissures more easily than water. The heavier petroleum fractions, such as diesel fuel and fuel oils, are more viscous than water and flow easily. Values of density and viscosity for several fuel products are listed in Table 6.2.

TABLE 6.2

Densities and Viscosities of Selected Fuels Compared to Water

Fluid	Density (g/mL)			Viscosity (centipoise)		
	0°C	15°C	25°C	0°C	15°C	25°C
Water	1.000	0.998	0.996	1.8	1.14	0.9
Automobile gasoline	0.76	0.73	0.68	0.8	0.62	—
Automotive diesel fuel	0.84	0.83	—	3.9	2.7	—
Kerosene	0.84	0.84	0.83	3.4	2.3	2.2
No. 5 jet fuel	0.84	—	—	—	—	—
No. 2 fuel oil	0.87	0.87	0.84	7.7	—	4.0
No. 4 fuel oil	0.91	0.90	0.90	—	47	23
No. 5 fuel oil	0.93	0.92	0.92	—	215	122
No. 6 fuel oil or Bunker C	0.99	0.97	0.96	7.4×10^7	—	3200

- Capillary attraction for the liquid contaminant to soil particles.
- Soil zone (vadose, capillary, or saturated), in which LNAPL is present, determines whether the pore space contains air, water, or contaminant.
- Magnitude of pressure and concentration gradients acting on the liquid free product.

LNAPL solubility in water is variable and depends on the chemical mixture. Literature data for solubility of pure compounds can be misleading because the solubility of a specific compound decreases when it is part of a blend (Table 6.3).

TABLE 6.3

Solubility Variability of Gasoline Components from Different Fuel Mixtures

	Concentration Dissolved in Water (mg/L)			
Compound	Regular Leaded	Regular Unleaded	Super Unleaded	Pure Compound
Benzene	30.5	28.1	67.0	1,740–1,860
Toluene	31.4	31.1	107.0	500–627
Ethylbenzene	4.0	2.4	7.4	131–208
1,2-Dichloroethane	1.3	—	—	8,524
Methyl-*tert*-butyl ether (MTBE)	43.7	35.1	966.0	48,000
t-Butyl alcohol	22.3	15.9	933.0	Miscible
m-Xylene	13.9	10.9	11.5	134–196
o-, *p*-Xylene	6.1	4.8	5.7	157–213
1,2-Dibromoethane	0.58			4,300

RULE OF THUMB

The aqueous solubility of a particular compound in a multicomponent NAPL can be approximated by multiplying the mole fraction of the compound in the NAPL mixture by the aqueous solubility of the pure compound.
Solubility of component i in an NAPL mixture is

$$S_i = X_i S_i^0 \tag{6.1}$$

where
S_i is the solubility of component i in the mixture
X_i is the mole fraction of component i in the mixture
S_i^0 is the solubility of pure component i

6.4 FORMATION OF PETROLEUM CONTAMINATION PLUMES

In the subsurface soil environment, petroleum compounds can be present in four phases, each of which can create its own contaminant plume. The four phases are

1. Liquid petroleum free product (LNAPL)
2. Petroleum compounds adsorbed to soil particles
3. Dissolved petroleum components
4. Vaporized petroleum components

Each phase behaves differently and poses different remediation problems. The liquid free product originates directly from the contamination source and initially has the same composition. The adsorbed, dissolved, and vapor phases are extracted from the liquid free product as it contacts soil, water, and air in the soil pore spaces. Each phase moves independently in its own distinct contaminant plume.

Generally, the vapor contaminant plume moves most rapidly. The dissolved plume moves more slowly, at groundwater velocity or less, depending on its retardation factor. Depending on whether its viscosity is greater or less than water, the free product plume may move slower or faster than the dissolved plume. The adsorbed plume may be immobilized or, in the saturated zone, part of it may be sorbed to mobile colloids and move at approximately the groundwater velocity.

RULES OF THUMB

1. With respect to the total mass of fuel contaminant in the subsurface soil environment, the floating free product and immobilized oil (trapped by capillary forces and adsorbed to soil particles) are generally more than 99% of the total mass.
2. When free product is present, the dissolved phase in the groundwater is generally less than 1% of the total mass. The dissolved plume is just the tip of the contamination iceberg.

EXAMPLE 1

COMPARING DISSOLVED AND LNAPL FREE PRODUCT MASSES

A leaking underground storage tank (LUST) released 1000 gal of gasoline (density about 0.7 g/mL) to the subsurface. After 1 year, the resulting dissolved-phase plume was about 1000 ft long, 100 ft wide, and averaged 10 ft deep. The average concentration of hydrocarbons in the plume was 2.20 mg/L (estimated by measuring and adding the total volatile hydrocarbon [TVH] and total extractable hydrocarbon [TEH] concentrations). The porosity of the aquifer was 0.30. If no hydrocarbon was lost due to volatilization or biodegradation, how much of the original release was in the dissolved phase and how much was in the LNAPL phase?

Answer:

Total mass released $= (1000 \text{ gal}) (3.78 \text{ L/gal}) (1000 \text{ mL/L}) (0.7 \text{ g/mL}) = 2.646 \times 10^6 \text{ g}$
Volume of contaminated groundwater $= (1000 \text{ ft}) (100 \text{ ft}) (10 \text{ ft}) (0.30) (28.3 \text{ L/ft}^3) = 8.490 \times 10^6 \text{ L}$
Mass of dissolved hydrocarbons $= (8.490 \times 10^6 \text{ L}) (2.20 \text{ mg/L}) (1 \text{ g/1,000 mg}) = 18,680 \text{ g}$
Mass of LNAPL free product $= 2.646 \times 10^6 \text{ g} - 18,680 \text{ g} = 2.627 \times 10^6 \text{ g}$
Percent of total mass that is dissolved $= (18,680 \text{ g}/2.646 \times 10^6 \text{ g}) \times 100 = 0.7\%$
Percent of total mass that is LNAPL $= (2.627 \times 10^6/2.646 \times 10^6) \times 100 = 99.3\%$

6.4.1 DISSOLVED CONTAMINANT PLUME

Water solubility is the most important chemical property for assessing the impact of a contaminant on the environment. Dissolved contaminants arise when the free product comes in contact with water. The water may be moisture retained in the soil, precipitation percolating downward through the soil, groundwater flowing through contaminated soil, or groundwater lying under a layer of free product. Both crude and refined petroleum products contain hundreds of different components with different water solubilities, ranging from slightly soluble to insoluble.

RULES OF THUMB

1. In general, lightweight aromatics such as the BTEX group (benzene, toluene, ethylbenzene, and xylenes) are the most soluble components of fuel mixtures. If MTBE or ethanol additives are present, they are the most soluble components by far.
2. The overall water solubility of commercial gasoline without additives is between 50 and 150 mg/L, depending on its exact composition. When free product gasoline is present, the dissolved portion generally accounts for less than 1% of the total contaminant mass present in the subsurface.
3. The overall solubility of fresh No. 2 diesel fuel in water is around 0.4–8.0 mg/L, again depending on its composition. When free product diesel fuel is present, the dissolved portion generally accounts for less than 0.1% of the total contaminant mass present in the subsurface.
4. Nevertheless, because of their typically high toxicity, dissolved contaminants can greatly exceed concentrations where water is regarded as seriously polluted.

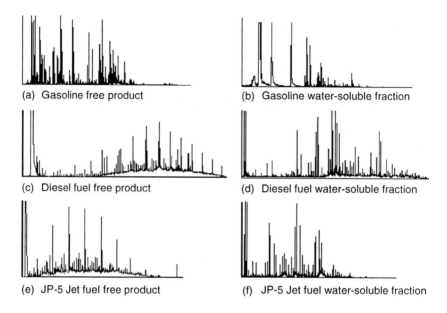

FIGURE 6.5 Gas chromatograph/flame ionization detector (GC/FID) chromatograms of gasoline, diesel, and JP-5 fuels and their respective water-soluble fractions. Time of elution, which corresponds roughly to the number of carbons in the eluted compound, increases from left to right. Thus, peaks corresponding to heavier compounds appear farther to the right in each figure. The composition of free product and dissolved fractions are very different in each type of product because the more soluble compounds become distributed preferentially into the water-soluble fraction. The water-soluble fractions are composed mainly of 1-, 2-, and 3-ring aromatic hydrocarbons.

The compositions of the free product and dissolved fractions are very different, as indicated in Figure 6.5, because the more soluble compounds partition out the free product and become concentrated in the water-soluble fraction.

Dissolved contaminants become a part of the water system and move with the groundwater, but they usually move at a lower velocity because of retardation by sorption processes. Sorption to soil and desorption back into the dissolved phase is a continual process that retards the movement of dissolved contaminants. The amount of retardation for any particular contaminant depends mainly on the organic content of the soil; retardation is greater in soils with more organic matter. Because their water solubilities are low, dissolved fuel contaminants continue to partition between the dissolved phase and soil particle surfaces, especially in soils with a high organic content.

RULE OF THUMB

Typical retardation factors for BTEX in sandy soil range from 2.4 for dissolved benzene (groundwater moves 2.4 times faster than benzene) to 6.2 for the dissolved xylene isomers.

6.4.2 VAPOR CONTAMINANT PLUME

Vapor phase contaminants arise from the volatile components of the free product escaping into adjacent air. Low mass hydrocarbon components commonly associated with the gasoline fraction are the most volatile. Vapor movement is not influenced by groundwater motion but is weakly influenced by gravity. It follows the most conductive pathways through the subsurface, from regions of higher to lower pressure. Vapors denser than air can collect in low spots like basements and sewers. Much of the vapor remains trapped in soil near its origin, slowly escaping to the surface atmosphere. A small portion of vapor phase contaminants may dissolve into soil–water, but it is generally insignificant.

RULES OF THUMB

1. A measurable vapor concentration will be produced if either:
   Henry's constant $(K_H = C_a/C_w) > 0.0005$ atm m^3 mol^{-1}
   (This produces significant partitioning from water to air.)
   or
   vapor pressure > 1.0 torr at 20°C
   (This results in significant diffusion upward through the vadose zone.)
2. Characteristic vapor pressures for gasoline:
   Fresh gasoline: 260 torr (0.34 atm)
   Weathered gasoline (2–5 years old): 15–40 torr (0.02–0.05 atm)

6.5 ESTIMATING THE AMOUNT OF LNAPL FREE PRODUCT IN THE SUBSURFACE

The initial steps in the remediation of a site where an LNAPL spill has occurred are

1. Try to limit the movement of contaminant plumes.
2. Remove from the subsurface as much free product as possible.

As long as free product is present, it continues to partition into the sorbed, dissolved, and vapor contaminant plumes, continually feeding their growth. Only after the mobile LNAPL has been removed from above the water table remediation of the contaminant plumes can be effective.

LNAPL in the subsurface is generally detected and measured by its accumulation in wells. To design a program for removing free product, one must obtain a reliable estimate of the volume of free product that is present. However, the relationship between the thickness of free product that accumulates in a well and the thickness of free product distributed above the water table is easily misinterpreted. This LNAPL thickness in a well is influenced not only by the thickness of LNAPL in the

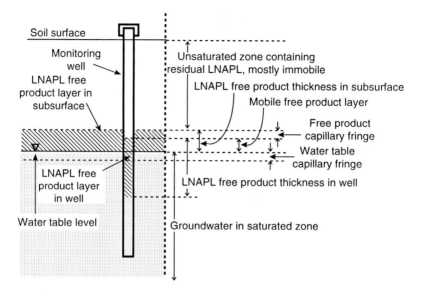

FIGURE 6.6 Thickness of LNAPL accumulated in a well compared to thickness in adjacent subsurface.

subsurface adjacent to the well but also by soil texture and fluctuations of the water table level.

Figure 6.6 illustrates some of the factors that affect free product accumulation in a well. In the soil subsurface away from a well, liquids are influenced by capillary attractions that draw them into small pore spaces and interstices. Where no LNAPL free product is present, three forces determine the aquifer water table elevation:

1. Gravity pulls water downward.
2. Water pressure in the aquifer acts upward against gravity.
3. Capillary forces at the interface between the saturated and unsaturated zones also act upward against gravity.
4. The water table rises to the level where the downward force of gravity is balanced by the two upward forces of water pressure and capillary attractions.

Within a well, there are no upward acting capillary forces affecting the liquid levels. Only the balance between gravity and water pressure in the aquifer determines the water level in a well. The result is that the water level in a well is lower than the top of the water table in the subsurface around the well, at the bottom of the capillary zone in the adjacent soil.

If LNAPL is present floating on the water table, it also develops a capillary zone at its interface with the unsaturated zone. Capillary zones occur at the upper boundaries of both the water table and the free product layer. In the capillary zones, liquids are drawn upward against gravity and are largely immobile, especially in horizontal directions. The thickness of the capillary zones depends on the soil texture. In coarse soils and sands with few capillary-size pore spaces and interstices, capillary zone layers may be only a few millimeters thick; in fine soils and sands, they may be several meters thick.

Where LNAPL free product lies on the water table, the water level is lowered by the weight of LNAPL (see Figure 6.7). Where the LNAPL free product layer is thin, it lies largely above the water capillary zone because the weight of LNAPL cannot easily displace water from this region. Where the LNAPL layer is thick, its excess weight makes it penetrate farther into, or even through, the water capillary zone, moving the free product–water interface still lower.

When an appropriately screened well passes through an LNAPL free product layer into the saturated zone, water and free product flow into the well from the surrounding subsurface soils. Liquid movement into the well occurs from the water and free product regions below their respective capillary zones, where liquid mobility exists. LNAPL from the mobile zone around the well flows into the well and, without any upward capillary forces within the well, the additional weight of LNAPL lowers the water level in the well to below the normal water table in the aquifer. LNAPL flows into the well until the top level of LNAPL in the well is the same as the top of the mobile zone in the surrounding soil.

Within a well, where no capillary forces exist, the weight of LNAPL lowers the LNAPL–water interface farther than in the surrounding subsurface. LNAPL will continue to flow into the well, lowering the water table, until the upward pressure of aquifer water balances the weight of LNAPL. The result is that LNAPL accumulates

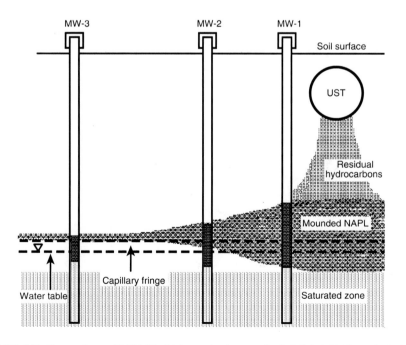

FIGURE 6.7 Comparison of LNAPL thickness in three wells (MW-1, MW-2, and MW-3).

in a well to a greater thickness than in the surrounding subsurface, where capillary forces buoy up both the water level and LNAPL free product layer. The upper level of LNAPL in the well is lower than the upper level in the surrounding subsurface by the thickness of the LNAPL capillary zone. The LNAPL–water interface in the well is lower than in the adjacent subsurface by an amount that depends on the soil texture and the thickness of the subsurface layer of mobile LNAPL. This behavior is illustrated in Figures 6.7 and 6.8.

Equation 6.2 may be used to calculate the water table level in the subsurface adjacent to a well from measurements in a well where LNAPL is present. Use of Equation 6.2 is useful for evaluating and plotting groundwater elevations when LNAPL is present in the wells.

$$\text{WTE} = \text{WE}_{\text{well}} + (\text{LNAPL density} \times \text{LNAPL thickness in well}) \qquad (6.2)$$

where
  WTE is the water table elevation in subsurface adjacent to the well
  WE_{well} is the water elevation at the water–LNAPL interface in the well

An estimate of LNAPL thickness in the adjacent subsurface, ignoring soil properties and capillarity, can be made from (Figure 6.6):

$$t_{\text{subsurface}} \approx \frac{t_{\text{well}}(\text{water density} - \text{LNAPL density})}{\text{LNAPL density}} \qquad (6.3)$$

where
  $t_{\text{subsurface}}$ is the thickness of LNAPL in the subsurface adjacent to the well (cm)
  t_{well} is the thickness of LNAPL in the well (cm)
  Water density is 1.0 g/cm^3
  LNAPL density is 0.7–0.8 g/cm^3 for gasoline and diesel fuels

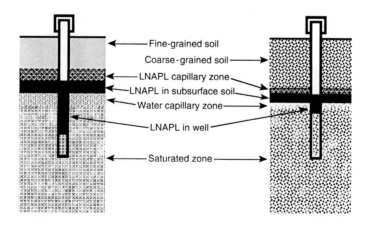

FIGURE 6.8 Effect of soil texture on LNAPL thickness in a well.

Equation 6.3 is useful for estimating the total volume of LNAPL in a plume whose area has been measured. Details for calculating more accurately the recoverable volume of LNAPL free product in the subsurface from well measurements have been published (Farr, et al., 1990; Lenhard and Parker, 1990; Parker, et al., 1996). Computer programs are also available for this and related calculations. However, Equation 6.3 is often sufficient for the initial evaluation of a remediation program.

6.5.1 How LNAPL Layer Thickness in the Subsurface Affects LNAPL Layer Thickness in a Well

The weight of LNAPL in the subsurface near wells MW-2 and MW-3 is not sufficient to force water downward through the capillary zone (see Figure 6.7). Near well MW-1, where the LNAPL is thicker because it has formed a dome, the weight of LNAPL is great enough to force water downward through the capillary zone and below the original water table. Within a well, there are no capillary forces acting upward on the water. In wells MW-2 and MW-3, the LNAPL thickness in the wells is greater than the LNAPL thickness in the surrounding soils because the absence of capillary forces in the wells reduces the net upward forces acting on the water so the water level is lower. In well MW-1, where LNAPL has pressed down through the capillary fringe, upward forces on the water are due only to aquifer pressure and are the same in the well and in the surrounding soil. Thus, the LNAPL thickness and the water level in well MW-1 are the same as in the surrounding subsurface.

6.5.1.1 Effect of Soil Texture on LNAPL in the Subsurface and in Wells

Soil texture determines the magnitude of the upward capillary forces that act on subsurface water and LNAPL. Capillary forces are much larger in fine-grained soil than in coarse-grained soil. Consequently, the difference between LNAPL thickness in a well and LNAPL thickness in the adjacent subsurface is greater in fine-grained soil. The effect of soil texture is illustrated in Figure 6.8.

RULES OF THUMB

1. Measured LNAPL thickness in a well often exceeds the corresponding LNAPL thickness in the surrounding subsurface by a factor of 2–10, because LNAPL above the water table flows into the well and depresses the well water level.
2. The LNAPL thickness ratio, $h_{well}/h_{subsurface}$, generally increases with decreasing soil particle size, increasing capillary zone thickness, and increasing LNAPL density.
3. A crude estimate, ignoring soil properties, of LNAPL thickness in the adjacent subsurface can be made with Equation 6.4, obtained by rearranging Equation 6.3.

(Continued)

RULES OF THUMB (Continued)

$$h_{\text{well}}/h_{\text{subsurface}} \approx \frac{(\text{LNAPL density})}{(\text{water density} - \text{LNAPL density})} \qquad (6.4)$$

4. Since fuel LNAPL (gasoline and diesel) generally has a density of 0.7–0.8 g/cm^3 and the density of water is 1.0 g/cm^3, Equation 6.4 gives, for fuel LNAPLs:

$$h_{\text{well}}/h_{\text{subsurface}} \approx 0.7/0.3{-}0.8/0.2 \text{ or } 2.3{-}4.0.$$

6.5.1.2 Effect of Water Table Fluctuations on LNAPL in the Subsurface and in Wells

Water table fluctuations promote vertical spreading of LNAPL in the subsurface and can influence the thickness of LNAPL that collects in monitoring wells.

When the groundwater table rises

- Some floating LNAPL free product is driven up into the unsaturated zone. Sorbed residual LNAPL in the formerly unsaturated zone can be remobilized by dissolving into the free product, causing further lateral spreading.
- Some free product remains trapped in pore spaces below the water level within the saturated zone.
- Mobile free product layer above the water table becomes thinner.
- When the groundwater table falls.
- LNAPL free product floating on the water table moves downward as the water table drops, leaving behind an immobilized fraction of free product as sorbed residual LNAPL retained in the newly unsaturated zone above the lowered water table.
- Mobile free product layer floating on the water table may become thicker because free product formerly trapped below the water table is free to migrate downward to the new water table.

The rising and falling of the water table leaves behind a "smear zone" of contamination that lies partially in the saturated zone and partially in the unsaturated zone. This behavior is illustrated in Figure 6.9.

6.5.1.3 Effect of Water Table Fluctuations on LNAPL Measurements in Wells

LNAPL spills are often first detected by the appearance of a free product layer above the water in downgradient wells. If the well free product layer diminishes during a remediation program, it is tempting to believe that the cleanup effort is working successfully. An increase in the well free product thickness may initiate a search for new LNAPL sources. However, unless fluctuations in groundwater depth are taken

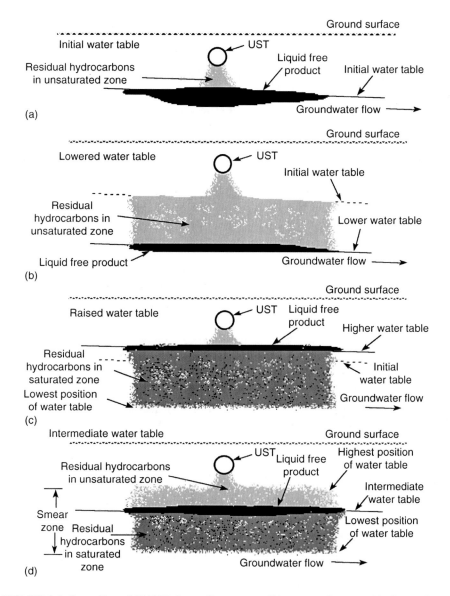

FIGURE 6.9 Spreading of LNAPL into a "smear zone" because of water table fluctuations.

into account, basing such conclusions on changes in the free product layer thickness in wells can lead to serious errors.

When groundwater rises, the thickness of the free product layer in wells generally decreases because a portion of the mobile free product becomes trapped below the water table and becomes immobile, thinning the mobile free product layer. When the water table falls, free product formerly trapped in the saturated zone

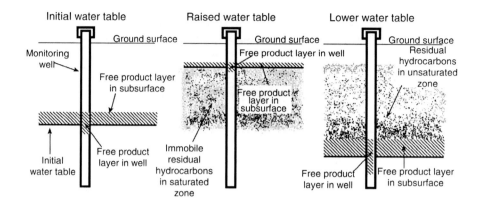

FIGURE 6.10 Effect of fluctuating water table on LNAPL accumulation in a well.

becomes mobile again and can accumulate in the free product layer over the water table, where it is free to flow into wells. This behavior is illustrated in Figure 6.10.

6.6 ESTIMATING THE AMOUNT OF RESIDUAL LNAPL IMMOBILIZED IN THE SUBSURFACE

Residual LNAPL in the subsurface is the portion that will not flow into a well. It is the part of an LNAPL spill that cannot be removed by pumping to the surface. Residual LNAPL must be remediated by processes such as biodegradation, chemical degradation, soil washing, volatilization with heat, vapor extraction, or excavation. Residual LNAPL is retained in the unsaturated zone by adsorption and capillary forces. Therefore, small soil particles and large surface area both increase the amount of residual LNAPL retained. The soil retention factor (volume of LNAPL per volume of soil) depends mainly on the soil pore size distribution, soil wettability, LNAPL viscosity, and LNAPL density.

Usually more LNAPL is immobilized in the saturated zone than in the unsaturated zone because part of the residual LNAPL in larger pores of the unsaturated zone eventually drains down to the water table. In the unsaturated zone, LNAPL is the wetting fluid and tends to spread into the smaller pores, where it can be retained by capillary forces. However, it can drain downward from the larger pores. In the saturated zone, water is the wetting fluid and LNAPL the nonwetting fluid. Here, if LNAPL were mobile, its buoyancy would drive it upward to the water table, but, instead, it is trapped in larger soil pore spaces by immobile water.

Figure 6.11 shows soil retention factors for several kinds of LNAPL in soils of different textures. The retention of LNAPL in soils above the water table usually ranges between about 80 L/m^3 of soil for fuel oil in silt, to about 2.5 L/m^3 for gasoline in coarse gravel. LNAPL in the unsaturated zone can often be remediated without excavation by some combination of soil washing, volatilization, or bioremediation.

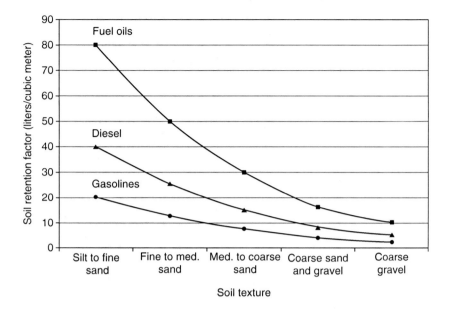

FIGURE 6.11 Soil retention factors for LNAPL fuels in different soils above the water table. Calculations assume a soil bulk density of 1.85 g/cm^3 and LNAPL densities of 0.7, 0.8, and 0.9 g/cm^3 for gasolines, diesel fuel, and fuel oils, respectively. (From Mercer, J. and Cohen, R.A., 1990, *J. Contamin. Hydrol.*, 6, 107–163.)

EXAMPLE 2

USING SOIL RETENTION FACTORS

One thousand gallons of fuel oil were spilled on a soil consisting mostly of medium to coarse sand. What volume of soil is required to immobilize the 1000 gal? If the spill area was confined by a berm to 100 ft^2, how deep into the soil will the oil penetrate? Could it endanger a shallow aquifer 35 ft below the surface?

Answer:

From Figure 6.11, the soil retention factor is about 30 L/m^3 for fuel oil in medium to coarse sand. The volume of soil needed to contain the entire spill is

$$V_{soil} = \left(\frac{1000\,\text{gal}}{30\,\text{L/m}^3}\right)\left(\frac{3.785\,\text{L}}{1\,\text{gal}}\right)\left(\frac{35.3\,\text{ft}^3}{1\,\text{m}^3}\right) = 4454\,\text{ft}^3$$

Assume the oil plume travels downward without spreading, so that its cross section is 100 ft^2. Then a volume of 4454 ft^3 will extend downward by

Depth of oil penetration until it is all retained and immobilized $= \dfrac{4454\,\text{ft}^3}{100\,\text{ft}^2} = 44.5\,\text{ft}$

Oil is likely to reach the aquifer at 35 ft.

6.6.1 SUBSURFACE PARTITIONING LOCI OF LNAPL FUELS

As part of an effort to provide an authoritative, defensible engineering basis for predicting contaminant behavior in soils, EPA has identified 13 soil loci among which petroleum contaminants become partitioned (USEPA, 1991). Contaminants may move within a given locus under the influence of pressure and concentration gradients, or from one locus to another because of molecular attractions (for example, when soluble components in the free product layer dissolve into the groundwater).

In Table 6.4, the partitioning behavior of gasoline in sandy soils has been calculated, using the methods and data recommended by the USEPA (1991). The table shows how much gasoline LNAPL is expected to be retained in the 13 different partitioning loci.

TABLE 6.4
Relative Importance of Different Subsurface Loci in Sandy Soils for Retention of Gasoline Contamination

	Loci of Subsurface LNAPL Retention	Average Gasoline Retention in Sandy Soils (mg/cm^3)	Percent of Total Retention in Sandy Soils
1	Gasoline vapors in soil pores in the unsaturated zone.	0.095	<0.1%
2	Liquid gasoline sorbed to dry soil particles in the unsaturated zone. Locus 2 is especially important in the soil volume immediately below a spill, but not downgradient of the spill.	36	9.88
3	Gasoline dissolved in water on wet soil particles in the unsaturated zone.	0.0010	<0.1%
4	Liquid gasoline sorbed to wet soil particles in the saturated and unsaturated zones.	0.076	<0.1%
5	Liquid gasoline in soil pore spaces within the saturated zone. Locus 5 contaminants may generally be regarded as immobile.	38	10.4
6	Liquid gasoline in soil pore spaces in the unsaturated zone. Contaminants enter locus 6 mainly from free product floating on the groundwater table when the table rises and then falls.	110	30.2
7	LNAPL gasoline free product floating on top of the groundwater table. The most important loss mechanism from locus 7 occurs when a fluctuating water table moves contaminant into loci 5 and 6, where some of it remains trapped.	180	49.4
8	Gasoline dissolved in groundwater.	0.020	<0.1%
9	Gasoline sorbed to colloidal particles in water in the saturated and unsaturated zones.	0.00013	<0.1%
10	Liquid gasoline diffused into mineral grains in the saturated and unsaturated zones.	0.000060	<0.1%
11	Gasoline sorbed onto or into microbiota in the saturated and unsaturated zones.	0.010	<0.1%
12	Gasoline dissolved into the mobile pore water of the unsaturated zone.	0.030	<0.1%
13	Liquid gasoline in rock fractures in the saturated and unsaturated zones.	0.21	0.17

Source: Calculated from data in USEPA (1991).

It is evident from Table 6.4 that when free product is present in locus 7 (gasoline LNAPL floating on top of the groundwater table), this condition is the controlling factor for the distribution of contaminants in other zones. With free product present above the water table, loci 2, 5, 6, and 7 are by far the most important in terms of mass and account for 99.9% of the total soil and groundwater contamination.

The mass of gasoline in the vapor and dissolved states (loci 1, 3, 8, and 12) is insignificant compared to that in the free product above the water table and in the smear zone created by a rising and falling water table (loci 2, 5, 6, and 7). Of course for remediation purposes, all loci are important and those with relatively small amounts of contaminant may be the most difficult to remediate to a regulated level. Because it is the most mobile and can flow into wells, the first goal of remediation should be to remove the LNAPL floating on the water table.

EXAMPLE 3

CALCULATION OF THE CONTAMINANT PLUME VOLUME REQUIRED TO IMMOBILIZE ONE MILLION GALLONS OF GASOLINE

Consider a site where LNAPL from a point source of contamination has leaked downward into sandy soil to the water table, where it has spread out above the saturated zone, moving downgradient in the direction of groundwater flow, as in Figure 6.4 for the left tank. For this case, we can assume that only loci 5 and 6, which comprise the smear zone caused by a fluctuating water table, are effective for immobilizing LNAPL contaminants. Locus 2 lies primarily under the area of the initial leak and probably is small compared to the total free product plume volume. LNAPL in locus 7 may be regarded as mobile. As LNAPL in locus 7 flows downgradient and is subjected to vertical movement caused by a fluctuating water table, it continually loses mass into loci 5 and 6 where it becomes immobile.

Assume an average sandy soil with a hydraulic gradient of 0.009 ft/ft. Using the methods laid out by the USEPA (1991), the estimated flow velocity for LNAPL gasoline in locus 7 is about 1.3 ft/day.* This may be compared with a groundwater velocity of about 10 ft/day for the same conditions. Further assume that the seasonal groundwater table fluctuations average around ±1.5 ft, giving a free product smear zone 3 ft in thickness. Taking the smear zone to be of uniform thickness and assuming that, on average, half of the smear zone is in the saturated zone and half in the unsaturated zone, we can say loci 5 and 6 are each 1.5 ft thick everywhere adjacent to the free product plume.

Answer:

Locus 5 retention

Take the density of aged gasoline LNAPL to be 0.74 g/cm^3 or 2800 g/gal. From Table 6.4, locus 5 retains 38 mg of gasoline per cubic meter of soil, or about

$$\frac{38\,mg}{cm^3} \times \frac{1\,g}{10^3\,mg} \times \frac{2.83 \times 10^4\,cm^3}{ft^3} = 1.08 \times 10^3\,g \text{ of gasoline per cubic foot of soil}$$

---

* Because of its greater viscosity, diesel fuel would move about one-third as fast.

The gallons of gasoline stored per cubic foot in locus 5 is

$$\frac{1.08 \text{ g}}{\text{ft}^3} \times \frac{1 \text{ gal}}{2.78 \times 10^3 \text{ g}} = 0.387 \text{ gal/ft}^3$$

The quantity of gasoline stored in locus 5 (1.5 ft thick) below 1 ft^2 of surface area is

$$\frac{0.387 \text{ gal}}{\text{ft}^3} \times 1.5 \text{ ft}^2 = 0.58 \text{ gal/ft}^2$$

Locus 6 retention

From Table 6.4, locus 6 retains 110 mg of gasoline per cm^3 of soil, or about 3.11×10^3 g of gasoline per ft^3 of soil. Calculations parallel to those above give a result of 1.11 gal of gasoline per ft^3 of soil, and 1.67 gal/ft^2 of surface area in locus 6.

The total quantity of gasoline retained in loci 5 and 6 below 1 ft^2 of surface area is $(0.58 + 1.67)$ gal/ft$^2 = 2.25$ gal/ft^2, or about 98,000 gal per acre of surface area.

Thus, the soil contaminant plume from a one million gallon gasoline spill in sandy soil, where the water table fluctuates ± 1.5 ft annually (creating a 3 ft smear zone), could become immobilized after it had spread into approximately 30 acre-ft, beneath a surface area of about 10 acres.

6.7 CHEMICAL FINGERPRINTING OF LNAPLS

Environmental professionals often need to do more than locate and cleanup pollutants. They may be asked to help identify the sources of the contamination. Someone always has to pay for a remediation effort. The high costs that are often involved mean that any prudent potentially responsible party (PRP) will want convincing proof that he must accept responsibility for the pollution. Polluted sites often have a history of different owners and uses, any or all of which may have contributed to the present site contamination. Off-site sources may also be responsible for all or part of the current pollution problems. Allocating responsibility for the expenses involved in remediation often becomes a legal issue. The methods used in building a legally defensible case that assigns responsibility for soil and groundwater pollutants have become part of a branch of environmental investigations known as environmental forensics.

One of the most common applications of environmental forensics is identifying the sources of fuel contamination. The use of petroleum fuels has been increasing for the past 100 years and fuel contaminants are pervasive wherever they have been used. There are many potential sources of fuel pollutants, from numerous small gasoline stations to large refinery operations, roadway accidents to criminal acts of disposal, pipeline breaks to tank corrosion. It is no wonder that the origin of any particular site contamination may often be legitimately questioned. Chemical fingerprinting of fuels is one tool in the toolbox of chemical forensics.

In chemical fingerprinting, the chemical characteristics of fuel contaminants are used to help distinguish among different possible sources of the pollutants. For example, knowing that only gasoline was ever used at a certain site, but that the groundwater hydrocarbon contaminants found there are from diesel fuel, clearly points to an off-site source.

6.7.1 First Steps in Chemical Fingerprinting of Fuel Hydrocarbons

Successful chemical fingerprinting of hydrocarbon contamination frequently consists of the sequential application of several investigative steps:

- Identify the fuel type of the contaminants. Are they gasoline, diesel fuel, or fuel oils? Each of these classes has unique chemical characteristics.
- If possible, distinguish between the ages of different contaminated samples. Weathering and aging processes often introduce predictable changes in the chemical makeup of fuels. If samples have weathered differently because of different exposure times, they will have different chemical profiles. Certain compounds in fuel mixtures are more readily biodegraded, solubilized, or volatilized than others. This causes changes in the overall fuel composition as time passes. The exact nature of the changes is always site-specific, but even if approximate ages cannot be assigned, sometimes the ages of different samples can be clearly demonstrated to be significantly different. In some cases, the presence of discontinued additives or certain degradation products may be a useful indicator of age.
- Also, fuel compositions have changed historically, as more efficient refinery production practices were developed and as clean air legislation and automotive engine development mandated changes. For example, in the United States gasoline containing organic lead-based octane boosters in gasolines was first marketed in 1923 and the elimination of lead compounds began in the early 1980s and was complete by 1995. The presence of lead in fuel-contaminated soil or water may indicate an old rather than recent spill, especially if accompanied by dichloroethane and/or dibromoethane, which were lead scavenger gasoline additives introduced around 1927. Table 6.5 chronicles the history of lead and lead scavenger additives in the United States.
- Look for unique chemical compounds that can serve as "markers," which might be present in contamination from one source but not from another. An example might be the presence of MTBE or ethylene dichloride (EDC) in some fuel-contaminated samples but not in others. Different production practices at different refineries sometimes produce subtle but distinctive differences in their fuel products, which can help to distinguish between possible sources. This approach generally requires extensive knowledge about the chemical composition of contamination at the site and, sometimes, knowledge about the chemicals previously and currently used at the site.

Chemical fingerprinting is sometimes fairly simple and sometimes very complicated, even impossible. A few approaches are described here to indicate the possibilities. Complicated cases will require the help of experienced forensic chemists.

TABLE 6.5
Timeline of Lead and Lead Scavenger Additives to Gasoline

1923—Leaded gasoline first marketed using tetraethyl lead (TEL) as an antiknock agent.

1926—U.S. Surgeon General recommends maximum lead content of 3.17 g of lead per U.S. gallon of gasoline (g Pb/gal).

1927/8—Lead scavengers 1,2-dibromoethane (EDB, ethylene dibromide,[a] 1,2-EDB, $C_2H_4Br_2$) and 1,2-dichloroethane (EDC, ethylene dichloride, 1,2-DCA, $C_2H_4Cl_2$) introduced for use with lead alkyl gasoline additives. Only EDB, not EDC, was used in piston engine aviation gasoline to avoid corrosion of aluminum parts. The amount of scavenger added was designed to react completely with all the lead to form volatile lead halides that would be removed with the exhaust gases. Ratios of EDC–EDB varied from 1:1 to about 2:1. Characteristic composition of a lead additive package with a 1:1 scavenger ratio during the 1980s was 62 wt% TEL, 18 wt% EDB, 18 wt% EDC, and 2 wt% of other ingredients such as dye, antioxidants, and stability improvers.

1950—Most gasolines in the United States were leaded by this time.

1959—Maximum permitted lead level peaked at 4.23 g Pb/gal. This value is well above the point of diminishing returns on the lead response curve and it is unlikely that much gasoline was ever produced with this much lead. More characteristic lead concentrations were between 2.25 (for regular) and 3 g Pb/gal (for premium).

1960—Alternative antiknock agent tetramethyl lead (TML) introduced by Chevron (then Standard Oil). After 1960, various mixtures of TEL, TML, and other organic lead additives, have been used.

1975—EPA proposed a scheduled reduction of lead to 1.7 g Pb/gal in 1975, 1.4 g Pb/gal in 1976, 1.0 g Pb/gal in 1977, 0.8 g Pb/gal in 1978, and 0.5 g Pb/gal in 1979.

1980—EPA set an overall maximum lead limit in all gasolines to 0.5 g Pb/gal, for large refiners.

1982—EPA set a maximum lead concentration for leaded gasoline, averaged over a 3 month manufacturing period (called a "pool standard"), at 1.10 g Pb/gal for large refiners. The actual level of lead in any two batches of gasoline could vary.

1985—Maximum permitted pool standard for lead in leaded gasoline, lowered to 0.5 g Pb/gal for all refiners. Lead credits were allowed. Many states began phasing out leaded gasoline during the middle to late 1980s.

1986—EPA scaled a pool standard decrease in lead to 0.1 g Pb/gal, to occur from 1986 to 1988.

1987—EPA eliminated lead credits.

1992—Manufacture of leaded gasoline eliminated in California.

1995—EPA eliminated lead in all U.S. gasoline produced after 1995, per Section 211(n) of the Clean Water Act.

---

[a] Despite their names, EDB (ethylene dibromide) and EDC (ethylene dichloride) do not contain double bonds. Their proper chemical names are 1,2-dibromoethane ($BrCH_2CH_2Br$) and 1,2-dichloroethane ($ClCH_2CH_2Cl$). Here, the use of the term "ethylene" is common in the petroleum trade and is based on the use of ethylene as a feed stock in their manufacture.

EXAMPLE 4

Estimating the Amounts of Lead, Ethylene Dichloride, and Ethylene Dibromide Contained in a Leaded Gasoline Release

Use Table 6.5 to estimate amounts of lead, EDC, and ethylene dibromide (EDB) contained in 1000 gal of leaded gasoline release believed to have occurred before 1980.

Answer:

A reasonable estimate for the concentration of lead in leaded gasoline before 1980s may be taken as around 2.0 g/gal. Assuming a typical additive package that was 62 wt% tetraethyl lead (C$_8$H$_{20}$Pb), 18 wt% EDC (C$_2$H$_4$Cl$_2$), and 18 wt% EDB (C$_2$H$_4$Br$_2$), the corresponding concentrations of EDC and EDB can be calculated as follows.

Assume the following:

Amount of lead in gasoline = 2.0 g/gal
Additive package contained 62% tetraethyl lead (TEL), 18% 1,2-dichloroethane, 18% 1,2-dibromoethane, and 2% other (EDC:EDB = 1:1)
Calculation: all percentages are weight percentages
Formula for TEL is C$_8$H$_{20}$Pb; MW = 323.45 g/mol
C: 29.70%; H: 6.23%; Pb: 64.06%
2.0 g Pb/gal = 0.64 × (grams of TEL)
TEL = 2.0/0.64 = 3.1 g TEL/gal
TEL is 62% of additive package.

Therefore, total grams of additive per gal = 3.1 g TEL/0.62 = 5.0 g additive/gal.

EDC = 18% of additive.

Therefore, EDC = 0.18 (5.0 g additive/gal) = 0.9 g EDC/gal.

EDB = 18% of additive.

Therefore, EDB = 0.18 (5.0 g additive/gal) = 0.9 g EDB/gal.

Using these figures, a spill of 1000 gal of leaded gasoline that occurred before about 1980 could release about 900 g (2.0 lb) each of EDC and EDB to the environment. Both EDC and EDB are moderately soluble and would become dissolved in groundwater, the water concentration depending on the length of contact time, extent of cosolvency with other contaminants, and dilution by water.

6.7.2 Identifying Fuel Types

The most useful analytical tool for identifying different fuel types is the gas chromatograph (GC), with a mass spectrometer detector (GCMS). When fuels are analyzed in a GC, retention times of different hydrocarbon compounds closely correspond to their boiling points; the higher the boiling point, the longer the retention time. Thus, lower boiling point gasoline components elute from the GC earlier than higher boiling point diesel components.

Physical and chemical characteristics of different fuels were described in Section 6.2. Table 6.1 and Figure 6.2 show that carbon number ranges characteristic of different fuel types correspond to different boiling point ranges; in general, the more carbons in a petroleum molecule, the higher is its boiling point (see the discussion of London forces in Chapter 2). In a gas chromatogram, the longer the retention time of a peak, the higher the boiling point of the corresponding compound and the more carbons in the molecule. Although the presence of structural differences, such as the presence of carbon side chains, introduces some variability, this general principle remains useful. Gas chromatograms of different types of fuel are different from one another and are useful for identifying fuel types.

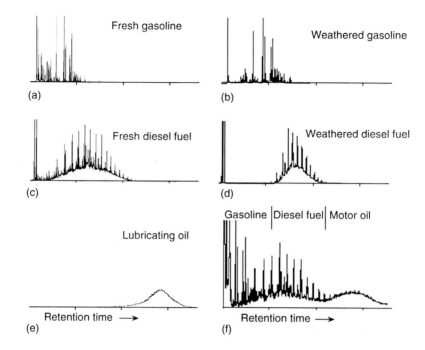

FIGURE 6.12 Gas chromatograms (GC) showing the differences in chromatographic signatures between different types of fresh and weathered petroleum hydrocarbon free product. Figure (f) at lower right is a GC of free product containing a mixture of gasoline, diesel fuel, and motor oil. Humps, where the chromatogram rises above the baseline, are due to hundreds of different hydrocarbon compounds that are not chromatographically resolved.

In Figure 6.12, note that the fresh gasoline GC signature contains more light-weight components (peaks farther to the left, indicating fewer carbon atoms) than do diesel fuel or lubricating oil. When fuel hydrocarbons are weathered by volatilization, dissolution, and biodegradation, the lighter components on the left side of the fresh gasoline and diesel fuel signatures are lost first from the free product LNAPL. Thus, the gas chromatogram of weathered diesel does not resemble that of gasoline, because it lacks the lightweight components that are characteristic of gasoline. Likewise, weathered gasoline does not resemble diesel fuel because it lacks the heavier components that are characteristic of diesel fuel. The components of lubricating oil are all heavier than those found in gasoline or diesel fuel.

6.7.3 Age-Dating Fuel Spills

6.7.3.1 Gasoline

As a gasoline additive, benzene increases the octane rating and reduces engine knocking. As a result, gasoline often contained 2%–5% by volume of benzene before the 1950s, when tetraethyl lead (TEL) replaced it as the most widely used antiknock additive. However, with the global phase-out of leaded gasoline, benzene has made a comeback as a gasoline additive in some nations. In the United States,

Canada, and Europe, concern over benzene's negative health effects and its contamination of groundwater have led to the regulation of gasoline's benzene content. In the United States, current gasolines are regulated to 1% benzene on a regional average.* Canadian and European gasoline specifications now contain a similar 1% limit on benzene content.

The gasoline content in fresh gasoline of the other BTEX compounds, toluene, ethyl benzene, and the xylene isomers (see Figure 6.1), is also variable, depending on the manufacturer, octane requirements, seasonal formulations, and geographic regions. Each BTEX compound also weathers at a different rate. Because benzene is more soluble and volatile than other common gasoline constituents, it is depleted more rapidly than other BTEX components in LNAPL samples of older gasoline contamination because of partitioning into air and groundwater.

Several simple, but approximate, indicators of the age of gasoline spills are related to the BTEX constituents. The concentrations of single BTEX compounds in a weathered sample are seldom useful for estimating how long ago the sample was released because gasolines have never been produced with consistent BTEX concentrations. Furthermore, after a spill or leak, the different components of gasoline are lost by volatilization, dissolution, and degradation at different rates. Thus, knowing the concentration of benzene in a gasoline sample is not particularly helpful by itself for estimating the age of the gasoline. However, certain ratios of the BTEX compounds have been shown empirically to be useful for estimating the age of gasoline releases.

6.7.3.2 Changes in BTEX Ratios Measured in Groundwater

Groundwater BTEX ratio age-dating techniques are based on the different loss rates by volatilization, dissolution, and biodegradation of BTEX compounds from gasoline LNAPL. In groundwater, the sequence of BTEX loss from LNAPL generally begins with benzene, because it is lost most rapidly from gasoline LNAPL by volatilization and partitioning into groundwater; benzene loss is followed by toluene, ethylbenzene, xylenes, in that order. Although biodegradation rates of the separate BTEX compounds are very site-specific, biodegradation of benzene and toluene is generally faster than that of ethylbenzene and the xylene isomers. Their different loss rates from LNAPL suggest that ratios of BTEX compounds might be useful for estimating the time that has elapsed since a gasoline release.

However, the uncertainty of initial BTEX content in gasoline-contaminated groundwater limits the usefulness of simple ratios such as B/T, B/E, T/X, etc. for contaminant age dating. Nevertheless, it has been found empirically from numerous site studies that a quantity known as the cumulative BTEX ratio (R_{BTEX}) can be a useful dating parameter (Kaplan et al., 1997). R_{BTEX} is defined as

---

* The 1% by volume limit for benzene in gasoline is a regional average. EPA has established a system of benzene credits that can be traded among regional refiners to allow some production flexibility while keeping the regional benzene average at or below 1%. The intent is to control regional benzene emissions from all sources. The result is that many individual gasolines will have levels less than 1% benzene and a few might exceed 1%.

$$R_{BTEX} = \frac{B + T}{E + X} \qquad (6.5)$$

where B, T, E, and X are the groundwater concentrations (mg/L) of benzene, toluene, ethylbenzene, and xylene, respectively.

Use of R_{BTEX} appears to minimize the uncertainties related to variations in initial BTEX concentrations. Field and laboratory measurements (Kaplan et al., 1997) show that R_{BTEX} values between 1.5 and 6.0 in gasoline-contaminated groundwater* generally indicate a release that occurred within the last 1–5 years. The ratio decreases exponentially with time with an estimated half-life of 2.3 years, and a value of R_{BTEX} less than 0.5 may be taken to indicate a release greater than 10 years old. These relations are more accurate for samples taken close to the release point because additional uncertainty is introduced by spatial separation of the BTEX components during migration through the subsurface (see Chapter 5, Section 5.7). Morrison (2000a, b) discusses additional sources of uncertainty in age dating with R_{BTEX}.

RULES OF THUMB

In the United States, older gasolines (before 1950) typically contained about 2%–5% by volume of benzene. Current gasolines are regulated to 1% on a regional average for health reasons. The ratio (by weight concentration) $R_{BTEX} = \frac{B+T}{E+X}$ can be used to obtain an approximate age of dissolved gasoline. Because of the large variability in initial gasoline composition and the site-specific nature of weathering processes, all dating methods using BTEX ratios should be considered only a first-order approximation. Whenever possible, additional lines of age-dating evidence should be compared.

Whenever BTEX ratios are used for age dating a fuel release, it is important to consider whether releases of other fuels or substances containing BTEX compounds might contribute to the samples collected.

In groundwater

1. If $R_{BTEX} = \dfrac{B + T}{E + X} = 1.5{-}6.0$, it generally indicates a release less than 5 years old.
2. If $R_{BTEX} = \dfrac{B + T}{E + X} = {<}0.5$, it generally indicates a release more than 10 years old.
3. The value of R_{BTEX} in groundwater is assumed to decrease exponentially with time after a gasoline release according to

$$R_{BTEX} = 6.0\,e^{-0.308t} \qquad (6.6)$$

where t is the time in years since initial release.

---

* The value of R_{BTEX} is also influenced by the amount of gasoline LNAPL in contact with groundwater.

RULES OF THUMB (Continued)

Equation 6.6 assumes that $R_{BTEX} = 6.0$ close to the source shortly after the release (whereas actual ratios immediately after a release were found to range between 1.5 and 6.0, (Kaplan et al., 1997)) and follows first-order decay kinetics with a half-life of 2.3 years.

Other clues to the age of gasoline contamination are based on the presence or absence of TEL, organic manganese compounds, EDB, and EDC, which were common additives to gasoline before 1980 (see Table 6.5). Therefore, the presence of these compounds is supporting evidence for gasoline contamination originating before 1980. However, EDB and EDC are present in some agricultural chemicals and must be used as an indicator with caution.

6.7.3.3 Diesel Fuel

Normal-alkanes (linear single-bonded hydrocarbons) in the approximate carbon number range C6–C24 are the most abundant components of fresh diesel oils, although many other types of hydrocarbons are also present (see Table 6.6). The table also shows that the different types of hydrocarbons in diesel oils biodegrade at different rates. Isoprenoids are branched, unsaturated hydrocarbons that biodegrade more slowly than linear alkanes with similar masses, because their chemical structure inhibits biodegradation.

The solubilities and volatilities of oil range *n*-alkanes and isoprenoids are quite low and similar for the two chemical classes. Therefore, they have similar rates of weathering by nonbiological processes. However, since isoprenoids biodegrade much more slowly than *n*-alkanes, the abundance ratio of *n*-alkanes to isoprenoids changes over time where biodegradation is occurring. Because the passive biodegradation rates of diesel in soils are fairly uniform at similar sites (Christensen and Larsen, 1993), the corresponding changes in composition can be used for estimating the age of the diesel contamination.

TABLE 6.6
Approximate Composition of Fresh Diesel Fuel (C6–C24)

Class of Compound	Percent
n-Alkanes (degrade fastest; smaller alkanes degrade faster than larger)	40
iso- and cyclo-alkanes (degrade slower)	36
isoprenoids (degrade very slowly)	3–4
Aromatics (most soluble; mainly parent and alkylated benzenes, naphthalenes, phenanthrenes, and fluorenes)	20
Polar (water-soluble sulfur, nitrogen, and oxygen compounds)	1

Christensen and Larsen (1993) found, as had others, that by comparing GC peak-height ratios of pristane (a C19 isoprenoid) with *n*-heptadecane (a linear C17 alkane) an estimate can be made of the degree of biodegradation that the sample has undergone. The ratios of phytane (C20 isoprenoid) and *n*-octadecane (linear C18 alkane) can be compared in a similar way. Using the relative extents of biodegradation, the relative ages of fuel contamination in different samples from similar sites may be estimated.

When the composition of fresh diesel fuel is compared with weathered fuels that have biodegraded in the subsurface environment, certain composition changes are apparent.

In particular, *n*-alkanes dominate the composition of fresh fuels and isoprenoids dominate the composition of highly degraded fuels. The C17/pristane peak-height ratio falls from about 2 (for fresh diesel) to 0 (after about 20 years). Figure 6.13 compares gas chromatograms of fresh and biodegraded No. 2 diesel oil. Changes in the relative peak heights of *n*-alkanes and isoprenoids are readily apparent.

Christensen and Larsen (1993) found a linear relation between the age of diesel oils and the C17/pristane peak-height ratio. From their data, it appears possible to determine the age of a diesel spill to within about 2 years, if it meets the following criteria:

- If it is between 5 and 20 years old.
- If it was created by a single sudden spill event.
- It has not been "weathered" significantly except by biodegradation.

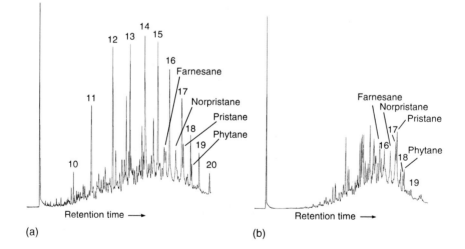

(a)   (b)

FIGURE 6.13 (a) A gas chromatogram of typical fresh No. 2 diesel oil. The numbered peaks are linear *n*-alkanes, where the number represents the carbon number (e.g., C17) of the alkane. The *n*-alkanes are the most abundant compounds in fresh diesel fuels and dominate the composition. The peaks of several isoprenoids are labeled. The peak-height ratio of C17/pristane is about 2:1. (b) A gas chromatogram of biodegraded No. 2 diesel oil. Isoprenoids are more abundant than *n*-alkanes. The peak-height ratio of C17/pristane is about 0.8, indicating an age of about 13 years according to Equation 6.7.

A linear best-fit to their data yields Equation 6.7:

$$\text{Age of diesel fuel (year)} = -8.3(\text{C}17/\text{pristane ratio}) + 19.5 \qquad (6.7)$$

RULES OF THUMB

1. Fresh diesel contains more n-alkanes than isoprenoids, such as pristane or phytane. Therefore, if relative depletion of the n-alkanes is observed, it indicates that biodegradation has taken place.
2. In a fresh diesel fuel, the C17/pristane and C18/phytane ratios are close to 2. Any ratio less than about 1.5 indicates that biodegradation has occurred.

EXAMPLE 5

FINGERPRINTING FUEL CONTAMINANTS AT AN INDUSTRIAL SITE

The following GC/MS data, taken from soils at an industrial site with contamination by gasoline, diesel, and heavy oil hydrocarbons, indicate that at least two different diesel spill events occurred, separated by approximately 6–10 years. The mass spectrometer detector was sometimes operated in the single-ion mode, which increases sensitivity. Single-ion monitoring consists of leaving the mass spectrometer tuned to a fixed mass number as the gas chromatogram peaks elute, rather than continually scanning the entire mass range for each GC peak. The difference between single-ion and full-range monitoring is shown in Figures 6.14a and 6.14b.

To estimate the age of the diesel fuel shown in Figure 6.14, the GC/MS spectra in the pristane/phytane region (15–17 min retention time) must be expanded. Expanded single-ion spectra are shown in Figure 6.15 for samples from Boreholes A and B, which are separated by about 200 yd.

Pristane (16.06 min in Figure 6.15a) biodegrades much more slowly than the C17 n-alkane (16 min). The same is true for phytane (16.92 min) and the C18 n-alkane (16.83 min). The retention times for these peaks are slightly different in Figure 6.15b. In the Borehole A sample, the C17/pristane peak-height ratio is 0.33 and the C18/phytane ratio is 0.26, indicating significant biodegradation. Although, site-specific conditions will determine how much time this amount of degradation would require, the diesel in this sample is at least 15–20 years old. Equation 6.7 suggests an age of about 19 years.

In the Borehole B sample, the C17/pristane ratio is 1.2, which indicates a more recent diesel spill. The diesel fuel from Borehole B is about 9 years old by Equation 6.7.

6.8 SIMULATED DISTILLATION CURVES AND CARBON NUMBER DISTRIBUTION CURVES

A simple approach to chemical fingerprinting is to generate simulated distillation curves (SDCs) and carbon number distribution curves (CNDCs). These plots allow a relative comparison of the concentrations of volatile and semivolatile compounds

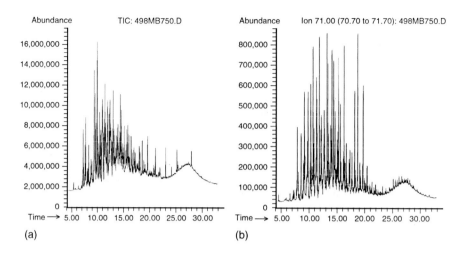

FIGURE 6.14 Full-range (a) and single-ion (b) GC/MS spectra of a contaminated soil sample from Borehole A. (a) GC/MS full-range spectrum of fuel contamination extracted from soil at Borehole A. The distribution of compounds indicates that the soil contains gasoline, diesel, and heavy oil hydrocarbons. (b) Single-ion spectrum at mass 71 of the same sample. Mass 71 is a fragment ion common to many hydrocarbon compounds of C7 and larger. It is useful for diesel fingerprinting because the mass 71 ion is produced abundantly in the fragmentation of pristane and phytane.

present in a sample, without requiring a detailed analysis of each compound present. The curves can be generated automatically from computerized gas chromatogram data without identifying the individual peaks.

A SDC shows the percentage of the sample that would be volatilized at various temperatures. Since the volatility of a petroleum compound is closely related to the size and number of carbon atoms in that compound (compounds with fewer carbon atoms are lighter in weight and boil at lower temperatures than compounds with more carbon atoms), one can estimate the general chemical makeup of petroleum contaminants from an SDC. A CNDC gives similar information but is often easier to interpret.

SDC and CNDC curves allow a visual comparison of the mass distribution of chemical compounds that are present in the analytical samples, based on their boiling points and masses. The shapes of these curves are distinctly different for different types of hydrocarbon mixtures. Gasoline, for example, contains relatively high concentrations of lightweight hydrocarbons such as benzene, while diesel fuel normally has very low concentrations of these lightweight compounds, but higher concentrations of heavier compounds. Weathering of organic compounds produces predictable changes in the shapes of SDCs and CNDCs.

When CNDCs and SDCs have distinctive shapes, they can be used as a "fingerprint" for determining whether contamination at one location is different from or similar to contamination at another location. Sometimes this preliminary analysis is sufficient for identification purposes, as in Example 5. In other cases, it serves to guide further study.

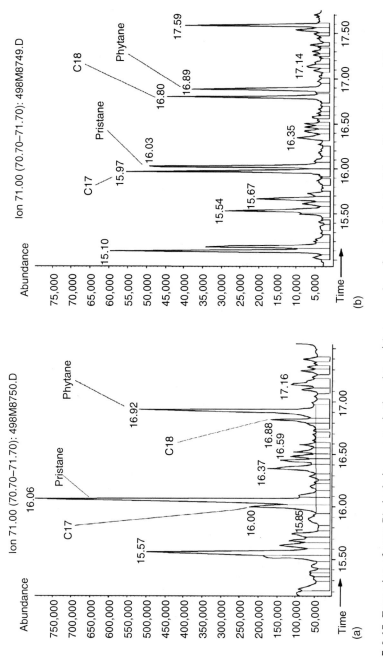

FIGURE 6.15 Expansion of mass 71 single-ion spectra in the pristane/phytane region of contaminated soil from Boreholes A and B.

EXAMPLE 6

Fingerprinting Diesel Contamination with Simulated Distillation and Carbon Number Curves

Mr. A, the owner of a newly purchased mountain home, frequently, but not always, detected strong fuel odors in his basement shortly after diesel fuel was delivered to a neighbor's underground storage tank (UST). Mr. B, the neighbor, was cooperative and allowed Mr. A to take samples from the UST for analysis and comparison with soil samples from around Mr. A's new home.

In Figure 6.16a, the SDC for No. 2 diesel fuel from Mr. B's UST lies somewhat to the left of the laboratory standard for diesel, showing that it contains a higher percentage of lower boiling (lighter weight) compounds. This could indicate that the UST fuel was contaminated slightly with gasoline (perhaps from a tanker truck used to carry both types of fuel) or that it was specially formulated for cold weather use. In Figure 6.16b, the CNDC for the UST fuel has a distinctive fingerprint that includes two prominent peaks at C11 and C13.

Figures 6.17a,b shows SDC and CNDCs for a soil sample collected from a free product seep adjacent to the foundation of Mr. A's home. Not only is the SDC for the

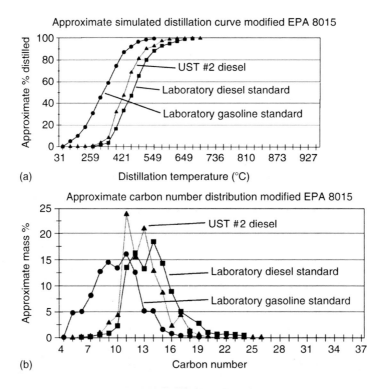

FIGURE 6.16 (a, b) Simulated distillation curves and carbon number curves comparing No. 2 diesel fuel from an underground storage tank with laboratory standards of diesel fuel and gasoline.

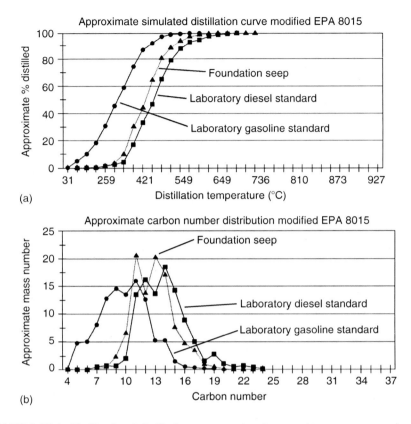

FIGURE 6.17 (a, b) Simulated distillation curves and carbon number curves comparing free product from a seep adjacent to a house foundation with laboratory standards of diesel fuel and gasoline.

foundation seep sample very similar to the SDC for the UST sample, its CNDC contains the two distinctive peaks at C11 and C13 that were also seen in the UST sample.

Mr. B acknowledged that this data strongly implicated his UST as the source of contamination at Mr. A's home, even if it did not explain why the leak appeared to be erratic. When his tank was leak tested, it was found that the filler pipe had cracked where it was fastened to the tank, below the soil surface. It leaked only when the tank was overfilled and diesel rose into the fill pipe above the crack.

REFERENCES

Bredehoeft, J., 1992, Much contaminated groundwater can't be cleaned up, *Groundwater*, 30, p. 30, November–December.

Christensen, L.B. and Larsen, T.H., Fall 1993, Method for determining the age of diesel oil spills in the soil, *Groundwater Monitoring and Remediation*, 13 (4), pp. 142–149.

Farr, A.M., Houghtalen, R.J., and McWhorter, D.B., 1990, Volume estimation of light nonaqueous phase liquids in porous media, *Groundwater*, 28 (1), 48–56.

Freeze, R.A. and Cherry, J.A., 1979, *Groundwater*, Prentice-Hall, Englewood Cliffs, NJ.

Kaplan, I.R., Yakov, G., Shan, T.L., and Ru, P.L., 1997, Forensic environmental geochemistry: Differentiation of fuel types, their sources and release time, *Organic Geochemistry*, 27 (5/6), 287–317.

Lenhard, R.J. and Parker, J.C., 1990, Estimation of free hydrocarbon volume from fluid levels in monitoring Wells, *Ground Water*, 28 (1), January–February, 57–67.

MacDonald, J.A. and Kavanaugh, M.C., 1994, Restoring contaminated groundwater: An achievable goal? *Environ. Sci. Technol.*, 13 (4), 142–149.

Mercer, J. and Cohen, R.A., 1990, A review of immiscible fluids in the subsurface: properties, models, characterization and remediation, *J. contamin. Hydrol.*, 6, 107–163.

Morrison, R.D., 2000a, Critical review of environmental forensic techniques: part I, *Environmental Forensics*, 1, 157–173.

Morrison, R.D., 2000b, Critical review of environmental forensic techniques: part II, *Environmental Forensics*, 1, 175–195.

Parker, J.C., Waddill, D.W., and Johnson, J.A., March 1996, UST corrective action technologies: Engineering design of free product recovery systems, EPA/600/SR-96/031, Order No. PB96-153556, National Technical Information Service, National Risk Management, Springfield, VA.

USEPA, 1991, Assessing UST corrective action technology, a scientific evaluation of the mobility and degradability of organic contaminants in subsurface environments, EPA/600/2–91/053, September, 1991.

USEPA, 2004, How to evaluate alternative cleanup technologies for underground storage tank sites, a guide for corrective action plan reviewers, United States Environmental Protection Agency, Solid Waste and Emergency Response 5401G, EPA 510-R-04–002, May 2004, www.epa.gov/oust/pubs/tums.htm.

7 Behavior of Dense Nonaqueous Phase Liquids in the Subsurface

7.1 DNAPL PROPERTIES

Dense nonaqueous phase liquids (DNAPLs) are liquids that are only slightly soluble in water and therefore exist in the subsurface as a separate fluid phase immiscible with both water and air.* The density of DNAPLs is greater than water (DNAPL density >1 g/cm^3 at 4°C) and their mobility in the subsurface is governed more by gravity and the properties of the DNAPL and surrounding soil than it is by ground-water movement.

Unlike light nonaqueous phase liquids (LNAPLs) such as gasoline, diesel fuel, and heating oil (which are less dense than water), DNAPLs released into soils can sink below the water table where their more-soluble components can slowly dissolve into flowing groundwater, giving rise to dissolved contaminant plumes. A release of DNAPL at the ground surface can therefore lead to long-term contamination of both the vadose and saturated zones at a site.

DNAPLs such as wood preservatives like creosote, transformer, and insulating oils containing polychlorinated biphenyls (PCBs), coal tar, and a variety of chlorinated solvents such as trichloroethene (TCE) and tetrachloroethene (PCE) have been widely used in industry since the beginning of the twentieth century. However, their importance as soil and groundwater contaminants was not recognized until the 1980s, mainly because of the limitations of early analytical methods. As a result, chemical material safety data sheets (MSDS) distributed as late as early 1970 sometimes recommended that waste chlorinated solvents be discarded by spreading them onto dry ground and allowing them to evaporate. These early MSDSs acknowledged the volatile nature of many DNAPL chemicals, but did not recognize their ability to infiltrate rapidly into the subsurface, causing soil and groundwater pollution. It is not surprising that DNAPLs are the contaminants of greatest concern at many Superfund and other hazardous waste sites.

Table 7.1 lists many of the DNAPLs commonly found at Superfund sites, along with their chemical formulas, some alternative names, and common abbreviations.

---

* See Chapter 6 to review the properties of nonaqueous phase liquids (NAPLs) in general and light nonaqueous phase liquids (LNAPLs).

TABLE 7.1
DNAPL Contaminants of Concern at Many Hazardous Waste Sites

Chemical Abstracts Service (CAS) Name	Abbreviation	CAS Number	Other Names	Molecular Formula	Structural Formula
Chloromethane	Artic; R40	74-87-3	Methyl chloride; monochloromethane	CH_3Cl	CH_3Cl
Dichloromethane	Methylene chloride; MC	75-09-2	Methylene dichloride	CH_2Cl_2	CH_2Cl_2
Trichloromethane	CF	67-66-3	Chloroform; methane trichloride	$CHCl_3$	$CHCl_3$
Tetrachloromethane	CT	56-23-5	Carbon tetrachloride	CCl_4	CCl_4
Chloroethane	CA	75-00-3	Ethyl chloride	C_2H_5Cl	$Cl_3C—CH_3$
1,1-Dichloroethane	1,1-DCA	75-34-3	Ethylidene dichloride	$C_2H_4Cl_2$	$Cl_3C—CH_3$
1,2-Dichloroethane	1,2-DCA, EDC	107-06-02	Ethylene dichloride	$C_2H_4Cl_2$	$Cl_3C—CH_3$
1,1,1-Trichloroethane	1,1,1-TCA	71-55-6	Methyl chloroform, chlorothene, methyltrichloromethane	$C_2H_3Cl_3$	$Cl_3C—CH_3$
1,1,2-Trichloroethane	1,1,2-TCA	79-00-5	Vinyl trichloride; β-trichloroethane	$C_2H_3Cl_3$	$Cl_2HC—CH_3$
Chloroethene	VC	75-01-4	Vinyl chloride; chloroethylene	C_2H_3Cl	$ClHC=CH_2$
1,1-Dichloroethene	1,1-DCE	75-35-4	1,1-Dichloroethylene; vinylidine chloride	$C_2H_2Cl_2$	$Cl_2C=CH_2$

Compound	Abbreviation	CAS Number	Synonyms	Formula	Structure
(E)-1,2-Dichloroethene	trans-1,2-DCE	156-60-5	trans-1,2-Dichloroethene; trans-1,2-dichloroethylene; acetylene dichloride	$C_2H_2Cl_2$	$t\text{-}ClHC=CHCl$
(Z)-1,2-Dichloroethene	cis-1,2-DCE	156-59-2	cis-1,2-Dichloroethene; cis-1,2-dichloroethylene; acetylene dichloride	$C_2H_2Cl_2$	$c\text{-}ClHC=CHCl$
Trichloroethene	TCE	79-01-6	Trichloroethylene	C_2HCl_3	$Cl_2C=CHCl$
Tetrachloroethene	PCE	127-18-4	Perchloroethylene; tetrachloroethylene	C_2Cl_4	$Cl_2C=CCl_2$
Chlorobenzene	CB	108-90-7	Monochlorobenzene, benzene chloride, phenyl chloride	C_6H_5Cl	C_6H_5Cl
1,2-Dichlorobenzene	1,2-DCB	95-50-1	o-Dichlorobenzene	$C_6H_4Cl_2$	$C_6H_4Cl_2$
1,3-Dichlorobenzene	1,3-DCB	541-73-1	m-Dichlorobenzene	$C_6H_4Cl_2$	$C_6H_4Cl_2$
1,4-Dichlorobenzene	1,4-DCB	106-46-7	p-Dichlorobenzene	$C_6H_4Cl_2$	$C_6H_4Cl_2$
1,2,3-Trichlorobenzene	1,2,3-TCB	87-61-6	vic-Trichlorobenzene	$C_6H_3Cl_3$	$C_6H_3Cl_3$
1,2,4-Trichlorobenzene	1,2,4-TCB	120-82-1	Trichlorobenzol	$C_6H_3Cl_3$	$C_6H_3Cl_3$
1,3,5-Trichlorobenzene	1,3,5-TCB	108-70-3	sym-Trichlorobenzene	$C_6H_3Cl_3$	$C_6H_3Cl_3$
1,2,3,5-Tetrachlorobenzene	1,2,3,5-TECB	634-90-2	1,2,3,5-TCB	$C_6H_2Cl_4$	$C_6H_2Cl_4$
1,2,4,5-Tetrachlorobenzene	1,2,4,5-TECB	95-94-3	s-Tetrachlorobenzene, sym-tetrachlorobenzene	$C_6H_2Cl_4$	$C_6H_2Cl_4$
Hexachlorobenzene	HCB	118-74-1	Perchlorobenzene	C_6Cl_6	C_6Cl_6
1,2-Dibromoethane	EDB	106-93-4	Ethylene dibromide; dibromoethane	$C_2H_4Br_2$	$C_2H_4Br_2$
Polychlorinated biphenyls	PCBs	—	Aroclor; Phenoclor; Pyralene; Clophen; Kaneclor	—	See Section 7.4

Properties of DNAPL liquids and surrounding soils that are useful for predicting DNAPL mobility are described in Table 7.2. Some important DNAPL compounds and their properties found at these sites are included in Table 7.3.

7.2 DNAPL FREE PRODUCT MOBILITY

In a DNAPL release, the free product sinks vertically downward through the vadose zone under gravitational forces, spreading laterally under capillary forces and leaving behind a trail of residual soil-sorbed DNAPL. In the vadose zone, DNAPL behaves similarly to LNAPL, moving downward while spreading laterally and leaving a trail of soil-sorbed and immobile liquid NAPL in the form of disconnected blobs and ganglia of free product that remain behind the trailing end of the downward-moving DNAPL body.

7.2.1 DNAPL in the Vadose Zone

Like LNAPL, DNAPL in the vadose zone will partition into solid, liquid, and vapor phases so that different portions are present as free product, pore space vapor, dissolved in water, and sorbed to soil (see Figure 6.3). Because of continual losses to other phases, the downward-moving free product is continually diminished in mass and volume. It also undergoes changes in composition as the more volatile and soluble components preferentially leave the free product mixture. A point may be reached at which the remaining DNAPL free product no longer holds together as a continuous phase, but rather is present as immobile isolated globules and ganglia, held in place by capillary forces. Only DNAPL present as a continuous, immiscible, liquid phase is mobile. If sufficient DNAPL was originally present, liquid free product will eventually reach the water table interface between the vadose and saturated zones.

The fraction of liquid hydrocarbon that is retained by sorption and capillary forces in the pores of soils is referred to as residual saturation and is relatively immobile.* Percent residual saturation (%RS) is defined by Equation 7.1.

$$\%RS = 100 \times \left(\frac{\text{volume of NAPL trapped in subsurface pore spaces}}{\text{total volume of pore spaces}} \right) \quad (7.1)$$

The amount of residual DNAPL retained in a typical soil such as silt, sand, or gravel is generally between 5% and 20% of the soil pore space.

In the vadose zone, only DNAPL in the vapor, dissolved, and liquid free product phases has significant mobility; DNAPL sorbed to soil surfaces or trapped in pores is immobile unless it partitions again into one of the three mobile phases. DNAPL in the vapor phase is generally denser than air and tends to sink. However, it spreads laterally wherever the subsurface is least permeable, often moving far beyond the region of residual saturation, where the vapors can contaminate soils and ground-water distant from the region of the spill.

---

* A common operational definition of NAPL mobility is that mobile NAPL can drain under gravity into a monitoring well, while immobile NAPL (residual saturation) cannot.

TABLE 7.2
DNAPL Properties Important for Predicting Mobility in Environment

Properties of DNAPL/Soil	Definition/Typical Units	Comments
Density (d)	d = mass/volume d = g·cm^{-3}; lb·ft^{-3}	Density distinguishes between LNAPLs ($d_{DNAPL} < d_{water}$) and DNAPLs ($d_{DNAPL} < d_{water}$). It depends on temperature, pressure, molecular weights of components, intermolecular forces, and bulk liquid structure.
Dynamic viscosity (μ)	μ = fluid internal resistance to flow or shear. The CGS unit is poise (P); SI unit is N·s·m^{-2}. 1 P = 100 centipoise = 1 g/cm·s = 0.1 Pa·s	Dynamic viscosity is a measure of the force required to move a liquid at a constant velocity. The common unit of μ is the centipoise (cP) because water at 20.2°C has a convenient viscosity of 1.000 cP. Viscosity decreases with increasing temperature (note water in Table 7.2). Intermolecular attractions are the main cause of viscosity. The lower the viscosity, the more fluid the liquid and the more easily it will flow through soils. The reciprocal of dynamic viscosity is called fluidity.
Kinematic viscosity (ν)	ν = dynamic viscosity/density The CGS unit is stokes (St) or centistokes (cSt); SI units are m^2·s^{-1}; stokes = poises/density 1 St = 100 cSt = 10^{-4} m^2·s^{-1}	When the force causing a liquid to move is only due to gravity, as in NAPL movement in the environment, the fluid density, as well as the dynamic viscosity, affects the rate of movement. Using kinematic viscosity includes density in its definition and eliminates the force term (N or Pa). Kinematic viscosity is convenient for calculating hydraulic conductivity, which is inversely proportional to ν. Since the density of water at 20.2°C is 0.998 g/cm^3, the kinematic viscosity of water at 20.2°C is, for most practical purposes, equal to 1.0 cSt.
Solubility in water (S)	S = mass of dissolved substance per unit volume of water, in equilibrium with the undissolved substance. For environmental pollutants in water, the common units are mg/L or μg/L.	Solubility measures a compound's tendency to partition from the bulk compound into water. For a single-component NAPL, the solubility is the concentration of dissolved component in equilibrium with the NAPL. For NAPLs that are mixtures, each component of the mixture has its own characteristic solubility, which is generally lower than the solubility of the pure component (see Section 6.3.8). Thus, the overall solubility of an NAPL mixture is variable, depending on its composition, and changes with time as the more-soluble components leave the NAPL by partitioning into the water. Solubility can vary with temperature, pH, TDS, and the presence of cosolvents (e.g., detergents, EDTA, etc.). In general, the greater the molecular weight (high polarizability) and symmetry (low polarity) and the fewer hydrogen-bonding atoms, the lower the solubility, see Section 2.9.

(Continued)

TABLE 7.2 (Continued)
DNAPL Properties Important for Predicting Mobility in Environment

Properties of DNAPL/Soil	Definition/Typical Units	Comments
Vapor pressure (P_v)	P_v = pressure exerted by a vapor in equilibrium with the liquid or solid phase of the same substance. There are many different units for pressure. The more common units are millimeters of mercury (mm Hg), torr, and atmosphere (atm). The SI unit is pascal (Pa). $1 \text{ mm Hg} = 1 \text{ torr} = 760^{-1} \text{ atm}$ $= 1.333 \text{ mbar} = 133.3 \text{ Pa}$ $= 1.934 \times 10^{-2} \text{ psi } 1 \text{ Pa}$ $= 1 \text{ N/m}^2 = 10^{-5} \text{ bar}$ $= 7.50 \times 10^{-3} \text{ torr}$ $= 1.450 \times 10^{-4} \text{ psi}$	Vapor pressure indicates an NAPL's volatility, or tendency to vaporize, at a given temperature. It depends only on the temperature and increases exponentially with increasing temperature. On a molecular level, vapor pressure is an indication of the strength of intermolecular attractive forces, see Section 2.8.6. The vapor pressure of DNAPLs ranges from very high to very low; for example, compare 1,1-dichloroethylene and chrysene in Table 7.2.
Henry's law volatility	The Henry's law volatility of a compound is a measure of the transfer of the compound from being dissolved in the aqueous phase to being a vapor in the gaseous phase.	The transfer process from water to the gaseous phase in the atmosphere is dependent on the chemical and physical properties of the compound, the presence of other compounds, and the physical properties (velocity, turbulence, depth) of the water body and atmosphere above it. The factors that control volatilization are the solubility, molecular weight, vapor pressure, and the nature of the air–water interface through which it must pass. The Henry's constant is a valuable parameter that can be used to help evaluate the propensity of an organic compound to volatilize from the water. The Henry's law constant is defined as the vapor pressure divided by the aqueous solubility. Therefore, the greater the Henry's law constant, the greater the tendency to volatilize from the aqueous phase, refer to Table 7.1.

TABLE 7.3
Values for Important Properties of DNAPL Contaminants Commonly Found at U.S. Superfund Sites

Chemical Compound	Density (g/cm³)	Water Solubility (mg/L)	Vapor Pressure (torr)	Henry's Law Constant (atm m³/mol)	Dynamic Viscosity[a] (centipoise)	Kinematic Viscosity[a] (centistokes)
Water	0.9991 (15°C)	—	12.8 (15°C)	—	1.145 (15°C)	1.146 (15°C)
(for comparison)	0.9982 (20°C)		17.5 (20°C)		1.009 (20°C)	1.011 (20°C)
Halogenated semivolatiles						
Aroclor[b] 1242	1.3850	0.45	4.06×10^{-4}	3.4×10^{-4}		
Aroclor[b] 1254	1.5380	0.012	7.71×10^{-5}	2.8×10^{-4}		
Aroclor[b] 1260	1.4400	0.0027	4.05×10^{-5}	3.4×10^{-4}		
Chlordane	1.6	0.056	1×10^{-5}	2.2×10^{-4}	1.104	0.69
1,4-Dichlorobenzene	1.2475	80	0.6	1.58×10^{-3}	1.258	1.008
1,2-Dichlorobenzene	1.3060	100	0.96	1.88×10^{-3}	1.302	0.997
Dieldrin	1.7500	0.186	1.78×10^{-7}	9.7×10^{-6}		
Pentachlorophenol	1.9780	14	1.1×10^{-4}	2.8×10^{-6}		
2,3,4,6-Tetrachlorophenol	1.8390	1,000				
Halogenated volatiles						
Carbon tetrachloride	1.5947	790	91.3	0.020	0.965	0.605
Chlorobenzene	1.1060	490	8.8	3.46×10^{-3}	0.756	0.683
Chloroform (trichloromethane)	1.4850	7,920	160	3.75×10^{-3}	0.563	0.379
1,1-Dichloroethane	1.1750	5,500	182	5.45×10^{-4}	0.377	0.321
1,2-Dichloroethane	1.2530	8,690	63.7	1.1×10^{-3}	0.840	0.67
cis-1,2-Dichloroethylene	1.2480	3,500	200	7.5×10^{-3}	0.467	0.364

(Continued)

TABLE 7.3 (Continued)
Values for Important Properties of DNAPL Contaminants Commonly Found at U.S. Superfund Sites

Chemical Compound	Density (g/cm^3)	Water Solubility (mg/L)	Vapor Pressure (torr)	Henry's Law Constant (atm m^3/mol)	Dynamic Viscosity[a] (centipoise)	Kinematic Viscosity[a] (centistokes)
trans-1,2-Dichloroethylene	1.2570	6,300	265	5.32×10^{-3}	0.404	0.321
1,1-Dichloroethylene	1.2140	400	500	1.49×10^{-3}	0.330	0.27
1,2-Dichloropropane	1.1580	2,700	39.5	3.6×10^{-3}	0.840	0.72
Ethylene dibromide	2.1720	3,400	11	3.18×10^{-4}	1.676	0.79
Methylene chloride	1.3250	13,200	350	2.57×10^{-3}	0.430	0.324
1,1,2,2-Tetrachloroethane	1.6	2,900	4.9	5.0×10^{-4}	1.770	1.10
1,1,2-Trichloroethane	1.4436	4,500	0.188	1.17×10^{-3}	0.119	0.824
1,1,1-Trichloroethane	1.3250	950	100	4.08×10^{-3}	0.858	0.647
Tetrachloroethylene (PCE)	1.620	200	14	0.0227	0.890	0.54
Trichloroethylene (TCE)	1.460	1,100	58.7	8.92×10^{-3}	0.570	0.390
Trichloromethane (chloroform)	1.4850	7,920	160	3.75×10^{-3}	0.563	0.379
Nonhalogenated semivolatiles						
2-Methyl naphthalene	1.0058	25.4	0.0680	0.0506		
o-Cresol	1.0273	31,000	2.45×10^{-1}	4.7×10^{-5}		
p-Cresol	1.0347	24,000	1.08×10^{-1}	3.5×10^{-4}		
2,4-Dimethylphenol	1.0360	6,200	0.098	2.5×10^{-6}		
m-Cresol	1.0380	23,500	1.53×10^{-1}	3.8×10^{-5}	21.0	20
Phenol	1.0576	84,000	5.293×10^{-1}	7.8×10^{-7}		
Naphthalene	1.1620	31	2.336×10^{-1}	1.27×10^{-3}		3.87

Benzo(a)Anthracene	1.1740	0.014	1.16×10^{-9}	4.5×10^{-6}
Fluorene	1.2030	1.9	6.67×10^{-4}	7.65×10^{-5}
Acenaphthene	1.2250	3.88	0.0231	1.2×10^{-3}
Anthracene	1.2500	0.075	1.08×10^{-5}	3.38×10^{-5}
Dibenzo(a,h)anthracene	1.2520	2.5×10^{-3}	1×10^{-10}	7.33×10^{-8}
Fluoranthene	1.252	0.27	7.2×10^{-5}	11×10^{-6}
Pyrene	1.2710	0.148	6.67×10^{-6}	1.2×10^{-5}
Chrysene	1.2740	6.0×10^{-3}	6.3×10^{-9}	1.05×10^{-6}
2,4-Dinitrophenol	1.6800	6.0×10^{-3}	1.49×10^{-5}	6.45×10^{-10}
Miscellaneous				
Coal tar (45°F)	1.028			18.98
Creosote	1.05			~1.08 (15°C)

Source: Adapted from USEPA, *Dense Nonaqueous Liquids*, S.G. Huling and J.W. Weaver, Ground Water Issue, Office of Research and Development, Office of Solid Waste and Emergency Response, Washington, DC, EPA/540/4-91-002, March 1991.

[a] Dynamic viscosity measures a liquid's resistance to flow. Kinematic viscosity is the ratio of dynamic viscosity to density, see Table 7.2.

[b] Aroclor is the trade name for polychlorinated biphenyls (PCBs) manufactured by Monsanto. See Section 7.3.4.

Because the vapor pressure of many DNAPL compounds is relatively high, the lifespan of residual DNAPL in the unsaturated zone, where vaporization occurs, can be much less than the lifespan of residual DNAPL below the water table, where vaporization cannot occur. The vaporization process can deplete residual DNAPLs having high vapor pressures, such as the solvents TCE and PCE, within 5–10 years in relatively warm and dry climates. This will not eliminate the presence of vapor phase, adsorbed phase, and aqueous phase contamination in the unsaturated zone, but it can lead to an absence of the DNAPL phase. The absence of DNAPL in the unsaturated zone at a site does not necessarily imply that no DNAPL was ever released at that site in the past, or that past releases of DNAPL have failed to reach the water table.

Water percolating downward through the vadose zone will preferentially leach the more-soluble components of DNAPL from the free product and residual saturation that it contacts; eventually carrying dissolved DNAPL to the saturated zone, contaminating groundwater there. Partitioning of residual saturation into the dissolved phase is facilitated further by the rise and fall of the water table.

7.2.2 DNAPL AT THE WATER TABLE

At the water table interface, DNAPL behaves very differently from LNAPL. Being denser than water, it does not float above the water table but tends to continue downward through the capillary zone of the water table into the saturated zone, where partitioning into the dissolved phase is maximized. To continue moving downward in the saturated zone, DNAPL must displace water held in the soil pore spaces by capillary forces. Consequently, at the water table interface, downward movement slows while DNAPL piles up and spreads laterally. If sufficient weight of DNAPL accumulates, it presses downward through the capillary zone and continues down through the saturated zone, see Figure 7.1. Because soil surfaces in the saturated zone are already wetted by water, DNAPL movement below the water table does not leave a trail of soil-sorbed DNAPL, although some DNAPL can become trapped as residual saturation where water is not readily displaced.

7.2.3 DNAPL IN THE SATURATED ZONE

In the saturated zone, DNAPL can exist only in three phases: the continuous liquid free product, dissolved, and residual saturation phases. The vapor phase is absent. In the saturated zone, residual saturation DNAPL is in continual contact with water and, therefore, continually partitions its more-soluble components into the dissolved phase. Thus, the properties of the DNAPL change progressively, generally toward greater density and higher viscosity. In most soils, hydraulic gradients large enough to mobilize horizontal movements of residual DNAPL are unrealistic. Therefore, investigation and remediation activities involving intensive well pumping are not likely to draw residual DNAPL into wells.

If the initial release was large enough, DNAPL will continue downward through the saturated zone to the bottom of the aquifer. Only an impermeable obstruction, such as bedrock, or complete depletion of mobile free product by sorption and capillary retention within the soil, stops the downward movement of DNAPL mobile

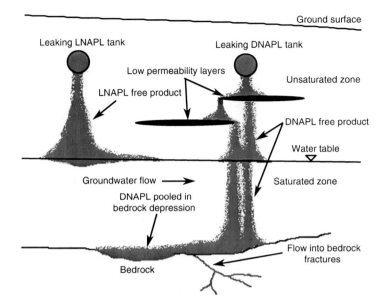

FIGURE 7.1 Comparison of dense nonaqueous phase liquids (DNAPLs) and light nonaqueous phase liquids (LNAPLs) movement in the subsurface after a spill. When mobile NAPL encounters stratigraphic units of low permeability, such as a clay lens or bedrock, it spreads out until it can enter a preferential pathway of greater permeability that allows it to continue downward. DNAPL entering fractured rock systems may follow a complex pattern of preferential pathways.

free product. A decrease in soil permeability, such as a clay layer,* whether in the unsaturated or saturated zone, affects DNAPL travel by slowing the downward movement and causing lateral spreading until soils that are more permeable are encountered. The lateral spreading is generally in the downward slope direction of the stratigraphic unit, but is influenced also by pool formation in depressions and penetration into cracks and fissures. This leads to the formation of DNAPL pools and fingerlike ganglia. Eventually, if sufficient DNAPL is present to move past the impermeable layers, bedrock is reached where DNAPL collects in pools and fractures. If the bedrock is slanted, DNAPL may migrate down the physical slope, even if the direction is opposite to the groundwater movement. Residual and pooled DNAPL together form what is called the DNAPL source zone. It is within the source zone that dissolution into groundwater occurs and aqueous phase plumes originate.

DNAPL solubilities are generally low, so DNAPL in the saturated zone will continue to dissolve slowly into the groundwater without significant diminution over many years. At typically slow groundwater velocities, even a small DNAPL release can persist for decades or longer under natural conditions before all the DNAPL has dissolved or degraded. Once in the subsurface, it is difficult or impossible to recover

---

* Also called a *clay lens* or *low permeability lens*.

all of the trapped residual DNAPL. DNAPL that remains trapped in the soil/aquifer matrix acts as a continuing source of groundwater contamination.

DNAPLs with low viscosity (e.g., methylene chloride, perchloroethylene, 1,1,1-TCA, TCE) can infiltrate into soil faster than water. The relative values of DNAPL viscosity and density, with respect to water, indicate how fast it will flow downgradient through the saturated zone compared to water. For example, several low-viscosity chlorinated DNAPLs (refer to Table 7.3) will flow 1.5–3.0 times faster than water, whereas higher viscosity compounds, including light heating oil, diesel fuel, jet fuel, and crude oil (i.e., LNAPLs) will flow 2–10 times slower than water. Both coal tar and creosote typically have a density greater than one and a viscosity greater than water. Note that the viscosity of NAPL changes with time as different components partition to other phases. As a fresh NAPL loses the lighter volatile components by evaporation, the NAPL becomes more viscous because the remaining heavier, more viscous components comprise a larger fraction of the NAPL mixture.

RULES OF THUMB FOR DNAPL

1. DNAPL movement is affected by gravity far more than groundwater movement. It moves with the slope of the bedrock below the aquifer, independently of the direction of groundwater movement, and forms pools in bedrock depressions.
2. Chlorinated hydrocarbons are generally denser than water (DNAPL). They sink to the bottom of the water table.
3. In Table 7.3, many chlorinated hydrocarbons, including TCE, tetra-chloroethylene, 1,1,1-TCA, methylene chloride, chloroform, and carbon tetrachloride, have viscosities less than water. They flow through the saturated zone 1.5–3.0 times faster than water and can penetrate small fractures and micropores, becoming inaccessible to in situ remediation.
4. The percent residual DNAPL retained as immobile liquid in a typical soil such as silt, sand, or gravel is generally between 5% and 20% of the soil pore space.
5. DNAPLs with high vapor pressure can totally evaporate from the DNAPL phase in the vadose zone in a relatively short time. Therefore, the absence of DNAPL in the unsaturated zone at a site does not necessarily imply that no DNAPL was ever released at that site in the past, or that past releases of DNAPL have failed to reach the water table. Vapor phase, sorbed phase, and dissolved phase contamination may still be present.
6. In most saturated zone soils, intensive well pumping cannot create a large enough hydraulic gradient to move residual DNAPL into the well.

7.3   TESTING FOR THE PRESENCE OF DNAPL

It is very difficult to locate DNAPL free product with monitoring wells. First, DNAPL remains at the bottom of the monitoring well and may go unnoticed. Second, DNAPL free product may be present in locations seemingly unrelated to the spill location, such as perched on low permeability layers in pools and cracks, or upgradient of the spill at the bottom of the aquifer in pools and fractures in the bedrock. There often are no obvious guidelines as to where a well should be placed or how it should be screened to collect free product.

In addition, there are risks of enlarging the contaminated volume when trying to locate and determine the extent of a DNAPL source zone. Unlike residual DNAPL, pooled DNAPL is relatively easy to mobilize by increasing the hydraulic gradient. An exploratory well can inadvertently be drilled through DNAPL perched on a clay lens or pooled on bedrock, resulting in vertical mobilization into previously uncontaminated regions. It often is prudent to use a "from outside toward inside" approach to delineating DNAPL sites, in order to minimize the chances of directly encountering pooled DNAPL during site characterization.

For these reasons, dissolved concentrations of DNAPL-related chemicals in groundwater wells distant from the source zone are often the only evidence that DNAPL free product is present at a site. The EPA has recommended an empirical approach for determining whether DNAPL free product is near a monitoring well where dissolved DNAPL-related compounds have been detected (USEPA, 1992). In order to use this approach, one must

1. Measure the concentrations of DNAPL-related compounds dissolved in groundwater.
2. Know the composition of the suspected DNAPL. See Example 3 for a useful procedure when the composition of the DNAPL is not known.
3. Calculate the effective solubility (S_{eff}) of the measured DNAPL components.
4. Apply the guidelines of Section 7.3.1.

The effective solubility is the theoretical solubility in water of a single component of a DNAPL mixture. It may be approximated by multiplying the component's mole fraction* in the mixture by the solubility of its pure phase.

$$S_{eff}(a) = X_a S_{pure}(a) \qquad (7.2)$$

---

* The mole fraction of compound a in a mixture of several compounds is written X_a.

$$X_a = \frac{\text{moles of } a}{\text{total moles of all compounds in mixture}}$$

For a mixture containing 1 mole of CCl_4 and 3 moles of $CHCl_3$, $X_{CCl_4} = 1/4 = 0.25$ and $X_{CHCl_3} = 3/4 = 0.75$. Note that the sum of all mole fractions must equal unity. The mole fraction of any pure substance equals unity.

where

$S_{eff}(a)$ = effective solubility, in mg/L, of component a in a DNAPL mixture
X_a = mole fraction of compound a in the mixture
$S_{pure}(a)$ = pure-phase solubility of compound a, in mg/L

7.3.1 CONTAMINANT CONCENTRATIONS IN GROUNDWATER AND SOIL THAT INDICATE THE PROXIMITY OF DNAPL

If any of the following conditions exist in groundwater, there is a high probability that DNAPL free product is near the sampling location.

- Groundwater concentrations of DNAPL-related chemicals are >1% of either their pure-phase solubility (S_{pure}) for a single component DNAPL or the effective solubilities (S_{eff}) for components of a DNAPL mixture. The factor of 1% of the solubility is intended to roughly account for the expected concentration decrease due to dilution, dispersion, and degradation of the DNAPL component while moving from the source zone to a monitoring well that is "near" the source. The higher the percentage factor, the closer the well is likely to be to the source zone.
- Soil concentrations of DNAPL-related chemicals are >10,000 mg/kg (1% of soil mass).
- Groundwater concentrations of DNAPL-related chemicals increase with depth or appear in anomalous upgradient/cross-gradient locations with respect to groundwater flow.
- Groundwater concentrations of DNAPL-related chemicals calculated from water–soil partitioning relationships are greater than their pure-phase solubility or effective solubility.

7.3.2 CALCULATION METHOD FOR ASSESSING RESIDUAL DNAPL IN SOIL

1. Measure the DNAPL compounds in the soil.
2. Calculate $S_{eff}(a)$ from Equation 7.2 for each compound.
3. Find K_{oc}, the organic carbon–water partition coefficient in Table 5.5, or from published literature. Otherwise estimate it from log K_{oc} = log K_{ow} − 0.21.
4. Determine f_{oc}, the fraction of organic carbon (oc) in the soil by lab analysis. Values for f_{oc} typically range from 0.03 to 0.00017 (mg oc)/(mg soil). Convert values reported in percent (mg oc/100 mg soil) to (mg oc)/(mg soil).
5. Determine or estimate the dry bulk density of the soil (d_b). Typical values range from 1.8 to 2.1 g/cm^3 (kg/L).
6. Determine or estimate the water-filled porosity (p_w) of the soil.
7. Determine K_d, the soil–water partition coefficient, from $K_d = K_{oc} \times f_{oc}$, Equation 4.16.
8. If the soil sample is collected from a source zone, DNAPL free product is present in the soil and the concentrations of DNAPL compounds dissolved in the pore water will be close to their calculated effective solubilities, S_{eff}.

Therefore, calculate from Equation 7.3 the minimum DNAPL concentration in soil, $C_{\text{soil}}^{\min}(a)$, that would result in a pore water concentration equal to S_{eff}.

9.
$$C_{\text{soil}}^{\min}(a) = \frac{S_{\text{eff}}(a) \times (K_d d_b + p_w)}{d_b} \tag{7.3}$$

10. If measured soil concentrations of compound $a > C_{\text{soil}}^{\min}(a)$, DNAPL free product was present in the soil sample.
11. If measured soil concentrations of compound $a < C_{\text{soil}}^{\min}(a)$, DNAPL free product was not present in the soil sample.

EXAMPLE 1

USING GROUNDWATER CONCENTRATIONS TO ESTIMATE THE PROXIMITY OF RESIDUAL SINGLE-COMPONENT DNAPL

Analysis of a water sample from a monitoring well indicated 6.4 mg/L of tetrachloro-ethene (PERC). Tetrachloroethene was a target contaminant because a dry cleaning establishment had once been on the site near the well. Is residual tetrachloroethene DNAPL likely to be in the subsurface upgradient near the well? Use data from Table 7.3.

Answer:

Since the observed DNAPL is a pure solvent (tetrachloroethene) and not a mixture, its mole fraction, X, equals unity and $S_{\text{eff}} = S_{\text{pure}}$. From Table 7.3, the solubility of pure tetrachloroethene is 200 mg/L. By the guideline in Section 7.3.1, if the measured concentration of a single-component DNAPL in a well is 1% or more of its pure-phase solubility, it is likely that a DNAPL source zone is near the well.

One percent of 200 mg/L is 2.0 mg/L. The measured concentration of tetrachloroethene in the well is 6.4 mg/L. Because this is significantly larger than 2.0 mg/L, it is likely that a source zone of tetrachloroethene DNAPL is quite close to the well.

EXAMPLE 2

USING GROUNDWATER CONCENTRATIONS TO ESTIMATE THE PROXIMITY OF RESIDUAL MULTICOMPONENT DNAPL MIXTURES, WHERE THE INITIAL COMPOSITION IS KNOWN

A remediation project was being planned for a site that had contained a metal degreasing facility. The degreaser solution that was used consisted of 70 wt% trichloromethane, 15 wt% trichloroethylene, and 15 wt% tetrachloroethylene. A matrix of monitoring wells was drilled to try to locate subsurface source zones of DNAPL releases. A water sample from well SW-4 contained 88 mg/L trichloromethane (MW = 119.37), 1.6 mg/L tetra-chloroethylene (MW = 165.82), and 4.2 mg/L trichloroethylene (MW = 131.37). Is this well likely to be close to an upgradient DNAPL source zone? Use data from Table 7.3.

Answer:

1. Convert the weight-percentages of each DNAPL component to mole fractions.
   a. 100 g of solvent contains 70 g trichloromethane, 15 g trichloroethylene, and 15 g tetrachloroethylene.
   b. 70 g trichloromethane $= \left(\frac{70\,\text{g}}{119.37\,\text{g/mol}}\right) = 0.586$ mol

c. 15 g trichloroethylene $= \left(\frac{15\,\text{g}}{131.38\,\text{g/mol}}\right) = 0.038$ mol

d. 15 g tetrachloroethylene $= \left(15\,\frac{\text{g}}{165.82\,\text{g/mol}}\right) = 0.090$ mol

e. Total moles DNAPL $= 0.586 + 0.038 + 0.090 = 0.714$ mol

f. Mole fractions: X(trichloromethane) $= \left(\frac{0.586}{0.714}\right) = 0.821$

g. X(trichloroethylene) $= \left(\frac{0.038}{0.714}\right) = 0.053$

h. X(tetrachloroethylene) $= \left(\frac{0.090}{0.714}\right) = 0.126$

i. Sum of mole fractions $= 0.821 + 0.053 + 0.126 = 1.000$

Calculate S_{eff} from Equation 7.2 and Table 7.3.

- S_{eff} (trichloromethane) $= 0.821 \times 7920$ mg/L $= 6502$ mg/L
- S_{eff} (trichloroethylene) $= 0.053 \times 1100$ mg/L $= 58.3$ mg/L
- S_{eff} (tetrachloroethylene) $= 0.126 \times 200$ mg/L $= 25.2$ mg/L

By the guideline in Section 7.3.1, if the measured concentration in a well of a multi-component DNAPL mixture is 1% or more of the effective solubilities of its components, it is likely that a DNAPL source zone is near the well. The measured concentrations

- C_{meas} (trichloromethane) $= 88$ mg/L
- C_{meas} (trichloroethylene) $= 4.2$ mg/L
- C_{meas} (tetrachloroethylene) $= 1.6$ mg/L

are all greater than 1% of their respective effective solubilities. Therefore, the sampled well is likely to be close to an upgradient DNAPL source zone.

EXAMPLE 3

USING GROUNDWATER CONCENTRATIONS TO ESTIMATE THE PROXIMITY OF RESIDUAL MLTICOMPONENT DNAPL MIXTURES, WHERE THE INITIAL COMPOSITION IS NOT KNOWN

When the composition of the source DNAPL mixture is not known, a variation of the method used in Example 2 can be applied. From Equation 7.2, the mole fraction of component a is

$$X_a = \frac{S_{\text{eff}}(a)}{S_{\text{pure}}(a)}$$

and the sum of mole fractions of all components of the mixture must equal unity,

$$\sum_i X_a = \sum_i \frac{S_{\text{eff}}(a)}{S_{\text{pure}}(a)} = 1$$

In the absence of any dilution, dispersion, or degradation effects, $S_{\text{eff}}(a)$ will be equal to the measured well concentration of component a, $C_{\text{meas}}(a)$. Using the EPA 1% guideline to account for loss effects, $C_{\text{meas}}(a) = 0.01\,S_{\text{eff}}$. This gives Equation 7.4, which uses

TABLE 7.4

Estimating the Proximity of Residual Multicomponent DNAPL Mixtures of Unknown Composition, Using Equation 7.4

Compound	Measured Concentration in Monitoring Well, C_{meas} (mg/L)	Solubility of Pure Compound, S_{pure} (mg/L)	$\dfrac{C_{meas}}{S_{pure}}$
Trichloromethane	88	7920	0.0111
Trichloroethylene	4.2	1100	0.00382
Tetrachloroethylene	1.6	200	0.008
$\Sigma_i \dfrac{C_{meas}}{S_{pure}}$			0.02292

only the measured concentrations of DNAPL components in a well and their pure compound solubilities to describe the conditions where a DNAPL source zone is likely to be near the monitoring well.

$$\sum_i \frac{(0.01)S_{eff}(a)}{S_{pure}(a)} = \sum_i \frac{C_{meas}(a)}{S_{pure}(a)} \geq 0.01 \qquad (7.4)$$

Suppose that, in Example 2, we did not know the initial composition of the DNAPL solvent. Assume that the only data available were the measured well concentrations:

- C_{meas} (trichloromethane) $= 288$ mg/L
- C_{meas} (trichloroethylene) $= 4.2$ mg/L
- C_{meas} (tetrachloroethylene) $= 1.6$ mg/L

The use of Equation 7.4 is illustrated in Table 7.4.

The sum of column 4 is greater than 0.01 and, therefore, it is likely that a source zone of a DNAPL mixture is upgradient near the monitoring well.

EXAMPLE 4

USING SOIL CONCENTRATIONS BELOW THE WATER TABLE TO ESTIMATE THE PROXIMITY OF RESIDUAL SINGLE-COMPONENT DNAPL

Trichloroethene (TCE) was measured to be 452 mg/kg in a soil sample from the saturated zone. No other DNAPL compounds were detected. Measured soil data are

Porosity, $p_w = 0.27$
Dry bulk density, $d_b = 1.9$ kg/L
Fraction of organic carbon, $f_{oc} = 0.003$
Is a TCE DNAPL free product phase likely to be present in the soil sample?

Answer:

Since only TCE was present, $S_{eff} = S$, the pure compound water solubility. Use Table 5.5 to obtain the following data for TCE:

$S = 1100$ mg/L
$K_{oc} = 166$ L/kg

Calculate C_{soil}^{min} (TCE) from Equation 7.3.

$$C_{soil}^{min}(a) = \frac{S_{eff}(a) \times (K_d d_b + p_w)}{d_b}$$

$$= \frac{1100\,mg/L \times ((166\,L/kg)(0.003)(1.9\,kg/L) + 0.27)}{1.9\,kg/L}$$

$$C_{soil}^{min}(a) = 704 \text{ mg/kg}$$

Since the measured TCE soil concentration of 452 mg/kg < 704 mg/kg, it most likely is residual TCE rather than free product DNAPL.

EXAMPLE 5

USING SOIL CONCENTRATIONS BELOW THE WATER TABLE TO ESTIMATE THE PROXIMITY OF RESIDUAL MULTICOMPONENT DNAPL, WHERE THE INITIAL COMPOSITION IS NOT KNOWN

Even though the initial composition of a DNAPL mixture is not known, the sum of its mole fractions must equal unity. Under conditions of DNAPL saturation, the sum of measured soil concentrations of the DNAPL components divided by their saturated values, C_{soil}^{min}, is equivalent to the sum of their mole fractions. Therefore,

$$\sum_a \frac{C_{meas}(a)}{C_{soil}^{min}(a)} \geq 1$$

Table 7.5 demonstrates an example calculation for determining if DNAPL free product was present in a soil sample found to contain four different DNAPL compounds. The soil had the following properties:

Porosity, $p_w = 32\%$
Dry bulk density, $d_b = 1.7$ g/cm^3

TABLE 7.5
Example Soil Concentration Calculation for Multicomponent DNAPL

Compound	C_{meas} (mg/kg)	Pure Compound Solubility, S (mg/L)	K_{oc} (L/kg)	K_d	C_{soil}^{min} (mg/kg)	$\frac{C_{meas}}{C_{soil}^{min}}$
Tetrachloroethene (PCE)	109	200	155	0.465	131	0.834
Trichloroethene (TCE)	368	1100	166	0.498	755	0.487
Chlorobenzene (CB)	84	472	219	0.657	399	0.211
1,1,1-Trichloroethane (111-TCA)	188	1330	110	0.330	689	0.273
					$\sum \frac{C_{meas}}{C_{soil}^{min}} = 1.805$	

Organic carbon, $f_{oc} = 2.7\%$

K_d is calculated from Equation 5.17: $K_d = K_{oc} f_{oc}$

C_{soil}^{min} is calculated from Equation 7.3: $C_{soil}^{min}(a) = \dfrac{S_{eff}(a) \times (K_d d_b + p_w)}{d_b}$

Since the sum of the estimated mole fractions, $\sum\limits_{a} \dfrac{C_{meas}(a)}{C_{soil}^{min}(a)}$, is greater than unity, DNAPL was present in the soil sample.

7.4 POLYCHLORINATED BIPHENYLS

7.4.1 BACKGROUND

Polychlorinated biphenyls (PCBs) are a family of stable man-made organic compounds produced commercially by direct chlorination of biphenyl. PCBs were manufactured and sold under various trade names (Aroclor, Pyranol, Phenoclor, Pyralene, Clophen, and Kaneclor) as complex mixtures differing in their average chlorination level. In the United States, production of PCBs stopped in 1977.

Since 1929, about 1.4 billion pounds of PCBs have been commercially produced, the majority in the United States. It is estimated that several hundred million pounds have been released to the environment. The world's primary producer was Monsanto, who produced PCBs under the trade name Aroclor from 1930 to 1977. General Electric had a rival product called Pyranol. As shown in Figure 7.2, individual PCB compounds are formed by substituting between 1 and 10 chlorine atoms onto the biphenyl aromatic structure. These substitutions can produce 209 different congeners (homologues and isomers).

PCBs have many desirable properties for commercial applications: very high chemical, thermal, and biological stability; low water solubility; low vapor pressure; high dielectric constant; and high flame resistance. It is not surprising that PCBs found wide application as coolant and insulation fluids in transformers and capacitors, and as flame retardants, plasticizers, solvent extenders, organic diluents,

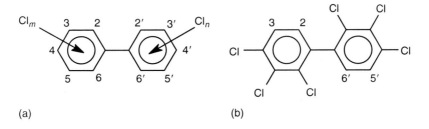

(a)   (b)

FIGURE 7.2 (a) General structure of polychlorinated biphenyls (PCBs). (b) One particular 6-chlorine PCB congener out of the 209 different types possible. The general formula for a PCB is $C_6Cl_mH_{5-m}C_6Cl_nH_{5-n}$, where m and n each can be any integer between 1 and 5. Individual PCB compounds are formed by substituting between 1 and 10 chlorine atoms onto the biphenyl aromatic structure. This substitution can produce 209 different congeners (homologues and isomers).

additives to epoxy paints, heat transfer fluids, in hydraulic fluids, in pesticides, and in printing inks. PCB's are also by-products of many industrial processes, such as the manufacturing of chlorinated solvents and chlorinated benzenes.

Industrial grade PCBs are mixtures of PCB compounds blended to give particular overall properties, such as viscosity, electrical resistance, boiling point, etc. For example, Aroclor-1242, see Table 7.1 (also called PCB-1242), is actually a mixture of more than 60 different PCB congeners with varying degrees of chlorination. It naturally has a complicated gas chromatogram. A four-digit numbering system was assigned to the mixtures. The first two numbers indicate the number of carbon atoms and the third and fourth numbers give the weight percent of chlorine.

7.4.2 ENVIRONMENTAL BEHAVIOR

PCBs are very stable species and do not degrade readily in the environment. Most of the released PCBs are believed to remain in mobile environmental reservoirs (Alder et al., 1993). They even survive ordinary incineration and can escape as vapors up the smokestack.

The wide use of PCBs has resulted in their common presence in soil, water, and air. PCB dispersion from source regions to global distribution occurs mainly through atmospheric transport and subsequent deposition. Because of their low vapor pressure and water solubility, PCBs typically have very high partition coefficients to abiotic and biotic particles. In aquatic systems, sediments are an important reservoir.

Environmental contamination was first reported in 1966, when high levels of PCBs were found in fish. PCBs in wastes dumped into Lake Michigan accumulated in the fatty tissue of fish and were subsequently found in the breast milk of nursing mothers who ate the fish. Children nursed by these mothers showed higher rates of development and learning disorders than those of nursing women in the same region who had not eaten the fish. Similar developmental effects were reported from Japan and Taiwan, where children of women who had eaten PCB contaminated rice products were underdeveloped physically and mentally. Adults working with PCBs were susceptible to a skin condition called chloracne, which produces pustules and cysts.

Eventually it was discovered that PCBs did not easily biodegrade and their use was restricted. In 1976, PCBs became regulated under the Toxic Substances Control Act and safe disposal became a major concern. Between 1974 and 1979, PCBs were used only in the production of capacitors and transformers. Monsanto stopped producing Aroclors in October 1977. In 1986, an international agreement was signed to ban most uses of PCBs and phase out the rest. Nevertheless, although they are no longer manufactured, they still leak from old electrical devices, including power transformers, capacitors, television sets, and fluorescent lights, and can be released from hazardous waste sites and historic and illegal refuse dumps. They also persist in fatty foods, such as certain fish, meat, and dairy products.

The toxicity of PCBs is a complicated issue since each congener differs in its toxicity. All PCBs are listed by EPA as known carcinogens and priority pollutants. When they are incinerated, they can produce dioxins, which are rated by EPA among the most toxic substances.

7.4.3   ANALYSIS OF PCBS

The gas chromatograph/mass spectroscopy (GC/MS) patterns of the different PCB mixtures show considerable overlap and common petroleum products, such as motor oil, also generate peaks in the PCB region. For these reasons, unambiguous identification of particular PCBs requires meticulous laboratory technique, especially if other organic compounds are present that have peaks in the PCB GC/MS regions.

7.4.4   CASE STUDY: MISTAKEN IDENTIFICATION OF PCB COMPOUNDS

A metal recycling company shredded automobile bodies, large appliances, industrial components such as power transformers and manufacturing equipment, etc. The nonmetallic residue from the shredding operation is called fluff, and consists of shredded solid plastics, foamed plastic, rubber, glass, wood, etc. The fluff was oily, having absorbed much of the residual oil remaining in the original metal components.

Fluff was disposed off by transport to a landfill. Acceptance by the landfill operators was conditional on a chemical analysis that showed the fluff did not contain excessive levels of toxic materials. High toxicity would require the fluff to be classified as hazardous waste, with more stringent disposal conditions. PCBs were a toxic substance of concern. If PCB levels exceeded 50 mg/kg, the fluff would be classified as a hazardous waste, requiring special and expensive disposal methods.

For several years, the recycling had never had PCB analyses from their fluff that were higher than about 15 mg/kg. Then, although they had no reason to believe their mix of shredded materials had changed significantly, the laboratory analyses were suddenly showing greater than 50 mg/kg of PCBs. Were these results accurate or not? Because PCB mass spectra overlapped the motor-oil GC/MS spectral range, it was possible that oil compounds were being mistaken for PCBs.

Arrangements were made with a knowledgeable laboratory director to be especially careful in sample cleanup and preparation. PCBs are very stable and can withstand strong acid and base extractions that will decompose most oils. Figure 7.3a and b compares the GC/MS spectra from an inadequate and a satisfactory cleanup procedure on similar fluff samples containing PCBs. By modifying the sample cleanup to more completely decomposed petroleum oils, it was shown that the PCB concentrations were well below the hazardous waste threshold and the fluff need not be treated as a hazardous waste.

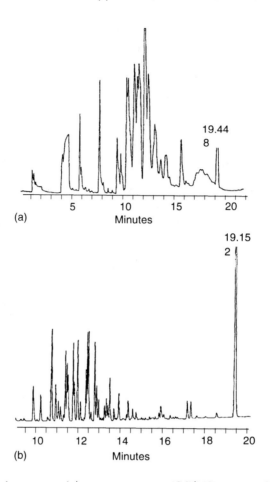

FIGURE 7.3 Gas chromatograph/mass spectroscopy (GC/MS) spectra of oily waste samples containing PCBs. (a) Incomplete removal of oils results in an overlap of oil and PCB spectra, causing poor resolution of the PCB components and an overestimation of PCB concentrations in the sample. (b) Better cleanup preparation of the sample decomposes most of the oil contaminants. This spectrum shows an expanded portion of (a) in the 10–20 min region. Individual PCB compounds show much better separation and the measured PCB concentration is much lower than in (a).

REFERENCES

Alder, A.C., Haggblom, M.M., Oppenheimer, S.R., and Young, L.Y., 1993, Reductive dechlorination of polychlorinated biphenyls in anaerobic sediments, *Environ. Sci. Technol.* 27, 530–538.

USEPA, 1991, *Dense Nonaqueous Liquids*, Scott G. Huling and James W. Weaver, Ground Water Issue, Office of Research and Development, Office of Solid Waste and Emergency Response, Washington, DC, EPA/540/4-91-002, March.

USEPA, 1992, *Estimating Potential for Occurrence of DNAPL at Superfund Sites*, Publication 9355.4-07FS, Office of Emergency and Remedial Response, Washington, DC, NTIS PB92-963338, January.

8 Biodegradation and Bioremediation of LNAPLs and DNAPLs

8.1 BIODEGRADATION AND BIOREMEDIATION

Biodegradation is the chemical breakdown of organic contaminants into smaller compounds through metabolic or enzymatic processes of living organisms in the environment, primarily bacteria, yeast, and fungi. It differs from chemical and physical degradation processes (e.g., chlorine oxidation, reduction by metallic iron particles, hydrolysis, photolysis, catalysis on reactive surfaces, etc.) in being caused by the action of living organisms. Some chemical structures are more susceptible to microbial breakdown than are others; vegetable oils, for example, will biodegrade more readily than petroleum oils, which, in turn, biodegrade more readily than polycyclic aromatic hydrocarbons (PAHs). Bioremediation* of soil and groundwater at sites contaminated with total petroleum hydrocarbons (TPH) and benzene, toluene, ethylbenzene, and xylenes (BTEX) is a well-established technology compared to sites contaminated with PAHs, chlorinated volatile organic compounds (VOCs), pesticides and herbicides, and explosives, which are more difficult to biodegrade.

When biodegradation converts hazardous contaminants into less hazardous or benign substances (not always), it is called bioremediation. Microbial metabolism is a series of biological reactions, predominantly oxidation–reduction (also called "redox") reactions, which convert organic compounds into energy and carbon to sustain microbial growth. In a typical metabolic redox reaction, an organic carbon compound serves as an electron donor that microbes use as a food source. The transfer of an electron from the donor to an electron acceptor proceeds through multiple reaction steps that generate energy and materials, carbon and other elements, for microbial cell growth.

---

* Bioremediation is one of several approaches for the remediation of sites contaminated with nonaqueous phase liquids. It is, however, the only remediation method treated with some depth in this book. Other methods, such as soil vapor extraction, air sparging, chemical oxidation, low-temperature thermal desorption, etc. are more appropriate for specialized reports. However, the fact that bioremediation processes are occurring does not preclude the additional use of more "active" remediation methods. A good starting reference for other remediation options is EPA's manual (USEPA, 2004).

Organic pollutants are often toxic because of their chemical structure that allows them to interfere with the normal functions of living organisms. Changing their structure in any way, which always occurs when an electron is lost or gained, will change their properties and may make them less toxic or even, in a few cases, more toxic. Eventually, usually with many reaction steps, biodegradation breaks organic pollutants into smaller and smaller molecules, finally ending with carbon dioxide, water, and mineral salts if the process is not interrupted. Although these final products represent the destruction of the original pollutant, some of the intermediate steps may temporarily produce compounds that are also pollutants, sometimes more toxic than the original. An example of a more toxic biodegradation product is vinyl chloride formed as an intermediate in the anaerobic biodegradation chain of the common solvent trichloroethylene (TCE, see Section 8.6).

For organic contaminants of low solubility and low volatility, for which most of the remediation processes described in Chapter 2 (e.g., chemical oxidation and reduction, volatilization, sorption, coagulation, precipitation, filtration, photolysis, etc.), are inefficient or impossible, biodegradation is the ultimate form of contaminant removal for soil and groundwater cleanup. Because some fraction of contaminants always becomes nearly irreversibly sorbed to soil and cannot readily be removed by desorption treatments (see Section 5.7.3), pollutants can remain in place for many years, serving as a continual source of groundwater contamination. In such cases, there are only three practical approaches to achieving a complete remediation:

1. Excavating the contaminated soil and sending it to a landfill or treating it on the surface (e.g., by incineration).
2. Isolating the contaminated soil by capping its surface with a membrane or other impervious cover and diverting surface and groundwater away from the area.
3. Allowing or assisting biodegradation to transform the contaminants into nonhazardous substances.

Biodegradation is often the most economical and practical approach to remediation. When biodegradation becomes bioremediation, i.e., when it is used as the main site cleanup process, it may require some assistance. If the rate of natural bioremediation (intrinsic bioremediation) is too slow, adding nutrients, oxygen, and appropriate microbes (engineered bioremediation) can often accelerate it. Much progress has been made in recent years in understanding the many different processes of biodegradation and adapting them to successfully degrading types of organic compounds once thought resistant to biodegradation, such as chlorinated hydrocarbons (National Research Council, 2000). EPA publishes many reports related to this new technology. The EPA Internet web page is a good place to find the latest references.

8.2   BASIC REQUIREMENTS FOR BIODEGRADATION

Biodegradation is a redox process. This means that energy for biodegradation arises from electron transfer reactions. There are six basic requirements that must be present for biodegradation to occur:*

1. Suitable degrading organisms, generally bacteria or fungi.
2. Electron donors that are the energy source (food) for the organisms. The electron donors are generally organic carbon compounds, which are mostly converted to CO_2, water, and mineral salts in the redox reactions that comprise the metabolism of the organisms.
3. Electron acceptors, generally O_2, NO_3^-, SO_4^{2-}, Fe^{3+}, and CO_2.
4. Carbon for cell growth, which comes from organic carbon. About 50% of bacterial dry mass is carbon.
5. Nutrients, including nitrogen, phosphorus, calcium, magnesium, iron, and trace elements. Of these, nitrogen and phosphorus are needed in the largest quantities. Bacterial dry mass is about 12% nitrogen and about 2%–3% phosphorus.
6. Acceptable environmental conditions: pH, salinity, hydrostatic pressure, solar radiation, toxic substances, oxygen, etc., must all be within acceptable limits for the particular bioprocesses. Microbes capable of degrading petroleum hydrocarbons generally prefer a pH between 6 and 8, and temperatures between 5°C and 25°C. Temperatures lower than 5°C tend to inhibit biodegradation in general.

Microbial reactions can be grouped into two classes, aerobic and anaerobic. Aerobic microbial reactions, such as the breakdown of fuel oils to carbon dioxide and water, require the presence of oxygen as an electron acceptor. Anaerobic microbial reactions, such as the reductive dechlorination of chlorinated solvents, utilize electron acceptors other than oxygen. When sufficient oxygen is present, it will be utilized in preference to alternative electron acceptors and only aerobic reactions will occur. When sufficient oxygen is absent, only anaerobic reactions can occur. Aerobic and anaerobic reactions generally require different kinds of bacteria, with aerobic reactions generally being considerably faster than anaerobic reactions.

In the list of six basic requirements mentioned above, items 1–5 must all be present in sufficient quantities for biodegradation to proceed. However, different pollutants can have very different quantity requirements. For example, the redox processes involved in the biodegradation of fuel hydrocarbons are very different from the biodegradation processes of chlorinated aliphatic hydrocarbons. In the biodegradation of fuel hydrocarbons, especially BTEX, aerobic reactions are far

---

* These requirements are similar to those for metabolic processes in animal life forms. People, for example, degrade food (electron donors) to obtain energy, breathe air to obtain oxygen (the main electron acceptor), require carbon and other nutrients for cell growth, and produce waste (degraded food), cell structures and energy.

more efficient than anaerobic. Since carbon electron donors can include the contaminant fuel molecules themselves, electron donors will generally be in excess and these aerobic redox reactions are generally limited by electron acceptor availability. Under aerobic conditions where oxygen is the main electron acceptor, an adequate supply of oxygen generally means that biodegradation will proceed until all of the contaminants accessible to the microbes are degraded. As aerobic biodegradation is the only natural process that actually reduces the mass of petroleum hydrocarbon contamination, it is the most important (and preferred) remediation mechanism for petroleum hydrocarbons.

On the other hand, the most highly chlorinated aliphatic hydrocarbons, such as the solvents tetrachloroethene (perchloroethene, PCE) and trichloroethene (TCE), typically are biodegraded by reductive dechlorination, an anaerobic process where the electron acceptors are the chlorinated solvent molecules themselves. Thus, as long as chlorinated contaminant remains there are electron acceptors available and the abundance of electron donors (e.g., fuel hydrocarbons, landfill leachate, or natural organic carbon) will always be the limiting factor. If the subsurface environment is depleted of electron donors before all the chlorinated aliphatic hydrocarbons are biodegraded, biological reductive dechlorination will cease.

8.3 BIODEGRADATION PROCESSES

The most important metabolic processes in biodegradation are redox reactions, classified according to the species serving as the electron acceptor:

Aerobic respiration

- Microbes use oxygen electron as acceptors to transform organic carbon into carbon dioxide.
- Electrons are transferred from the contaminant to oxygen. The contaminant is oxidized and the oxygen is reduced, forming water and carbon dioxide.
- The key requirement is adequate oxygen.

Anaerobic respiration

- Microbes use an "oxygen substitute" to serve as electron acceptors (usually nitrate, sulfate, Mn^{4+}, Fe^{3+}, or CO_2). Organic carbon is chemically transformed, often to carbon dioxide but sometimes to methane.
- Electrons are transferred from the contaminant (oxidizing it) to an electron acceptor (reducing it). The products formed by the anaerobic electron acceptors (CO_2, NO_3^-, SO_4^{2-}, and Fe^{3+}) are shown in Table 8.1 (Equations 8.1 through 8.6).
- The key requirement is an adequate supply of electron acceptors and an absence of oxygen.

Cometabolism

- Enzymes produced by microbes during the degradation of organic matter fortuitously react chemically to transform a contaminant that resists

TABLE 8.1
Normal Sequence of Biodegradation Reactions

Oxygen reduction	$\{CH_2O\} + O_2 \rightarrow CO_2 + H_2O$	(Aerobic)	(8.1)
NO_3^- reduction to N_2 (denitrification)	$5\{CH_2O\} + 4NO_3^- + 4H^+ \rightarrow 5CO_2 + 2N_2 + 7H_2O$	(Anaerobic)	(8.2)
Mn^{4+} reduction	$\{CH_2O\} + 2MnO_2 + 4H^+ \rightarrow CO_2 + 2Mn^{2+} + 3H_2O$	(Anaerobic)	(8.3)
Fe^{3+} reduction	$\{CH_2O\} + 4Fe(OH)_3 + 8H^+ \rightarrow CO_2 + 4Fe^{2+} + 11H_2O$	(Anaerobic)	(8.4)
SO_4^{2-} reduction	$2\{CH_2O\} + SO_4^{2-} + H^+ \rightarrow 2CO_2 + HS^- + 2H_2O$	(Anaerobic)	(8.5)
Methane production	$2\{CH_2O\} \rightarrow CH_4 + CO_2$	(Anaerobic)	(8.6)
	For organic matter with element ratios different from the model compound $\{CH_2O\}$, especially those containing chlorine, reduction leads to the generation of H_2, which undergoes a redox reaction with dissolved CO_2 as follows:		
	$CO_2 + 4H_2 \rightarrow CH_4 + 2H_2O$	(Anaerobic)	(8.7)

Note: In the stoichiometric equations, organic pollutants being degrading are approximated by the model compound CH_2O, which has element ratios similar to typical hydrocarbon contaminants.

biodegradation. For example, during biodegradation of methane, some bacteria produce an enzyme that breaks down chlorinated solvents, such as TCE.

• The key requirement is the presence of a substance that, when metabolized by microbes, produces the right enzymes to transform the contaminants.

Each of these three biodegradation processes requires an electron donor that is oxidized and an electron acceptor that is reduced. The overall electron transfer process provides metabolic energy for the microbes. Favorable conditions for biodegrading various organic compounds are listed in Tables 8.2. and 8.3, which identifies some of the site-specific factors that favor successful bioremediation.

Upon accepting electrons from an energy source, electron acceptors are converted to the products indicated in Equations 8.1 through 8.7. Only Equation 8.1 is aerobic; all the others are anaerobic. Equations 8.1 through 8.7 are listed in the order of energy release; Equation 8.1 releases the most energy and Equation 8.7 the least. Thus, when oxygen is available, Equation 8.1 will occur before any of the others. If all the oxygen is consumed and nitrate is available, Equation 8.2 becomes the preferred process, and so on.

8.3.1 CASE STUDY

8.3.1.1 Passive (Intrinsic) Bioremediation of Fuel LNAPLs: California Survey

Sometimes the best approach to treating soil contaminated with fuel hydrocarbons is to do nothing. If passive biodegradation rates are high enough, a pollutant plume may shrink or become immobile within an acceptably short time frame.

TABLE 8.2

Biodegradation Processes for Some Organic Compounds

Hydrocarbons (HCs)

Gasoline, diesel, fuel oil	Readily biodegradable under aerobic conditions; more slowly degradable under anaerobic conditions.
PAHs	Aerobically biodegradable under a narrow range of conditions.
Creosote	Readily biodegradable under aerobic conditions.
Alcohols, ketones, esters	Readily biodegradable under aerobic conditions.
Ethers	Biodegradable under a narrow range of conditions using aerobic or nitrate-reducing microbes.

Chlorinated aliphatic HCs

Highly chlorinated (e.g., tetrachloroethene)	Biodegraded and cometabolized by anaerobic microbes, primarily by reductive dechlorination; cometabolized by aerobic microbes in special cases.
Less chlorinated (e.g., dichloroethene)	Aerobically biodegradable under a narrow range of conditions; cometabolized by anaerobic microbes.

Chlorinated aromatic HCs

Highly chlorinated (e.g., pentachlorophenol)	Aerobically biodegradable under a narrow range of conditions; cometabolized by anaerobic microbes.
Less chlorinated (e.g., chlorobenzene)	Readily biodegradable under aerobic conditions.

Polychlorinated Biphenyls (PCBs)

Highly chlorinated	Cometabolized by anaerobic microbes.
Less chlorinated	Aerobically biodegradable under a narrow range of conditions.

Nitroaromatics

Trinitrotoluene, nitrobenzene, etc.	Aerobically biodegradable; converted to innocuous volatile organic acids under anaerobic conditions.

Metals

Cr, Cu, Ni, Hg, Cd, Zn, etc.	Solubility and reactivity can be changed by a variety of microbial processes.

Passive, or intrinsic, bioremediation is becoming increasingly acceptable as a treatment alternative, especially for fuel hydrocarbons (light nonaqueous phase liquids [LNAPL]), if there is no immediate threat to water uses and if natural site conditions are favorable (see Table 5.12). With favorable site conditions, passive remediation may be expected to eventually stabilize a contaminant plume's length and mass, even if an active source such as free product is present in the subsurface, continually dissolving contaminants into the plume. If all mobile active sources are removed to the point of residual saturation, biodegradation will often reduce the plume mass to the point of completing the cleanup.

A survey of fuel hydrocarbon contamination in California by Lawrence Livermore National Laboratory, commissioned in 1994 by the Underground Storage Tank (UST) Program of the California State Water Resources Control Board (Rice et al., 1995), shows the value of a passive approach to remediation. A detailed analysis of 271 cases involving leaking underground fuel tank (LUFT) showed that

TABLE 8.3

Hydrogeologic Factors Favoring In Situ Bioremediation

Type of Bioremediation	Important Site Characteristic	Favorable Indicators
Engineered	Transmissivity of subsurface to fluids	Hydraulic conductivity $>10^{-4}$ cm/s (if system circulates water) Intrinsic permeability $>10^{-9}$ cm^2 (if system circulates air)
	Relative uniformity of subsurface medium	Common in river delta deposits, floodplains of large rivers, and glacial outwash aquifers
	Low residual concentrations of NAPL contaminants on subsurface solids	NAPL concentration $<10,000$ mg/kg
Intrinsic/passive	Consistent groundwater flow (velocity and direction)	Seasonal variation in depth to water table <1 m Seasonal variation in regional flow path $<25°$
	Presence of pH buffers	Carbonate minerals (limestone, dolomite, and shell material)
	High concentration of electron acceptors	Oxygen, nitrate, sulfate, Fe^{3+}
	Presence of elemental nutrients	Nitrogen and phosphorus

- In general, fuel hydrocarbon plume lengths change slowly and tend to stabilize at short distances from the source.
- Plume boundaries, defined by a concentration of 10 ppb, extended no farther than 76 m in 90% of the cases.
- Plume mass decreased more rapidly than plume length.
- Residual fuel in the soil degraded more slowly than dissolved fuel and continued to dissolve contaminants into the groundwater. Thus, the length of the plume is defined mainly by the extent of soils containing residual adsorbed fuel.
- In 50% of sites with no actively engineered remediation, groundwater benzene concentrations decreased about 70% as fast as where pump-and-treat plus excavation treatment was applied.
- After removing contaminant sources, degradation generally removed 50%–60% of the remaining pollutant mass per year.

RULE OF THUMB

Once a fuel hydrocarbon (e.g., gasoline, diesel, heating oil) source is removed, passive remediation requires 1–3 years to reduce the dissolved plume mass by a factor of 10.

8.4 NATURAL AEROBIC BIODEGRADATION OF NAPL HYDROCARBONS

As discussed in Chapters 6 and 7, liquid hydrocarbons of low water solubility (oils, many solvents, gasoline, etc.) are called "nonaqueous phase liquids (NAPLs)." NAPLs are further divided into hydrocarbons that are less dense than water, LNAPLs, and hydrocarbons that are denser than water, dense nonaqueous phase liquids (DNAPLs). If NAPLs are mixed with water, they separate from water into a separate immiscible liquid phase. LNAPL floats on the water surface and DNAPL sinks to the bottom of the water. Generally, most of the NAPL (>90%) is in the immiscible phase and a small fraction dissolves into the water (<10%).

When natural aerobic biodegradation of NAPLs occurs in the saturated zone, indigenous aerobic bacteria react with dissolved oxygen (DO) to consume some of the immiscible-phase NAPL directly. These bacteria also release a biosurfactant that helps to increase the rate of NAPL dissolution into groundwater, enhancing their food supply.

RULES OF THUMB

1. Within the unsaturated zone, pore space contains mostly air and in situ aerobic biodegradation of NAPL is often limited by the availability of nitrogen nutrients. Addition of nitrogen, as nitrate or ammonia, usually enhances biodegradation.
2. Within the saturated zone, DO is usually the limiting factor for in situ aerobic biodegradation of NAPL.
3. For gasoline, the contaminants of regulatory interest are usually the most toxic and most soluble components, known as BTEX (benzene, toluene, ethylbenzene, and the xylene isomers).
4. It takes about 1 mg of O_2 to biodegrade 0.32 mg of BTEX.
5. The most easily biodegraded hydrocarbons are low molecular weight unbranched alkanes (smaller than C30–C40) and aromatics (smaller than C10). About 95% of their mass is converted to CO_2 and water in a few months.
6. The remainder, unbranched alkanes larger than C40 and branched alkanes, alkenes, and aromatics larger than C10, can resist degradation for many years.
7. Polar hydrocarbons containing S, O, N, Cl, or Br are often resistant to biodegradation.
8. Highly water-soluble chemicals biodegrade more readily than do those with low water solubility.
9. Chemicals that sorb weakly to soils biodegrade more readily than do those that sorb strongly. Strongly sorbed chemicals are less available to microbes.
10. Chemicals with small K_{ow} values biodegrade more readily than do those with large K_{ow} values.

RULES OF THUMB (Continued)

11. Chemicals that leach easily from soils biodegrade more readily than those that are not easily leached. These chemicals are more soluble and less strongly sorbed.
12. In the saturated zone, rates of hydrocarbon biodegradation roughly double for every 10°C increase in groundwater temperature, in the range of 5°C–25°C.

EXAMPLE 1

How long will it take to naturally biodegrade the BTEX contained in 250 kg of gasoline (an LNAPL) immobilized within the saturated zone, given the following conditions?

Depth of LNAPL zone into saturated zone = 2 m
Width of LNAPL zone in saturated zone = 10 m
Groundwater Darcy velocity = 1 m/day
Background upgradient DO concentration = 5 mg/L
Oxygen–hydrocarbon consumption ratio = 1 mg O_2/0.32 mg BTEX (from Rules of Thumb).

Assume that gasoline LNAPL is immobilized by sorption in the soil matrix and that BTEX is 25% of the LNAPL weight. Also assume that aerobic biodegradation is essentially instantaneous compared to normal groundwater movement. In the presence of excess oxygen, aerobic bacteria can degrade 1 mg/L of BTEX in about 8 days, essentially instantaneous compared to the years often required for flowing groundwater to replenish a plume area with oxygen. Under these conditions, the rate of biodegradation is equal to the rate at which sufficient DO can be brought into the residual gasoline LNAPL zone by groundwater flow. Approach the problem by calculating how long it would take enough water to pass through the plume cross-sectional area to supply 1 mg O_2 per 0.32 mg of BTEX.

Answer:

$$\text{Time to degrade} = (250\,\text{kg NAPL}) \times \left(\frac{0.25\ \text{g BTEX}}{1\ \text{g NAPL}}\right) \times (10^6\ \text{mg/kg})$$
$$\times \left(\frac{1\ \text{mg } O_2}{0.32\ \text{mg BTEX}}\right) \times \left(\frac{1\ \text{L}}{5\ \text{mg } O_2}\right) \times \left(\frac{1}{(2 \times 10)\text{m}^2}\right)$$
$$\times \left(\frac{1\ \text{day}}{1\ \text{m}}\right) \times \left(\frac{1\ \text{m}^3}{10^3\ \text{L}}\right)$$

Time to degrade = 1950 days or about 5.4 years.

This approach of calculating how rapidly DO can be supplied to the plume area does not work for DNAPL composed of chlorinated compounds because they are resistant to aerobic biodegradation. It also neglects the fact that part of the LNAPL becomes increasingly less unavailable to microbes because of stronger sorption and penetration into cracks and soil pores with time, which increases the time needed for remediation.

8.5   DETERMINING THE EXTENT OF BIOREMEDIATION OF LNAPL

Spilled fuel LNAPL, a major type of subsurface contamination, is present in the subsurface as

- Mobile LNAPL free product, which will drain into a well under gravity
- Residual LNAPL held by adsorption and capillarity in soil pore spaces, which is immobile and unable to drain into a well under gravity
- Dissolved LNAPL compounds in water
- Volatile LNAPL vapors

In an LNAPL spill, the most soluble and volatile components are lost first from the LNAPL free product by water leaching (washing and volatilization). Nevertheless, months later, the remaining LNAPL still contains about

- 90% of the benzene
- 99% of the total BTEX
- 99.9% of the TPH (total petroleum hydrocarbons)

Since most of the LNAPL has low water solubility, it seems clear that the first remediation step should be to remove physically as much LNAPL as possible. However, frequently less than 10% of the total LNAPL can be removed by recovery of mobile LNAPL. The remaining part stays trapped in the soil by sorption and capillarity. For this remaining part of the contamination, the best choice for corrective action may be bioremediation, especially if excavation is not practical. The next remediation step should be to determine if intrinsic bioremediation is occurring at a sufficient rate that no other action is required.

The U.S. Air Force Center for Environmental Excellence has published a technical document that describes a protocol for data collection and analysis that can be used for judging whether intrinsic bioremediation is occurring at a useful rate (Wiedemeier et al., 1995). This report is notable for comprehensively discussing the current state of knowledge and for its thorough list of references. Much of the following material is adapted from the Wiedemeier report.

8.5.1   USING CHEMICAL INDICATORS OF THE RATE OF INTRINSIC BIOREMEDIATION

Certain water quality parameters change because of biodegradation. By measuring how these parameters change with time and location within a contaminant plume in the saturated zone, the occurrence and rate of active biodegradation can be determined. The most important parameters that change are presented in Table 8.3 and include the pollutants being degraded, the waste products that are formed, and the redox reactants. For aerobic respiration, the electron acceptor is oxygen. Anaerobic electron acceptors include nitrate, sulfate, ferric iron, manganese, and carbon dioxide.

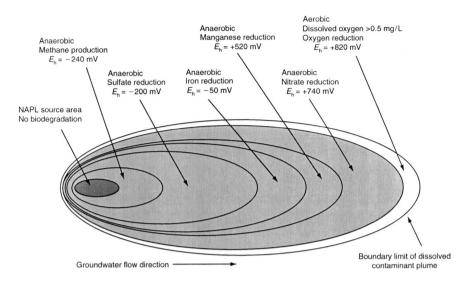

FIGURE 8.1 Idealized LNAPL plume in the saturated zone, showing aerobic (white) and anaerobic (shaded) zones of biodegradation in a dissolved NAPL plume. The distribution of different electron acceptors according to redox potential is also shown.

Figure 8.1 shows how the redox reactions associated with these electron acceptors are distributed in an idealized LNAPL plume undergoing active biodegradation. The processes that establish this distribution are

1. Aerobic biodegradation consumes available oxygen resulting in anaerobic conditions in the core of the plume and a zone of oxygen depletion along the outer margins.
2. The progressive lowering of the redox potential from the outer plume boundary toward the source zone, as preferred anaerobic electron acceptors become depleted according to Equations 8.1 through 8.6, establishes successive zones where different electron acceptors are dominant.
3. As aerobic biodegradation is relatively rapid and the rate of DO replacement is slow in the saturated zone, DO becomes depleted where LNAPL concentrations are high. Thus anaerobic redox reactions occur in most of the plume, where the abundance of anaerobic electron acceptors is large relative to DO. Since the anaerobic zone is typically more extensive than the aerobic zone, anaerobic biodegradation is usually the dominant process overall.
4. Water carrying DO diffuses into the plume from outside the plume boundary, enabling aerobic redox reactions to occur around the plume periphery.
5. It is likely that all biodegradation is inhibited in the LNAPL source area because of pollutant concentrations high enough to be toxic to microorganisms.
6. For both aerobic and anaerobic processes, the rate of contaminant degradation is controlled by the concentration of electron acceptors, not the rate that microorganisms consume the electron acceptors. As long as there is a

TABLE 8.4

Water Quality Parameters that Indicate Biodegradation Activity

Chemical Indicators in Groundwater for Biodegradation	Trend in Indicator Concentration During Biodegradation	Processes Responsible for Trend
Hydrocarbon concentrations	Decreases	Biodegradation
Dissolved oxygen	Decreases	Aerobic respiration
Nitrate	Decreases	Denitrification
Manganese (II)	Increases	Manganese (IV) reduction
Iron (II)	Increases	Iron (III) reduction
Sulfate	Decreases	Sulfate reduction
Methane	Increases	Methanogenesis
Alkalinity	Increases	Increased by aerobic respiration, denitrification, iron (III) reduction, and sulfate reduction; not affected much by methanogenesis
Oxidation/reduction Potential (pE)	Generally decreases toward plume center	Serves as a crude indicator of which redox reactions may be operating at a given time
Volatile fatty acids	Increases	Metabolic byproducts of biodegradation

sufficient supply of the electron acceptors, the rate of metabolism does not make any practical difference in the length of time required to achieve remediation objectives.

The first step in determining if biodegradation is occurring at a rate fast enough to be useful for remediation, is to look for appropriate changes in these chemical indicators (Table 8.4), which are discussed in more detail below.

8.5.2 HYDROCARBON CONTAMINANT INDICATOR

If significant LNAPL biodegradation is occurring, hydrocarbon concentrations will diminish with time and distance from the spill source. However, because this could occur due to dilution, confirmatory evidence is always required. Seek confirmatory evidence by measuring hydrocarbon concentrations in the groundwater plume close to the flow centerline and near the spill source, where dilution has less effect. Analyze the groundwater for hydrocarbon compounds of regulatory concern. These are usually BTEX and trimethylbenzenes, but total volatile hydrocarbons (TVH), total extractable hydrocarbons (TEH, also called semivolatiles), and polycyclic aromatic hydrocarbons (PAH) often should also be measured. The determination of whether biodegradation is occurring at a useful rate hinges on the rate at which these parameters are disappearing.

Based on solubilities, the highest combined dissolved concentrations of BTEX plus trimethylbenzenes should not be greater than about 30 mg/L for JP-4 jet or diesel fuel, or about 135 mg/L for gasoline. If these concentrations are exceeded, sampling errors have probably occurred, such as collecting some free product. This error is likely if emulsification of LNAPL has occurred in the water sample.

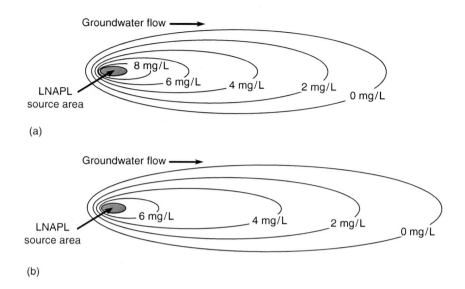

FIGURE 8.2 Idealized dissolved total BTEX isopleths (lines of equal concentration) for a groundwater plume. (a) Initial total BTEX measurement. Contour interval is 2 mg/L. Indicated isopleth values are mg/L of total BTEX. (b) Measurement of total BTEX isopleths 1 year later.

Figure 8.2a and b are idealized representations of a dissolved total BTEX groundwater plume caused by a gasoline spill, measured twice, 1 year apart. Comparing total BTEX plumes at the start of the study and 1 year later shows that, although plume area has increased a little, total mass of BTEX in the plume has decreased. Data from other chemical indicators may provide evidence that the decrease is primarily due to intrinsic bioremediation and not dilution.

8.5.3 ELECTRON ACCEPTOR INDICATORS

Subsurface microbes utilize different redox reactions in the order of decreasing energy-yielding value (from top to bottom in Figure 8.3). If oxygen is available, using it as an electron acceptor will always yield the greatest energy. Therefore, aerobic biodegradation reactions always occur first whenever sufficient DO is available. The DO level within the plume will be below background levels outside the plume wherever aerobic degradation is occurring.

Anaerobic processes will begin when DO has been depleted sufficiently, depending on which electron acceptors are available. Reduction of nitrate is the second highest energy-yielding process and occurs next if nitrate is present. However, if DO is present in concentrations greater than 0.5 mg/L, it is toxic to anaerobic-only (obligate) bacteria. Therefore, nitrate denitrification cannot begin until most of the DO has been consumed.

Electron acceptors available in the groundwater determine the sequence in which biodegradation reactions will occur. As the redox potential changes from positive to increasingly negative values, different electron acceptors are used in the

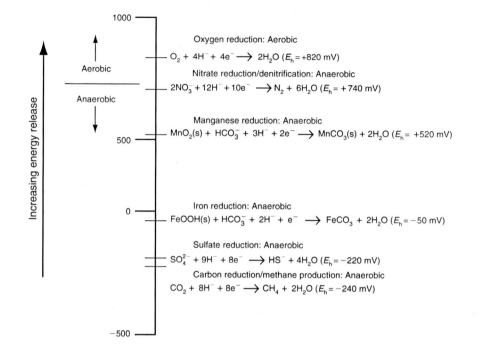

FIGURE 8.3 Order of successive microbially mediated redox reactions, is from top to bottom, from higher to lower redox potential. The more positive the redox potential is, the more energy is released per electron transferred. (Adapted from Wiedemeier, T.H., Wilson, J.T., Kampbell, D.H., Miller, R.N., and Hansen, J.E., 1995, *Technical Protocol for Implementing Intrinsic Remediation with Long-Term Monitoring for Natural Attenuation of Fuel Contamination Dissolved in Groundwater*, U.S. Air Force Center for Environmental Excellence, San Antonio, 1995.)

biodegradation process. As reduction of electron acceptors progresses, the redox potential (E_h) of the groundwater becomes increasingly negative and the energy obtained per electron transfer decreases.

Figure 8.1 shows how aerobic and anaerobic biodegradation zones develop in an LNAPL plume undergoing active biodegradation. Figure 8.3 illustrates the normal sequence of the most important biodegradation redox reactions and the potential at which they are initiated. Table 8.1 gives the stoichiometry of the biodegradation reactions. Each electron acceptor indicator is discussed in detail below.

8.5.4 DISSOLVED OXYGEN INDICATOR

If significant aerobic biodegradation is occurring, dissolved oxygen (DO) will diminish with time in the plume zone. DO will be consumed first before other electron acceptors can be used. Each 1.0 mg/L of O_2 consumed by microbes will destroy approximately 0.32 mg/L of BTEX, based only on the production of CO_2 and H_2O. This is a conservative estimate that ignores the conversion of carbon to cell mass. If cell mass production is included, each 1.0 mg/L of O_2 consumed by microbes can destroy as much as 0.97 mg/L of BTEX under ideal conditions. Because other factors such as nutrient availability affect respiration, it is best to use the more conservative 0.32 mg/L value as the amount of BTEX destroyed per 1.0 mg/L of O_2 consumed. The overall redox reaction of oxygen with benzene is

$$C_6H_6 + 7.5O_2 \rightarrow 6CO_2 + 3H_2O \qquad (8.8)$$

EXAMPLE 2

Suppose the background DO in groundwater at a remediation site is 6 mg/L. Excluding cell mass production, the groundwater conservatively has the capacity to biodegrade:

$$6 \text{ mg/L DO} \times \frac{0.32 \text{ mg/L BTEX}}{1.0 \text{ mg/L DO}} = 1.9 \text{ mg/L BTEX}$$

FIELD TEST FOR DO CONSUMPTION BY MICROBES

- Analyze groundwater for DO at different locations in the BTEX plume.
- Areas with elevated BTEX concentrations should have depleted (relative to background) or zero DO concentrations.
- This is a strong evidence for the occurrence of aerobic biodegradation.

8.5.5 NITRATE PLUS NITRITE DENITRIFICATION INDICATOR

If significant biodegradation is occurring and DO has been depleted to less than 0.5 mg/L, anaerobic reactions that utilize nitrate (NO_3^-) and nitrite (NO_2^-) as electron acceptors can commence. If nitrate/nitrite is present in the plume, their concentrations will diminish with time in the plume zone. Each 1.0 mg/L of dissolved NO_3^- will destroy about 0.21 mg/L of BTEX. The final reaction products are CO_2, H_2O, and N_2.

The denitrification steps are

$$NO_3^- \rightarrow NO_2^- \rightarrow NO \rightarrow N_2O \rightarrow NH_4^+ \rightarrow N_2 \qquad (8.9)$$

The overall redox reaction with benzene is

$$6NO_3^- + 6H^+ + C_6H_6 \rightarrow 6CO_2 + 6H_2O + 3N_2 \qquad (8.10)$$

Requirements for denitrification are

- Dissolved nitrate/nitrite
- Organic carbon
- Denitrifying bacteria
- Reducing conditions (DO < 0.5 mg/L)

Denitrification is favored when

> $6.2 < pH < 10.2$
> -200 mV $<$ redox potential (E_h) $< +665$ mV

Nitrate reduction is rapid. The rate at which nitrate and nitrite are supplied by groundwater to the reduction zone limits the reaction rate. Under denitrifying conditions, biodegradation of BTEX occurs in the following order:

> Toluene $> p$-xylene $> m$-xylene $>$ ethylbenzene $> o$-xylene $\gg$ benzene

---

FIELD TEST FOR NITRATE $\pm$ NITRITE CONSUMPTION BY MICROBES

- Analyze groundwater for nitrate plus nitrite.
- Areas with elevated BTEX concentrations should have depleted (relative to background) or zero nitrate plus nitrite concentrations.

8.5.6 METAL REDUCTION INDICATORS: MANGANESE (IV) TO MANGANESE (II) AND IRON (III) TO IRON (II)

The reduction of oxidized forms of iron and manganese (Fe^{3+} and Mn^{4+}) results in the production of reduced species that are water soluble. Elevated levels of these reduced metals (Fe^{2+} and Mn^{2+}) in the plume relative to background is indicative of anaerobic biodegradation.

If significant biodegradation is occurring and DO and nitrate/nitrite have been depleted, reduction of manganese from Mn(IV) to Mn(II) and iron from Fe(III) to Fe(II) will be initiated. Figure 8.3 shows that as the redox potential decreases, manganese will be reduced first, followed by iron. As the reduced forms of both metals are more soluble than the oxidized forms, dissolved manganese (II) and iron (II) concentrations will increase with time in the plume zone if there are available forms of manganese (IV) and iron (III) minerals to be used as electron acceptors. Aquifer sediments often contain large quantities of manganese (IV) and iron (III), frequently in the form of amorphous iron and manganese oxyhydroxides, so the reduction reaction, Equation 8.11, is a common occurrence.

$$\text{Mn(IV), Fe(III)(insoluble)} \rightarrow Mn^{2+}, Fe^{2+} \text{ (soluble)} \qquad (8.11)$$

The presence of increasing concentrations of Mn^{2+} and Fe^{2+} within the BTEX plume is a strong evidence for anaerobic biodegradation reactions.

FIELD TEST FOR METAL REDUCTION BY MICROBES

- Analyze groundwater for Mn^{2+} and Fe^{2+}.
- Areas with elevated BTEX concentrations should have elevated (relative to background) Mn^{2+} and Fe^{2+} concentrations.

8.5.7 SULFATE REDUCTION INDICATOR

After available DO, nitrate, manganese (IV), and iron (III) are consumed, available sulfate can be used as an electron acceptor. Sulfate reduction to sulfide is favored at pH 7 and E_h −200 mV. Sulfate-reducing microorganisms are sensitive to temperature, inorganic nutrients, pH, and redox potential. Small imbalances in environmental conditions can severely limit the rate of BTEX degradation via sulfate reduction. Each 1.0 mg/L of BTEX that is biodegraded requires the reduction of about 4.7 mg/L of sulfate. The overall reaction for benzene oxidation by sulfate reduction is

$$7.5H^+ + 3.75SO_4^{2-} + C_6H_6 \rightarrow 6CO_2 + 3.75H_2S + 3H_2O \qquad (8.12)$$

FIELD TEST FOR SULFATE REDUCTION BY MICROBES

- Analyze groundwater for sulfate (SO_4^{2-}), and perhaps sulfide (S^{2-}).
- Depleted sulfate concentrations and increased sulfide concentrations (relative to background) within the BTEX plume indicates active biodegradation by sulfate-reducing bacteria.

8.5.8 METHANOGENESIS (METHANE FORMATION) INDICATOR

After available DO, nitrate, manganese (IV), iron (III), and sulfate are consumed, the redox reactions of organic carbon to form CH_4 (C^{4-}) can be used to biodegrade BTEX, resulting in an increase in CH_4 concentrations in the plume zone.

Methanogenesis generates less energy for microbes than the other reducing reactions, and always occurs last, after other electron acceptors have been depleted. Methanogenesis causes the redox potential to fall below −200 mV at pH 7. The presence of elevated levels of methane in the presence of elevated levels of BTEX indicates that BTEX biodegradation is occurring as a result of methanogenesis.

Because methane is not present in LNAPL fuels, the presence of methane above background in groundwater adjoining LNAPL fuels indicates microbial degradation of fuel hydrocarbons. The overall reaction for benzene oxidation by methanogenesis is

$$C_6H_6 + 4.5H_2O \rightarrow 2.25CO_2 + 3.75CH_4 \qquad (8.13)$$

This reaction occurs in a minimum of four steps, at least one of which involves CO_2 accepting electrons and reacting with H^+ to form CH_4. In the process, C_6H_6 is oxidized to form additional CO_2. The biodegradation of 1 mg/L of BTEX by methanogenesis produces about 0.78 mg/L of methane.

FIELD TEST FOR METHANOGENESIS

- Analyze groundwater for methane (CH_4).
- High methane concentrations (relative to background) within the BTEX plume indicate active biodegradation by methanogenesis.

8.5.9   REDOX POTENTIAL AND ALKALINITY AS BIODEGRADATION INDICATORS

Groundwater redox potential and alkalinity also undergo measurable changes in regions where significant biodegradation of fuel hydrocarbons is occurring.

8.5.9.1   Using Redox Potentials to Locate Anaerobic Biodegradation within the Plume

Each successive biodegradation redox reaction involving the electron acceptors in Table 8.1 from DO to methane lowers the redox potential of the groundwater in which it occurs. Thus, a decrease with time of groundwater redox potential should serve as an indication of biodegradation activity.

FIELD TEST FOR OCCURRENCE OF REDOX REACTIONS

1. Map groundwater redox potentials at the site. Include at least one location upgradient of the plume. It is important to avoid aeration of well samples for these measurements.
2. Locations where groundwater redox potentials are lower than background are where electron acceptor species are being reduced, a sign of biodegradation.
3. Redox potentials within the plume can help to indicate which electron acceptors are active in different locations.
4. In regions of biodegradation activity, the zone of low redox potential (reducing zone) will become larger with time, as diminishing DO concentrations move farther and farther from the spill region.

8.5.9.2   Using Alkalinity to Locate Anaerobic Biodegradation within the Plume

Groundwater alkalinity increases during aerobic respiration, denitrification, iron (III) reduction, and sulfate reduction, and is unchanged during methanogenesis. The two main processes that increase alkalinity are

1. All of the biodegradation redox reactions produce carbon dioxide (see Table 8.1; Equations 8.1 through 8.7). Addition of CO_2 increases the total carbonate and alkalinity of the groundwater.

2. Redox reactions involving nitrate, manganese (IV), iron (III), and sulfate as electron acceptors all consume acidity as H^+ (see Table 8.1; Equations 8.2 through 8.5). This also increases alkalinity in groundwater where these reactions take place.

A measurement of alkalinity within a hydrocarbon plume can be used to infer the amount of petroleum hydrocarbons destroyed. For every 1 mg/L of alkalinity (as $CaCO_3$) produced, 0.13 mg/L of BTEX is destroyed (Wiedemeier et al., 1995).

FIELD TEST FOR USING ALKALINITY AS AN INDICATOR OF BIODEGRADATION

1. Map groundwater alkalinity concentrations at the site. Include at least one location upgradient of the plume.
2. Locations where groundwater alkalinity is higher than background are where CO_2 is being produced and H^+ is being consumed, which are signs of biodegradation.
3. Alkalinity levels within the plume can help to indicate the amount of petroleum hydrocarbons that have been destroyed.
4. In regions of biodegradation activity, the zone of higher alkalinity will become larger with time.
5. Increased alkalinity levels within the plume can be used to infer the extent of biodegradation occurring.

8.6   BIOREMEDIATION OF CHLORINATED DNAPLs

Chapter 7 contains a general discussion of chlorinated DNAPLs. Chlorinated solvents are among the most frequently encountered organic groundwater contaminants in the United States (Morrison, 2000; USEPA, 2000). Although these compounds have been known since before 1900, large-scale production and use of chlorinated solvents began around 1950 and continued until around 1978, when regulations were enacted controlling their production and use. The largest uses of chlorinated solvents are vapor degreasing of metal parts and dry cleaning of clothing (Morrison, 2000).

The most important process for natural biodegradation of highly chlorinated solvents such as PCE, TCE, and 1,1,1-TCA is reductive dechlorination, where a chlorine atom is removed from the solvent molecule and replaced with a hydrogen atom. The chlorine atom acquires an electron from the hydrogen atom and enters solution as a chloride anion (Wiedemeier et al., 1996, 1998). The major requirements for reductive dechlorination are the presence of electron donors to serve as a source of reducing power and metabolic energy, and a population of dehalorespiring microorganisms. The required electron acceptor is the chlorinated solvent molecule itself. Electron donors commonly found in contaminated groundwater include petroleum-derived aromatic hydrocarbons found in fuels (BTEX and the trimethyl-benzene isomers) or other types of organic carbon (landfill leachate, aliphatic fuel hydrocarbons, or natural organ carbon) (Wiedemeier et al., 1996, 1998; Wilson et al., 2001).

8.6.1 REDUCTIVE DECHLORINATION OF CHLORINATED ETHENES

There are six chlorinated ethenes:

1. Tetrachloroethene (PCE, $CCl_2 = CCl_2$)
2. Trichloroethene (TCE, $CCl_2 = CHCl$)
3. *cis*-1,2-Dichloroethene (*cis*-1,2-DCE, $CHCl = CHCl$)
4. *trans*-1,2-Dichloroethene (*trans*-1,2-DCE, $CHCl = CHCl$)
5. 1,1-Dichloroethene ($CCl_2 = CH_2$)
6. Vinyl chloride ($CHCl = CH_2$)

Starting with PCE, reductive dechlorination proceeds by sequential removal of one chlorine atom after another, replacing each with a hydrogen atom, in the following series of steps:

$$PCE \ (CCl_2 = CCl_2) \rightarrow TCE \ (CCl_2 = CHCl) \qquad (8.14)$$

$$TCE \rightarrow cis\text{-}1,2\text{-}DCE \ (CHCl = CHCl);$$
$$trans\text{-}1,2\text{-}DCE \ (CHCl = CHCl); \ 1,1\text{-}DCE \ (CCl_2 = CH_2) \qquad (8.15)$$

$$DCE \rightarrow vinyl \ chloride \ (CHCl = CH_2) \qquad (8.16)$$

$$Vinyl \ chloride \rightarrow ethene \ (CH_2 = CH_2) \qquad (8.17)$$

Equations 8.14 through 8.17 are included as part of Figure 8.4. Of the three DCE isomers, *cis*-1,2-DCE is usually the most abundant product, while *trans*-1,2-DCE and 1,1-DCE are formed in lesser amounts (Wiedemeier et al., 1996, 1998; Wilson et al., 2001). The presence of *cis*-1,2-DCE as a reaction product is an indication that reducing conditions exist. Even if aerobic conditions prevail, detection of 1,1-DCE indicates that microsites of anaerobic activity are present where reductive dechlorination is biodegrading PCE and TCE.

Each successive product in the reductive dechlorination sequence is less oxidized than its precursor and, accordingly, is less susceptible to further reduction. Consequently, the rate of reductive dechlorination decreases as the degree of chlorination decreases. For example, Equation 8.16 (the formation of vinyl chloride from the DCE isomers) requires more strongly reducing site conditions than does Equation 8.15 (the formation of the DCE isomers from TCE). The degradation of maximally chlorinated PCE to TCE, Equation 8.14, is the most rapid step in the sequence (Wiedemeier et al., 1996, 1998).

It has been observed that the degradation of *cis*-DCE is inhibited at some sites, apparently because of some combination of the following (Newell, 2001; Shim et al., 2001):

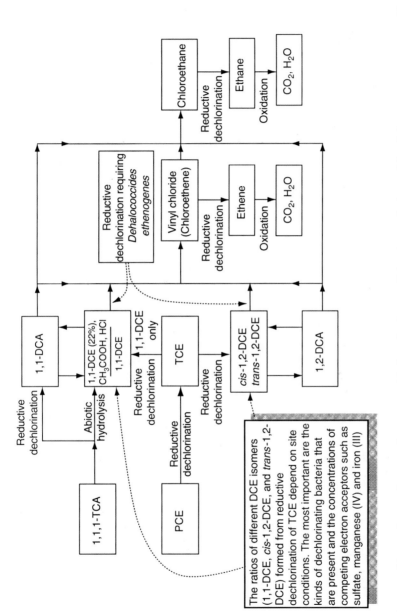

FIGURE 8.4 Decomposition pathways for several common chlorinated solvents. Solvents are designated by their common names or abbreviations. Chemical names and formulas may be found in Chapter 7 (Table 7.1).

- Particular bacteria were absent.
- Inhibiting enzymes were produced by the degradation of other organics at the site.
- Redox conditions were not sufficiently reducing.

8.6.2 REDUCTIVE DECHLORINATION OF CHLORINATED ETHANES

There are nine chlorinated ethanes:

1. Chloroethane (CA)
2. 1,1-Dichloroethane (1,1-DCA)
3. 1,2-Dichloroethane (1,2-DCA)
4. 1,1,1-Trichloroethane (1,1,1-TCA)
5. 1,1,2-Trichloroethane (1,1,2-TCA)
6. 1,1,1,2-Tetrachloroethane (1,1,1,2-TeCA)
7. 1,1,2,2-Tetrachloroethane (1,1,2,2-TeCA)
8. Pentachloroethane (PCA)
9. Hexachloroethane (HCA)

In general, both water solubility and vapor pressure decrease with increasing chlorination, while density and melting point increase. At room temperature, chloroethane is a gas, hexachloroethane a solid, and the others are liquids. Chloroethane is an LNAPL and the others are DNAPLs. All are sufficiently soluble to be a concern as water pollutants. Toxicity to aquatic life increases with increasing chlorination. All of the chlorinated ethanes are found as contaminants in soil and groundwater.

Although chlorinated ethanes (see Table 7.1) are similar to chlorinated ethenes in that they biodegrade by reductive dechlorination, two additional characteristics of chloroethanes are important to note:

1. They also degrade abiotically at environmentally significant rates, often at rates much greater than biodegradation.
2. Both biological and abiotic degradation of chloroethanes can produce chloroethenes (Wiedemeier et al., 1996, 1998).

In the case of 1,1,1-TCA, a common industrial solvent and degreaser, abiotic chemical transformation is the most prevalent environmental degradation process, producing 1,1-DCE (about 22%) by elimination reactions and acetic acid (about 78%) by hydrolysis (Smith, 1999).

8.6.3 CASE STUDY: USING BIODEGRADATION PATHWAYS FOR SOURCE IDENTIFICATION

Chlorinated contaminants found in soils and groundwater beneath and in the subsurface vicinity of an urban Superfund site included PCE, TCE, 1,1-DCE, cis-1,2-DCE, trans-1,2-DCE, and chloroethane.

From Figure 8.4, it is evident that the presence of *cis*-1,2-DCE, *trans*-1,2-DCE, 1,1-DCE, and 1,2-DCA are indicative of reductive dechlorination reactions of PCE and TCE occurring in groundwater beneath and in the subsurface vicinity of the site. However, 1,1-DCE can result from both the reductive dechlorination of PCE and TCE and the abiotic dechlorination of 1,1,1-TCA. Although 1,1,1-TCA was not detected, it could be possible that it had been present but was fully degraded, leaving only the degradation product 1,1-DCE.

The presence of chloroethane, which also is a degradation product of 1,1,1-TCA but not of PCE and TCE, is a supporting evidence that 1,1,1-TCA was also once present at the site. This result might possibly implicate a new potentially responsible party (PRP) for sharing in the remediation costs if they were known to have used 1,1,1-TCA as a solvent in their activities.

REFERENCES

Morrison, R.D., 2000, *Environmental Forensics: Principles and Applications*, CRC Press, Boca Raton, FL, p. 351.

National Research Council, 2000, *Natural Attenuation for Groundwater Remediation*, Chapter 3, The National Academy of Sciences Press, Washington, DC.

Newell, C.J., 2001, *The BIOCHLOR Natural Attenuation Model*, Powerpoint presentation, Headquarters U.S. Air Force, January 31, 2001.

Rice, D.W. et al., 1995, California leaking under-ground fuel tank (LUFT) historical case analyses, California State Water Resources Control Board.

Shim, H. et al., 2001, Aerobic degradation of mixtures of tetrachloroethylene, trichloroethylene, dichloroethylenes, and vinyl chloride by toluene-o-xylene monooxygenase of *Pseudomonas stutzeri* OX1, *Appl. Microbiol. Biotechnol.*, 56, 265–269.

Smith, J., 1999, The determination of the age of 1,1,1-trichloroethane in groundwater, in: *Conference Abstracts, Second Executive Forum on Environmental Forensics*, International Business Communications, Southboro, MA, p. 1.

USEPA, 2000, Engineered approaches to in situ bioremediation of chlorinated solvents: fundamentals and field applications, solid waste and emergency response (5102G), EPA 542-R-00–008, July 2000 (revised), http://clu-in.org/download/remed/engappinsitbio.pdf.

USEPA, 2004, How to evaluate alternative cleanup technologies for underground storage tank sites: A guide for corrective action plan reviewers, EPA 510-R-04–002, May 2004, www.epa.gov/oust/pubs/tums.htm.

Wiedemeier, T.H. et al., 1995, *Technical Protocol for Implementing Intrinsic Remediation with Long-Term Monitoring for Natural Attenuation of Fuel Contamination Dissolved in Groundwater*, U.S. Air Force Center for Environmental Excellence, San Antonio, TX.

Wiedemeier, T.H. et al., 1996, Overview of the technical protocol for natural attenuation of chlorinated aliphatic hydrocarbons in ground water under development for the U.S. Air Force Center for Environmental Excellence, in: *Symposium on Natural Attenuation of Chlorinated Organics in Ground Water*, Dallas, TX, September 11–13, EPA/540/R-96/509.

Wiedemeier, T.H. et al., 1998, Technical protocol for evaluating natural attenuation of chlorinated solvents in groundwater, USEPA, EPA/600/R-98/128.

Wilson, J.T. et al., 2001, Evaluation of the protocol for natural attenuation of Chlorinated solvents: Case Study at the Twin Cities Army Ammunition Plant, USEPA, EPA/600/R-01/025, March 2001.

9 Behavior of Radionuclides in the Water and Soil Environment

9.1 INTRODUCTION

This chapter is intended to give the nonspecialist a helpful understanding of how radionuclides behave in water and soil environments. Another purpose is to assemble information used for evaluating environmental radionuclide measurements into a form that is useful for a nonnuclear environmental professional. For example, the drinking water MCL for gross β emissions is 4 mrem/y, but laboratory results are generally given in terms of pCi/L. Tables and rules of thumb for many required conversions are found in this chapter. A third purpose, less important perhaps than the first two, is to offer a concise introduction to the basics of radioactivity and the properties of radiation. The nuclear processes of fission and fusion are not covered. Section 9.2, which comprises the introduction to nuclear structure, is not essential to using the rest of the chapter, but might help to remove some of the mystery that often surrounds a layman's perception of radionuclides and radioactivity.

9.2 RADIONUCLIDES

A radionuclide is an atom that has a radioactive nucleus. A radioactive nucleus is an atomic nucleus that emits radiation in the form of particles or photons, thereby losing mass and energy and changing its internal structure to become a different kind of nucleus, perhaps radioactive, perhaps a different element, perhaps neither. All radionuclides have finite lifetimes, ranging between billions of years to less than nanoseconds; each time a particle is emitted, the original radionuclide is transformed into a different species. The emitted particles can possess enough energy to penetrate into solid matter, altering and damaging the molecules with which they collide. Radionuclides cannot be neutralized by any chemical or physical treatment; they can only be confined and shielded until their activity dies to a negligible level. Radionuclides are unique in being the only pollutants that can act at a distance, harming life forms and the environment without physical contact.

This section addresses two basic questions about atomic nuclei:

- What are atomic nuclei made of?
- Why are some nuclei radioactive and some not radioactive?

The next sections discuss the conditions under which radiation is a hazard to the environment and to human health, and offer a guide through the "maze" of different units (becquerels, curies, rads, rems, and more) used to measure the amount of radiation and dose received.

Finally, the behavior of radionuclides in the environment is discussed. This behavior is mainly dependent on their chemical properties rather than their nuclear properties.

9.2.1 A Few Basic Principles of Chemistry

9.2.1.1 Matter and Atoms

All matter is composed of atoms. There are about 112 kinds of atoms, each with different chemical and physical properties. These are called the elements and comprise the entries in the periodic table, reproduced on the inside front cover of this book. For a useful working definition of an atom, regard it as the smallest bit of matter that can be identified as one of the elements by measuring its properties. Atoms can combine to form larger units of two or more atoms called molecules. A molecule is the smallest bit of matter that is recognizable as any chemical substance other than an element. Molecules can assemble into the still larger entities that make up the world we perceive with our normal five senses.

Atoms themselves have a substructure; they are assembled from subatomic articles called protons, electrons, and neutrons. This is the deepest level of subdivision needed for interpreting chemical behavior. However, a description of nuclear structure and radioactivity requires that we consider the next deeper level of substructures where still smaller units of matter, called quarks and leptons, combine to make protons, electrons, and neutrons. These are discussed in Section 9.2.4.

The structure of an atom is determined by the properties of their component parts, the protons, electrons, and neutrons.

- An electron carries a single negative charge and has insignificant mass (about 9.109×10^{-28} g) compared to a proton or neutron.
- A proton carries a single positive charge and is about 1800 times heavier than an electron (about 1.673×10^{-24} g).
- A neutron has no electric charge and a mass nearly the same as a proton, but just a little heavier (1.675×10^{-24} g).
- The electrical charges on a proton and an electron are equal in magnitude but opposite in sign. One positive charge can attract and neutralize one negative charge, resulting in zero net charge.

Every atom has a small positively charged nucleus in its center containing both protons and neutrons, which electrically attracts electrons until it is surrounded by a

Two protons give the nucleus a charge of +2

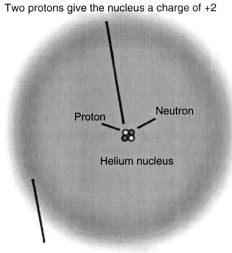

Two electrons form a spherical cloudlike orbital
around the nucleus with a charge of −2

FIGURE 9.1 Representation of a helium atom with two protons and two neutrons in its nucleus. The two positive charges of the protons attract and hold two negatively charged electrons depicted as a spherical cloud of negative charge surrounding the nucleus. In an actual helium atom, the diameter of the electron cloud is approximately 100,000 times larger than the diameter of the nucleus.

"cloud" of electrons (see Figure 9.1). The electrons are not drawn into the nucleus itself because of short-range repulsive forces. The number of electrons in the cloud is equal to the number of positive charges in the nucleus, so that the atom is electrically neutral overall.* The nucleus contains essentially all the mass of the atom in its protons and neutrons, while the electron cloud is virtually weightless by comparison.

9.2.1.2 Elements

Originally, a chemical element was defined as a substance that cannot be decomposed by chemical means into simpler substances. The test was whether any of its chemical properties could be changed by chemical decomposition processes. Elements identified by such tests could be combined into new substances called compounds having new properties, but decomposition of the compounds always brought back the original set of starting elements. However, there were many different elements, each with a unique set of properties. There had to be reasons for the differences among elements.

Eventually, in the early 1900s, the internal nuclear structure of atoms was revealed and the properties of elements were shown to depend on the number of protons in the nucleus. Each different element has a different number of protons in its nucleus. The periodic table arranges the elements from left to right in rows, so that each successive element contains one more proton in its nucleus than the preceding element, beginning with hydrogen, which has one proton. Each element is numbered with an atomic number equal to the number of protons in its nucleus and each element has a unique set of chemical and physical properties.

For example, the element carbon, with atomic number 6, has 6 protons in its nucleus and the element nitrogen, with atomic number 7, has 7 protons in its nucleus. The carbon nucleus can attract 6 electrons to itself before the atom becomes neutral and does not attract additional electrons. The electrical forces within the atom hold the electrons to the vicinity of the nucleus and, because the electrons repel one another, the 6 carbon electrons become distributed around the nucleus in a pattern unique to carbon atoms. The electron pattern around a nitrogen atom with 7 protons in its nucleus and 7 electrons distributed around it is unlike that of carbon.

One of the most obvious differences between carbon and nitrogen is that a large quantity of carbon atoms forms a solid at room temperature whereas a large quantity of nitrogen atoms forms a gas. The reasons for this have to do with their different electron distributions. When atoms come near one another, they interact with their electron clouds; the nuclei remain separated by relatively large distances. The attractions that form different compounds or cause a substance to be a gas, liquid, or solid at room temperature depend on the nature of the electron distributions around the interacting atoms.

In summary, the chemical properties of an element are primarily determined by the number of electrons it contains, and the number of electrons is equal to the number of positively charged protons in the nucleus.

9.2.2 PROPERTIES OF AN ATOMIC NUCLEUS

We have seen that an element is defined by the number of protons in its nucleus. What about the neutrons in the nucleus, what do they do? Since neutrons are not charged, they cannot attract or repel electrons and, therefore, do not affect the number of electrons around the nucleus. Neutrons in the nucleus of an element do not influence the chemical properties of the element. Thus, two different nuclei with the same number of protons but different numbers of neutrons have the same chemical properties and are the same element. However, they differ in their masses because of their different number of neutrons. Such atoms are called different isotopes of the same element. We will see that neutrons are needed to hold the protons together in a nucleus, against the repulsive forces between the positive electrical charges of protons.

Most of the naturally occurring elements are mixtures of several isotopes. The term nuclide refers to the nucleus of a particular isotope. Collectively, all the isotopes of all the elements form the set of nuclides. The distinction between the terms isotope and nuclide is somewhat blurred, and they are often used interchangeably. Isotope is best used when referring to several different nuclides of the same element and when the chemistry of the element is of interest as well as its isotope-specific nuclear properties. Nuclide is more generic and is used when referencing only one nucleus or several nuclei of different elements and the emphasis is mainly on nuclear properties.

9.2.2.1 Nuclear Notation

- The number of protons in a nucleus is called either the atomic number or the proton number and is designated by Z.
- The number of neutrons in a nucleus is called the neutron number and is designated by N.
- The sum of protons and neutrons in a nucleus is called the mass number and is designated by A.

The symbolic representation of the nucleus of an element is $^A_Z X$, where X is the chemical symbol of the element. The number of neutrons in X is found from $N = A - Z$. Note that there is some redundancy in this notation because only Z or X is needed to define an element, but not both. For this reason, Z is sometimes omitted and the nucleus may be written $^A X$. If needed, Z can be obtained from the periodic table.

EXAMPLES

$^4_2 He$ is the nucleus of the most abundant isotope of the element helium (He), with 2 protons and 2 neutrons ($N = A - Z = 4 - 2 = 2$).

$^{56}_{26} Fe$ is the nucleus of the most abundant* isotope of the element iron (Fe), with 26 protons and 30 neutrons ($N = A - Z = 56 - 26 = 30$).

$^{58}_{26} Fe$ is the nucleus of a less abundant isotope of the element iron (Fe), with 26 protons (Fe) and 32 neutrons ($N = A - Z = 58 - 26 = 30$).

RULES OF THUMB

1. All nuclei are composed of two types of particles: protons and neutrons.
2. The number of electrons in an atom equals the number of protons in its nucleus, making the atom electrically neutral.
3. The atomic number (or proton number), Z, equals the number of protons in the nucleus.
4. The neutron number, N, equals the number of neutrons in the nucleus.
5. The mass number, A, equals the total number of nucleons (protons plus neutrons) in the nucleus.
6. The nuclei of all atoms of a particular element must have the same number of protons but can contain different numbers of neutrons.
7. Isotopes are different forms of the same element, having the same number of protons (same atomic number: Z) but different numbers of neutrons.

---

* Percent natural abundance $= \dfrac{\text{number of atoms of a given isotope}}{\text{number of of all isotopes of that element}} \times 100\%$. The sample being measured must be a naturally occurring sample of the element as found on Earth. Natural abundances can vary over a wide range. For example, the natural abundances of the stable isotopes of oxygen are 99.759% for $^{16}_8 O$ (oxygen-16), 0.037% for $^{17}_8 O$ (oxygen-17), and 0.204% for $^{18}_8 O$ (oxygen-18).

9.2.3 Isotopes

Different isotopes of the same element have different mass numbers ($A = Z + N$) because they have the same number of protons but different numbers of neutrons. A stable isotope is one that does not spontaneously decompose into a different nuclide. With two exceptions, hydrogen-1 (^1_1H) and helium-3 (^3_2He), the number of neutrons is equal or greater than the number of protons in the stable nuclides.

For convenience, when comparing relative atomic masses, as when analyzing mass spectral data, an atomic mass unit, amu or u, is defined to be exactly one-twelfth of the mass of a single atom of the most abundant isotope of carbon, $^{12}_6C$. This definition is used because it results in the most precise mass spectrometer determination of the relative masses of other isotopes. Since carbon-12 contains 6 protons, 6 neutrons, and 6 electrons, the definition of an amu implies that protons and neutrons are considered to be of equal mass, 1 amu each, while the mass of the electrons is neglected. The mass of any nucleus is equal to its mass number, $A = N + P$, in amu.

There are about 112 different elements, while the number of different isotopes identified so far is about 3000, of which only about 265 are stable.* Clearly, most elements are a mixture of several isotopes. Most elements with proton numbers between 1 and 82 have at least two stable isotopes, a few have only one, and there are others with more than two (tin, e.g., has 10 stable isotopes). All isotopes with proton numbers greater than 82 are unstable and radioactive. If the number of neutrons N is plotted as a function of the number of protons Z in the nuclei of each of the approximately 266 stable isotopes, Figure 9.2 results. Thousands of unstable (radioactive) isotopes are not included in Figure 9.2.

Careful examination of Figure 9.2 suggests that the stability of a nucleus is dependent on the neutron to proton ratio (N/Z) in the nucleus. Figure 9.2 also reveals some interesting relationships between the numbers of protons and neutrons in a stable nucleus and the abundance of the corresponding isotope. This data has been used to develop theoretical models for the internal structure of nuclides.

1. There is a zone of stability within which all stable nuclei lie. If a nucleus has an N/Z ratio too large or too small and falls outside the stable zone, it will be unstable and radioactive.
2. For the lighter stable elements, from $Z = 1$ (hydrogen, ^1_1H) to about $Z = 20$ (calcium, $^{40}_{20}Ca$), the number of neutrons in the most abundant isotope is approximately equal to the number of protons, i.e., the slope of a best-fit line through the lighter stable isotopes is close to unity.

---

* The number of identified stable isotopes depends on how stability is defined, because experimental methods are currently capable of measuring radioactive decay half-lives as long as 10^{19} years, which is about a billion times longer than the current estimated age of the universe, 13.7×10^9 years. Several isotopes once thought to be completely stable have been shown in recent years to be slightly radioactive with very long half-lives. An example is $^{209}_{83}Bi$ (bismuth-209), traditionally regarded as the element with the heaviest stable isotope. However in 2003, bismuth-209 was shown to be an α emitter with a half-life of 19×10^{18} years (Marcillac et al., 2003). Although such long-lived isotopes may be regarded as stable for any practical purpose, their instability is of great theoretical interest. Bismuth-209 and several other isotopes with comparably long half-lives are often treated as stable and are still included in Figure 9.2.

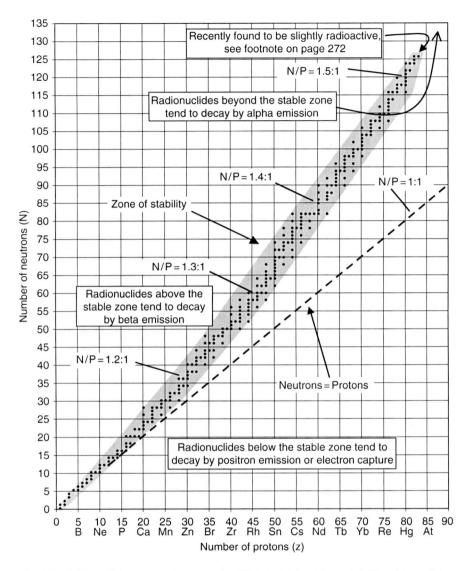

FIGURE 9.2 Plot of the neutron/proton ratio (N/Z) for 266 stable nuclei. Unstable nuclei are not shown. For nuclei with 20 or fewer protons, stable nuclei have N/Z close to unity. As Z increases beyond 20, nuclei require an increasing N/Z ratio to be stable. There are no stable nuclei with more than 82 protons (see footnote on page 272). For nuclei with 82 or fewer protons, the envelope of the dots (shaded area) represents a zone of stability. Nuclei with an N/Z ratio either too large or too small to lie within the zone of stability are unstable (radioactive).

3. For stable elements heavier than calcium, a best-fit line bends noticeably upward, away from the $N = Z$ line. As the number of protons increases, the ratio of neutrons to protons needed to produce a stable nucleus also increases, to a maximum of about 1.5 to 1.

4. The maximum number of protons in a stable nucleus appears to be 82 (three stable isotopes of lead, $^{206}_{82}Pb$, $^{207}_{82}Pb$, $^{208}_{82}Pb$, but see footnote on page 272). All nuclei with 83 or more protons are unstable (radioactive).

5. If we classify all nuclides by whether their numbers of protons and neutrons are even or odd, four groups are evident:

    a. Even Z and even N (e.g., $^{32}_{16}S$, $^{12}_{6}C$); this group contains more than half of all stable nuclides.

    b. Odd Z and odd N (e.g., $^{14}_{7}N$, $^{2}_{1}H$); this group contains the fewest stable nuclides, and $P = N$ in all of them.

    c. Even Z and odd N (e.g., $^{13}_{6}C$, $^{67}_{30}Zn$); this group contains about 20% of the stable nuclides.

    d. Odd Z and even N (e.g., $^{19}_{9}F$, $^{63}_{29}Cu$); this group contains about 16% of the stable nuclides.

6. The natural abundances of isotopes (see footnote on page 272) can vary over a wide range. Nuclides with an even number of protons, neutrons, or both are the most abundant, indicating that even numbers of nucleons impart an increased probability of formation.

7. Certain numbers of protons and neutrons are especially favored combinations for forming a nuclide. These numbers, all even, are called magic numbers. They are 2, 8, 20, 26, 28, 50, 82, and 126. The analogy between the nucleon magic numbers and the unusual stability of elements with filled electron shells (the noble gases with 2, 10, 18, 36, 54, and 86 electrons) has led to the development of theories of a nuclear shell structure for nuclides. The most abundant nuclides have Z or N numbers that correspond to the magic numbers. Nuclei that have both Z and N equal to one of the magic numbers are called "doubly magic", and are especially abundant. Some examples of doubly magic isotopes are helium-4 ($^{4}_{2}He$), oxygen-16 ($^{16}_{8}O$), calcium-40 ($^{40}_{20}Ca$), calcium-48 ($^{48}_{20}Ca$), tin-100 ($^{100}_{50}Sn$), and lead-208 ($^{208}_{82}Pb$). Helium-4 and oxygen-16 are the second and third most abundant isotopes in the universe, after hydrogen-1 ($^{1}_{1}H$).

8. The zone of stability seen in Figure 9.2 contains all of the stable nuclides. However, some nuclides that lie within the stable zone are not stable and these all have an odd number of protons, an odd number of neutrons, or both. Examples are technetium (Tc, $Z = 43$) and promethium (Pm, $Z = 61$), which have no stable isotopes. There also are nuclides like argon and potassium that have both stable and unstable nuclides within the zone of stability. Argon ($Z = 18$) has an even number of protons and has stable isotopes with even numbers of neutrons (Ar-36, Ar-38, and Ar-40), but its isotopes with odd numbers of neutrons (Ar-37 and Ar-39) are unstable. Potassium ($Z = 19$) has an odd number of protons and has stable isotopes with an even number of neutrons (K-39 and K-41), but its isotopes with odd numbers of neutrons (K-38, K-40, and K-42) are unstable. Two other potassium isotopes with an even number of neutrons (K-37 and K-45), which nevertheless are unstable, lie just outside the zone of stability.

9.2.4 Nuclear Forces

Why do nucleons stay assembled together in a nucleus at all? The existence of both stable and radioactive nuclides is evidence that sometimes they do and sometimes they do not. Some nuclides appear to be completely stable, some have very long half-lives (i.e., hold together for long periods of time; thousands to billions of years), and some have very short half-lives (hold together for very short periods of time; days to fractions of a second).

Protons are packed so closely together in an atomic nucleus that the coulombic repulsive force between them, which varies inversely with the square of the distance between charges of the same sign, is very strong. Another force must be present that is attractive and strong enough to hold the nucleus together. This very strong force is called the nuclear force. The nuclear force is a strong attraction between all nucleons, whether they are protons or neutrons. It is neither electrical nor gravitational in nature, is always attractive, and is a short-range force, acting only over very small distances (about 10^{-13} cm). When protons or neutrons are within about 10^{-13} cm of each other, the nuclear force binds them together strongly, overcoming the electrostatic repulsion between protons.

Nuclear forces have the following important properties:

1. They are extremely strong, much stronger than gravitational or electrical forces.
2. They have a very short range, about 10^{-13} cm and become saturated; one nucleon can only exert the nuclear force on a limited number of other nucleons.
3. They are always attractive and are charge independent; for nucleons within the 10^{-13} cm effective range, the force is just as strong between two neutrons, two protons, or a proton and a neutron. However, although two neutrons or a proton and a neutron can only attract each other because they experience only nuclear and not coulombic forces, two protons also have a coulombic repulsion, which can negate the attraction of their nuclear force under certain conditions.
4. Although nuclear forces are much stronger than electrostatic forces at very small distances, electrostatic forces are effective over much longer distances. In the case of two protons alone, the nuclear force does not hold them together against their coulombic repulsion. There are no stable nuclei consisting of two or more protons with no neutrons.
5. Because each neutron in a nucleus adds additional forces of attraction to every nucleon within its attractive range without adding any electrostatic repulsion, their presence in the nucleus is very important for holding the protons together. Add one neutron to the unstable two-proton nucleus of item 4 above and the stable helium-3 nuclide results, although in very low natural abundance ($1.3 \times 10^{-4}\%$). Add two neutrons and the very stable helium-4 nuclide result, with almost 100% abundance.
6. The fact that electrostatic proton–proton repulsive forces are long range and influence all the protons in the nucleus, while nuclear forces are short-range and saturate with only a few nucleons, gives rise to the important

observation that as the number of protons in a nucleus becomes greater, a relatively greater number of neutrons are needed to stabilize the nucleus. This can be seen in Figure 9.2, where the zone of stability curves upward with increasing proton number.

RULES OF THUMB

1. Light nuclei (up to about $Z = 20$, $^{40}_{20}Ca$) are stable with approximately an equal number of protons and neutrons.
2. Heavier nuclei require more neutrons than protons to be stable because the attractive nuclear force is short range and saturates, while repulsive coulombic force between protons is long range and does not saturate. As the number of protons increases, the coulombic repulsion increases rapidly and more and more neutrons are needed to hold the nucleus together.
3. At $Z = 83$, the repulsive force of 83 protons cannot be negated by adding more neutrons. All nuclei with $Z = 83$ or greater are unstable (radioactive). The maximum number of protons in a stable nucleus appears to be 82. The nuclide $^{208}_{82}Pb$ has the distinction of being the stable nuclide with the largest mass number and the largest atomic number. All nuclei with $Z \geq 83$ or $N \geq 126$ are unstable (radioactive).
4. All elements with atomic numbers between 83 (bismuth) and 92 (uranium) are naturally occurring unstable radionuclides (on Earth).
5. There are three causes of radioactivity related to the neutron/proton ratio in an atomic nucleus:
   a. There are more than 82 protons or more than 126 neutrons in the nucleus.
   b. There are 82 or fewer protons in the nucleus but the neutron/proton ratio is too low or too high to lie within the zone of stability.
   c. A few isotopes have neutron/proton ratios within the zone of stability but are unstable because they contain odd numbers of both neutrons and protons.

9.2.5 QUARKS, LEPTONS, AND GLUONS

For about 30 years after their discovery in the early 1900s, protons and neutrons, along with electrons, were believed to be the fundamental particles of matter. However, studies of radioactivity and high-energy particle physics soon revealed that matter could be subdivided still further. Two early observations started the search for an inner structure of protons and neutrons.

1. Certain radioactive nuclides were observed to emit negatively charged particles identical to electrons, called β particles. How could negative particles come from a nucleus consisting of protons and neutrons?

2. Neutrons in the free state, after being emitted from nuclei in nuclear reactions, are not stable. They decay into electrons and protons with a half-life of about 13 min.

The emission of an electron from a nucleus effectively changes a nuclear neutron into a proton, making the nucleus more positive by one charge unit. This actually adds a proton to the nucleus, increasing its atomic number by one unit and not changing its mass number. For a time, it was proposed that the neutron was actually a combination of an electron and a proton. However, further studies revealed greater complexity.

There appear to be three families of more fundamental particles, called quarks (six different kinds), leptons (six different kinds), and force carriers (the photon is the force carrier particle that carries electromagnetic forces). The strong nuclear force is carried by a force carrier particle called a gluon. The electron is the most familiar lepton. Protons and neutrons contain three quarks each, of just two different kinds (named up quark, or u, and down quark, or d). The three quarks in a neutron are a udd combination and the three quarks in a proton are a uud combination. When a neutron decays, one of its down quarks is transformed into an up quark. In this process, the neutron becomes a proton and conservation of charge and momentum is preserved by the creation of an electron and an antineutrino.

9.2.6 RADIOACTIVITY

An unstable nuclide cannot hold all of its nucleons together indefinitely. Eventually, the nucleus will change its internal structure by losing energy in the form of high-energy photon radiation or losing mass and energy by releasing one or more nucleons as energetic particles. Its new structure may or may not be stable; if not, the release of photons and nucleons will continue. This process is called radioactivity or radioactive decay. The photons and nucleons released are collectively called radiation or emissions. Radioactivity is the result of an unstable nucleus rearranging its nucleons by emitting radiation, a process that continues until a stable nuclear configuration is achieved.

The most common forms of radionuclide emissions are named after the first three letters of the Greek alphabet—alpha (α), beta (β), and gamma (γ).* These emissions accompany the process of transmutation, a nuclear reaction in which an unstable isotope of one element, called the parent isotope, is transformed into an isotope of

---

* Other less common forms of radioactivity are not regulated by EPA, because they normally are not hazards to health or the environment. They include

- Electron capture, where a parent nucleus captures one of its orbital electrons and emits a neutrino. This converts a nuclear proton to a neutron and lowers the nuclide's atomic number by one unit.
- Positron emission, also called positive β decay, where a positron (antielectron) is emitted. Positron emission has the same effect on the nucleus as electron capture, decreasing the atomic number by one unit.
- Internal conversion, where electric fields within the nucleus interact with orbital electrons, resulting in the ejection of an orbital electron from an outer shell of the atom.

a different element (possibly still radioactive, possibly not), called a daughter isotope.

All nuclides with $Z > 82$ (Pb) are radioactive and most of these undergo α decay, which is the radiation that lowers the value of Z most efficiently. Radioactive nuclides with $Z \leq 82$ are those with a neutron/proton ratio that does not fall within the zone of stability (Figure 9.2) or, in a few cases, nuclides with an odd number of both neutrons and protons. Isotopes with too many neutrons decay by β emission, which converts a neutron into a proton, decreasing the N/Z ratio. Isotopes with too few neutrons decay by positron emission or electron capture, which changes a proton into a neutron, increasing the N/Z ratio.

9.2.6.1 α Emission

α Emission is the ejection of an α particle, which is a nuclide unit consisting of two protons and two neutrons, from a nucleus. After emitting an α particle, the nucleus lowers its atomic number by two units and its mass number by four units. For example, the nuclide uranium-238, $^{238}_{92}U$, becomes thorium-234, $^{234}_{90}Th$.

An α particle is identical to a helium-4 nucleus, $^{4}_{2}He$, and will become a helium atom when it comes to rest and acquires two electrons from its surroundings. It carries two positive charges and has a mass number of four, the largest and heaviest particle emitted by natural radioactivity. The fact that these four nucleons are emitted as a single unit from a radioactive nucleus testifies to the unusual stability of the combination of two protons and two neutrons.

9.2.6.2 β Emission

β Emission is the ejection of a β particle (an electron) and an antineutrino from a nucleus. β Decay changes a neutron into a proton. The term "beta particle" is an historical term used in the early description of radioactivity. A nucleus that has too many neutrons can decrease the N/Z ratio by emitting an electron in β decay.

9.2.6.3 γ Emission

γ Emission usually occurs after a prior emission of an α or β particle leaves the nucleus in an excited energy state. It can then relax to the more stable ground state by emitting a high-energy γ photon (see Table 9.1). γ Radiation is the highest energy form of electromagnetic radiation because it results from transitions between widely spaced nuclear energy levels. Next on the electromagnetic energy scale are x-rays and ultraviolet light, which result from transitions between more closely spaced energy levels of the orbital electrons.

9.2.7   BALANCING NUCLEAR EQUATIONS

A nuclear equation describes the nuclear changes that occur because of radioactivity. The examples in Table 9.1 are all balanced nuclear equations. The rules for balancing nuclear equations are very simple:

TABLE 9.1
Nuclear Changes Caused by Radioactive Emissions

Type of Emission	Symbol	Mass Number (A) Change	Atomic Number (Z) Change	Example
Alpha, α	^4_2He	Decreases by 4	Decreases by 2	$^{226}_{88}\text{Ra} \rightarrow {}^{222}_{86}\text{Rn} + {}^4_2\text{He}$
Beta, β	$^0_{-1}\text{e}$	No change	Increases by 1	$^{14}_6\text{C} \rightarrow {}^{14}_7\text{N} + {}^0_{-1}\text{e}$
Gamma, γ	$^0_0\gamma$	No change	No change	$^{12}_5\text{B} \rightarrow {}^{12}_6\text{C}^* + {}^0_{-1}\text{e}$, followed by $^{12}_6\text{C}^* \rightarrow {}^{12}_6\text{C} + {}^0_0\gamma$
Positron	$^0_{+1}\text{e}$	No change	Decreases by 1	$^{38}_{19}\text{K} \rightarrow {}^{38}_{18}\text{Ar} + {}^0_{+1}\text{e}$
Electron capture	EC	No change	Decreases by 1	$^{123}_{52}\text{Te} + {}^0_{-1}\text{e} \rightarrow {}^{123}_{51}\text{Sb}$
Neutron	^1_0n	Decreases by 1	No change	$^{13}_4\text{Be} \rightarrow {}^{12}_4\text{Be} + {}^1_0\text{n}$

Notes:
1. C* signifies a carbon nucleus in an excited energy state.
2. The principles of balancing nuclear equations, as in the Example column, are explained in Section 9.2.7.
3. Electron capture creates a vacancy in the inner orbital electron orbital, leaving the orbital electrons in an excited energy state. As orbital electrons cascade downward to return to the ground energy state, electromagnetic photons, including x-rays, are emitted.

1. In a nuclear equation, the sum of the mass numbers (A) on both sides of the equation must be equal, to establish conservation of mass.
2. The sum of the charge numbers for nuclides, electrons, neutrons, and gammas must be equal on both sides of the equation to establish conservation of charge. Note that the atomic number (Z) is actually the nuclear charge number, so that balancing the atomic numbers is the same as balancing the nuclear charges.
3. For each particle in the equation, the chemical symbol (X), mass number (A), and atomic number (Z) are used in the form $^A_Z X$. For non-nuclides like electrons, protons, neutrons, gammas, etc., the charge number is the same as the atomic number for nuclides and is the lower left subscript.

EXAMPLE 1

Write the balanced equation for a nuclear reaction in which uranium-238 emits an α particle to form thorium-234.

Answer:

Uranium has atomic number 92 and thorium has atomic number 90. Therefore,
$$^{238}_{92}\text{U} \rightarrow {}^{234}_{90}\text{Th} + {}^4_2\text{He}$$
The mass numbers balance: 238 on the left $= 234 + 4$ on the right.
The charge numbers balance: 92 on the left $= 90 + 2$ on the right.

EXAMPLE 2

When nitrogen-14 in the upper atmosphere absorbs a neutron that enters the atmosphere from outer space, the nuclear reaction forms carbon-14. What other particle must also be formed?

Answer:

$$^{14}_{7}N + ^{1}_{0}n \rightarrow ^{14}_{6}C + ?$$

The unknown particle must have a mass number of 1 so that the mass numbers on both sides of the equation are equal to 15, and a charge number of 1 so that the charge numbers on both sides of the equation are equal to 7. The particle with a mass number of 1 and a charge number of 1 is the proton, written $^{1}_{1}H$ (or $^{1}_{1}p$). The balanced equation is

$$^{14}_{7}N + ^{1}_{0}n \rightarrow ^{14}_{6}C + ^{1}_{1}H$$

EXAMPLE 3

The carbon-14 formed in Example 2 is radioactive, emitting a β particle. Its use in carbon age dating is discussed in Example 7. Write the balanced equation that identifies the product nuclide of carbon-14 decay.

Answer:

$$^{14}_{6}C \rightarrow ^{14}_{7}? + ^{0}_{-1}e$$

The unknown product must have an atomic number of 7, so that $-1 + 7$ (on the right) $= 6$ (on the left). The element with atomic number 7 is nitrogen and the balanced equation is

$$^{14}_{6}C \rightarrow ^{14}_{7}N + ^{0}_{-1}e$$

EXAMPLE 4

Complete and balance the following nuclear reaction, which could occur in a particle accelerator.

$$^{250}_{98}Cf + ^{11}_{5}B \rightarrow ? + 5\,^{1}_{0}n$$

Answer:

The sum of mass numbers on the right must equal 261, to be the same as the sum of mass numbers on the left. The sum of charge numbers on the right must equal 103, to be the same as the sum of charge numbers on the left. Since the 5 neutrons on the right have a total mass number of 5 and a total charge number of 0, the unknown nuclide must have a mass number of 256 and a charge, or atomic, number of 103. The element with an atomic number of 103 is Lawrencium, so that the unknown nuclide is $^{256}_{103}Lr$. The balanced equation is

$$^{250}_{98}Cf + ^{11}_{5}B \rightarrow ^{256}_{103}Lr + 5\,^{1}_{0}n$$

EXAMPLE 5

Write a nuclear reaction for the α decay of polonium-210.

Answer:

The charge number is 84 for polonium and 2 for an α particle ($^4_2$He). The mass number is 210 for the polonium isotope and 4 for an α particle. Therefore, the other product nucleus has a mass number of $210 - 4 = 206$ and a charge number of $84 - 2 = 82$. The product is lead-206. The balanced equation is

$$^{210}_{84}\text{Po} \rightarrow {}^{206}_{82}\text{Pb} + {}^4_2\text{He}$$

9.2.8 RATES OF RADIOACTIVE DECAY

The rate of radioactive decay of a nuclide can only be determined by counting the emitted particles. Decay rates are most easily measured with pure samples, to insure that only one kind of emitting nuclide is present. However, even a sample that begins pure can become contaminated by radioactive daughter products. When more than one kind of emitter is present, their total emissions have to be sorted out by identifying the different kinds of particles, their characteristic energies, and their different rates of emission.

In the discussion that follows, it is assumed that a pure nuclide is the source of all emissions. Under these conditions, the rate of radioactive decay is the number of disintegrations per unit time, which is equal to the decrease in the number of parent radioactive nuclei per unit time.

Radioactive decay follows a first-order rate law, which means that the rate of decay of a given radionuclide at any time is directly proportional to the number of radioactive nuclei remaining at that time. Mathematically, the rate equation is written

$$\text{Rate} = \frac{d\mathcal{N}}{dt} = -k\mathcal{N} \tag{9.1}$$

where

$\mathcal{N}$ = the number of radioactive nuclei present
t = time
k = the proportionality constant, known as the decay rate constant. The minus sign indicates that N decreases with time

Integrating the rate equation gives

$$\ln\left(\frac{\mathcal{N}}{\mathcal{N}_0}\right) = -kt \tag{9.2}$$

where

$\mathcal{N}_0$ = initial number of radioactive nuclei at the start of a measurement
$\mathcal{N}$ = the number of radioactive nuclei remaining after a time t

9.2.8.1 Half-Life

For radioactive decay, it is usual to express the rate in terms of the half-life. The half-life is the time required for ½ of the radioactive nuclei initially present at any time to undergo disintegration.

For example, if there were 10,000 radioactive nuclei present in a sample at the start of a measurement, the half-life is the time it takes for the sample to decay until there are only 5,000 of the original radioactive nuclei remaining. The integrated equation, $\ln\left(\frac{N}{N_0}\right) = -kt$, can be used to derive a simple expression for the rate constant in terms of half-lives. Let $t_{1/2}$ be the time required for $1/2$ of the initial nuclei to decay, i.e., the half-life. When one half-life of the nuclide being measured has elapsed, the number of remaining nuclei, N, will equal $1/2$ of the initial number of nuclei present at the start of the measurement, N_0, or $N = (1/2)N_0$.

The integrated Equation 9.2 then becomes, $\ln\left(\frac{1/2N_0}{N_0}\right) = \ln \frac{1}{2} = -0.693 = -kt_{1/2}$, so that

$$k = \frac{0.693}{t_{1/2}} \tag{9.3}$$

and

$$\ln\left(\frac{N}{N_0}\right) = -\left(\frac{0.693}{t_{1/2}}\right)t \tag{9.4}$$

The use of Equations 9.3 and 9.4 is shown in the examples that follow.

EXAMPLE 6

Measurements on a sample containing iodine-131 indicate an initial activity of 3153 disintegrations per minute (dpm). After 52.5 h, the activity has fallen to 2613 dpm. What is the half-life of $^{131}_{53}I$? The activity of a sample is directly proportional to N_0, the number of radioactive nuclei present.

Answer:

Insert the values into Equation 9.3:

$$\ln\left(\frac{N}{N_0}\right) = -\left(\frac{0.693}{t_{1/2}}\right)t$$

$$\ln\left(\frac{2613}{3153}\right) = \ln(0.8287) = -\left(\frac{0.693}{t_{1/2}}\right)(52.5\ h)$$

$$t_{1/2} = -\frac{0.693 \times 52.5\ h}{\ln(0.8287)} = 194\ h$$

EXAMPLE 7

$^{14}_{6}C$ is a β emitter with a half-life of 5730 years. It is formed when $^{14}_{7}N$ in the upper atmosphere absorbs a neutron from outer space. Radioactive carbon dating assumes that formation and decay of carbon-14 are in equilibrium and that the concentration of carbon-14 in the atmosphere has been constant, proportional to 15.3 dpm, for the past several hundred thousand years. Plants and animals incorporate carbon from atmospheric CO_2 into their body structures. While they are alive, the $^{14}C/^{12}C$ ratio in their bodies remains constant, but when they die, there is no further carbon exchange with the atmosphere and the $^{14}C/^{12}C$ ratio begins to decrease because of ^{14}C decay.

A piece of wood found in an archeological digging site had a count rate of 9.8 dpm. Estimate the age of this wooden sample.

Answer:

Assuming that the wood originally had a ^{14}C count rate of 15.3 dpm, the age of the sample is found from Equation 9.3:

$$\ln\left(\frac{N}{N_0}\right) = -\left(\frac{0.693}{t_{1/2}}\right)t$$

$$\ln\left(\frac{9.8}{15.3}\right) = -\left(\frac{0.693}{5730 \text{ y}}\right)t$$

$$t = -\frac{5370 \text{ y} \times \ln(0.641)}{0.693} = 3446 \text{ y}$$

EXAMPLE 8

It is a rule of thumb that a radioactive sample is effectively "gone" after about 15 half-lives. What fraction of the original activity is left after 15 half-lives?

Answer:

Let $t = 15\ t_{1/2}$. Then, $\ln\left(\frac{N}{N_0}\right) = -\left(\frac{0.693}{t_{1/2}}\right)t = -\left(\frac{0.693}{t_{1/2}}\right)(15\ t_{1/2}) = -10.4$
$N/N_0 = \text{antiln}(-10.4) = e^{-10.4} = 3.06 \times 10^{-5}$, or about 0.003% of the original activity remains.

EXAMPLE 9

Calculate the weight in grams, W, of 1 mCi (3.700×10^7 dps) of ^{14}C from its half-life of 5720 years.

$$\text{Half-life} = t_{1/2} = \frac{0.693}{k}, \text{ where } k \text{ is a characteristic decay rate constant}$$

$$\text{for the radioisotope.}$$

The number of disintegrations per second = activity $= A = -\frac{dN}{dt} = kN$, where N is the number of ^{14}C atoms present.

$$N = \frac{\text{grams of } ^{14}C}{\text{mass number of } ^{14}C} \times \text{Avogadro's number} = \frac{W}{14} \times 6.02 \times 10^{23}$$

$$k = \frac{0.693}{t_{1/2}} = \frac{0.693}{5720 \text{ years}} = 1.211 \times 10^{-4}/\text{years}$$

$$A = \frac{1.211 \times 10^{-4} \text{ y}^{-1}}{3.156 \times 10^7 \text{ s y}^{-1}} \times \frac{W}{14} \times 6.02 \times 10^{23} = 1.65 \times 10^{12} \text{ s}^{-1} \times W$$

$$\text{Let } \mathcal{A} = 1 \text{ mCi} = 3.700 \times 10^7 \text{ s}^{-1}$$

$$W = \text{grams of }^{14}\text{C emitting 1 mCi} = \frac{3.700 \times 10^7 \text{ s}^{-1}}{1.65 \times 10^{12} \text{ s}^{-1} \text{g}^{-1}} = 2.24 \times 10^{-5} \text{ g}$$

9.2.9 RADIOACTIVE DECAY SERIES

A radioactive nucleus seeks greater stability by spontaneous emission of nuclear particles. After a particle is emitted, the original nuclide has changed the neutron and proton content of its nucleus and has become a different nuclide.

- If the new nuclide resulting from a radioactive emission has a stable neutron/proton ratio, see Section 9.2.4, the new nuclide is not radioactive and lies within the zone of stability in Figure 9.2.
- If the new nuclide resulting from a radioactive emission has an unstable neutron/proton ratio, it is radioactive and eventually will emit another nuclear particle. This will continue with each new nuclide until a stable nuclide is formed.
- The particle (α, β, neutron, etc.) with the highest probability of being emitted is one that changes the neutron/proton ratio in a way that moves the nuclide most efficiently toward the zone of stability (in Figure 9.2).
  - Nuclides with mass numbers that are too large ($A > 208$) can decrease the mass number most efficiently by emitting α particles, the heaviest emission. Therefore, most heavy radioisotopes ($A > 208$) are α emitters. Emission of an α particle (^4_2He) makes Z decrease by 2, N decrease by 2, and A decrease by 4 (e.g., $^{238}_{92}\text{U} \rightarrow \,^{234}_{90}\text{Th} + \,^4_2\text{He}$).
  - Nuclides with too many neutrons can decrease the N/Z ratio by emitting an electron in β decay ($^0_{-1}\text{e}$), which decreases N by 1 and increases Z by 1 without changing A (e.g., $^{14}_6\text{C} \rightarrow \,^{14}_7\text{N} + \,^0_{-1}\text{e}$).
  - Nuclides with too few neutrons can increase the N/Z ratio by emitting a positive electron, or positron, ($^0_{-1}\text{e}$), which increases N by 1 and decreases Z by 1 without changing A, (e.g., $^{20}_{11}\text{Na} \rightarrow \,^{20}_{10}\text{Ne} + \,^0_{+1}\text{e}$).
  - Often, after radioactive decay, a nucleus is left in an excited energy state. It can relax to the stable ground state by emitting a high-energy γ-ray photon, e.g., $^{12}_5\text{B} \rightarrow \,^{12}_6\text{C}^* + \,^0_{-1}\text{e}$, followed by, $^{12}_6\text{C}^* \rightarrow \,^{12}_6\text{C} + \gamma$.

9.2.10 NATURALLY OCCURRING RADIONUCLIDES

Before 1940, the periodic table ended with uranium, $^{238}_{92}\text{U}$. No elements with $Z > 92$ were known to exist because they are all radioactive with half-lives that are short compared with the lifetime of the earth, which is currently believed to be about 4.5×10^9 years. Although transuranic elements might have existed during the early years of earth's history, they have long since decayed away to stable elements with $Z \leq 82$. There are just three radioisotopes found naturally on earth with half-lives long enough to have persisted since earth's creation. They are uranium-238 ($^{238}_{92}\text{U}$, $t_{1/2} = 4.67 \times 10^9$ years), uranium-235 ($^{235}_{92}\text{U}$, $t_{1/2} = 7.13 \times 10^8$ years), and

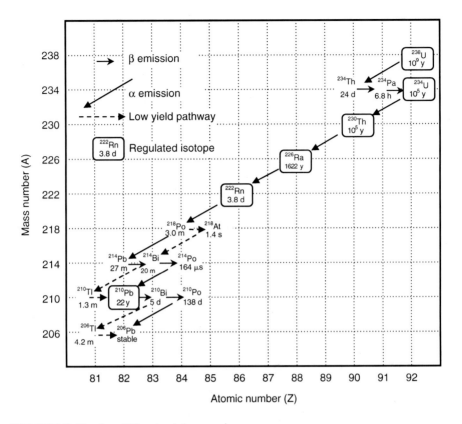

FIGURE 9.3 Uranium-238 natural decay series.

thorium-232 ($^{232}_{90}$Th, $t_{1/2} = 1.39 \times 10^{10}$ years). All the other naturally occurring radioisotopes found on earth today are daughter isotopes of these three parent radioisotopes.

Therefore, there are just three naturally occurring radioactive decay series* (Figures 9.3 through 9.5). Each series starts with one of the long-lived parent radioisotopes whose half-life exceeds that of any of its daughter products. A series continues forming one radioactive daughter after another by emitting α and β particles until a stable isotope is finally made. The half-life of each daughter is an indication of its stability; a longer half-life means a more stable nuclide. The three decay series, starting with $^{238}_{92}$U, $^{235}_{92}$U, and $^{232}_{90}$Th, terminate in the stable isotopes of lead $^{206}_{82}$Pb, $^{207}_{82}$Pb, and $^{208}_{82}$Pb, respectively.

A series sometimes follows alternative paths in which a main sequence of, say, α emission followed by β emission, is reversed to be a β emission followed by an α emission. If undisturbed by natural chemical separation processes, such as

---

* See footnote on page 272. Although bismuth-209 is naturally occurring and radioactive, its half-life is too long (19×10^{18} years) for it to produce a detectable series of daughter products. Therefore, it is not included among the naturally occurring decay series.

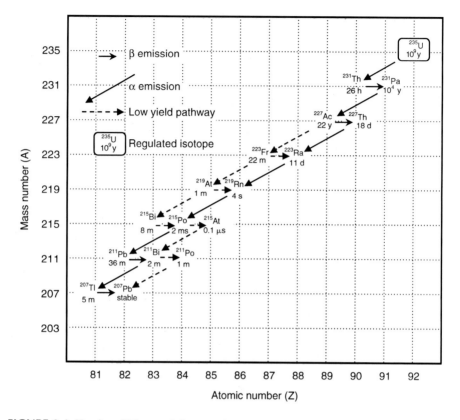

FIGURE 9.4 Uranium-235 natural decay series.

groundwater dissolving a soluble mineral containing a daughter isotope and carrying it away from the original rock formation, a parent radioisotope and its daughters reach a secular equilibrium. Secular equilibrium occurs when each radioisotope in the decay series is at a concentration where its rate of formation equals its rate of decay. This condition requires that the activity ($\mathcal{A}$, dpm) of each isotope in the series be identical.

9.3 EMISSIONS AND THEIR PROPERTIES

The health and environmental hazards associated with different radioisotopes are based primarily on the type and energy of their radiation emissions. Except for radon, most of the commonly encountered radioisotopes are heavy metals and these represent an additional risk based on their chemical toxicity.

When radioisotope radiation passes through human tissue, it interacts with the molecules it encounters, losing energy by forcing electrons out of their normal molecular orbitals, resulting in broken bonds, ionization, and configuration changes in the molecules. This can cause damage or death to cells in the tissue. Damage to a cell's reproductive mechanisms can result in abnormal reproductive behavior such as cancer.

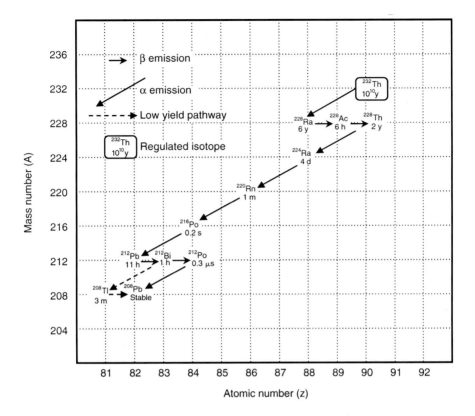

FIGURE 9.5 Thorium-232 natural decay series.

Three types of radiations present the greatest environmental concerns because they are produced by radionuclides that are commonly found in minerals and waste products and have a long enough half-life to allow dangerous quantities to accumulate.

1. Alpha particles: α Particles are doubly charged helium nuclei. Because they are charged and are relatively heavy (more than 7000 times heavier than β particles), they interact strongly with matter and lose their kinetic energy over very short distances of travel. In air, a typical α particle may travel about 10 cm, whereas in water or tissues its range is about 0.05 cm. They are completely stopped by a single sheet of paper or the outer layer of skin. In their passage through matter, α particles cause intense ionization of the molecules in their wake. As they lose energy and slow down, the degree of ionization decreases until they are moving slowly enough to trap electrons, which neutralizes them to harmless helium atoms.

   Since α particles do not penetrate through liquids or solids very far, α emitters (such as Ra, Th, U, and Pu) cause little radiation damage unless they are ingested or inhaled. Workers with α emitters are protected by special clothing and respirators designed primarily for preventing inhalation, ingestion, and transport of dust and small particles away from the site.

α-Emitting wastes require little or no shielding, although they must be contained to prevent their movement by wind or water.

If transported inside an organism by inhalation or ingestion, α emitters can cause profound damage to tissues around their immediate location, because all their energy is deposited within a very small volume. Ra, ^{90}Sr, and ^{133}Ba (all α emitters) substitute for calcium in bone tissue and the intense localized α radiation can destroy the tissue's ability to produce blood cells by causing ionization and bond-breaking in the DNA and RNA molecules of the cells.

2. Beta particles: β Particles are negatively charged electrons emitted from nuclei. They are singly charged and very much lighter than α particles. Thus, they interact less strongly with matter and have greater penetrating power and less ionizing capacity. Higher energy β particles can penetrate skin and travel up to 9 m in air. Although their damage is less localized, their cumulative effects can be as serious as those of α particles. Because their range in matter is so long, β-emitting wastes require shielding by a minimum of 5 mm of aluminum or 2 mm of lead.

3. Gamma rays and x-rays: γ- and x-rays are photons (electromagnetic radiation), are uncharged, and have no mass. They interact with matter relatively weakly by quantum mechanical processes rather than by collisional impact. They have a much greater penetrating power than α or β particles and deposit their energy over much longer path lengths. γ- and x-ray sources require extensive shields to block their emissions. Unlike charged particle radiation, the attenuation of γ radiation by shielding is exponential and, in principle, there always is some probability that a percentage of γ particles penetrate any thickness of shielding. For this reason, γ shielding is usually described in terms of the thickness required to attenuate γ radiation by a certain factor. For example, 4 in. of lead shielding will attenuate 1 MeV (million electron-volts) γ-rays by a factor of 3200, and 2 MeV γ-rays by a factor of 175.

The damage produced in matter by any of these particles depends on the activity (number of particles emitted per second) and on their energy. A given activity of tritium causes less damage than the same activity of ^{14}C because the β particles from tritium are lower energy than those from ^{14}C. Tritium β particles have a maximum energy of 0.0179 MeV, whereas those from ^{14}C are 0.156 MeV. In a sufficient mass of tissue to absorb all the β particles completely, ^{14}C will deposit almost 10 times more energy than the same activity of tritium.

9.4 UNITS OF RADIOACTIVITY AND ABSORBED RADIATION

Concentrations of radionuclides in the environment are typically expressed in terms of activity of the radionuclide per unit of volume of water (e.g., picocuries per liter, or pCi/L), per unit volume of air (e.g., picocuries per cubic meter, or pCi/m^3), or per unit mass of soil of solid (e.g., picocuries per kilogram, or pCi/kg). Activity measures the rate of disintegration of a radionuclide per unit mass or volume. Because the carcinogenic effect of a radionuclide is due to its emissions, concentrations of radionuclides are generally measured in terms of activity for health evaluation purposes.

TABLE 9.2
Radioactivity Parameters

Quantity	SI Units	Special SI Name/Symbol	Conventional Name/Symbol	Conversion: From Conventional to SI Units
Activity (dps)	s^{-1}	becquerel (Bq)	curie (Ci)	3.7×10^{10} Bq/Ci
Absorbed dose (kinetic energy absorbed per unit mass of absorbing matter)	J/kg	gray (Gy)	rad (rad)	0.01 Gy/rad
Dose equivalent	J/kg	sievert (Sv)	rem (rem)	0.01 Sv/rem
Photon exposure	C/kg	C/kg	roentgen (r, R)	2.58×10^{-4} C/kg/r

This section summarizes the three basic kinds of radioactivity measurements: (1) emission rate or activity, (2) absorbed dose, and (3) dose delivering a given biological effect. Tables 9.2 through 9.4 provide conversion factors among the most commonly used quantities. Table 9.6 gives calculated conversion factors for many β-and photon-emitting radionuclides between the emission rate (becquerels, Curies) and the dose delivering a given biological effect (sieverts, rems), for which there are no direct conversions. Table 9.7 helps to compare measurements commonly reported in picocuries per liter with drinking water standards for β and photon emitters commonly given in terms of rems.

TABLE 9.3
Unit Conversions That Apply to Disintegration Rate (Activity)

To Convert From	To	Multiply by
Curies (Ci)	Picocuries (pCi)	10^{12}
Curies (Ci)	Microcuries (μCi)	10^6
Curies (Ci)	Becquerels (Bq)	3.7×10^{10}
Curies (Ci)	Disintegrations per second (dps)	3.7×10^{10}
Picocuries (pCi)	Microcuries (μCi)	10^{-6}
Picocuries (pCi)	Millicuries (mCi)	10^{-6}
Picocuries (pCi)	Curies (Ci)	10^{-12}
Picocuries (pCi)	Becquerels (Bq)	0.037
Picocuries (pCi)	Disintegrations per second (dps)	0.037
Picocuries (pCi)	Disintegrations per minute (dpm)	2.22
Becquerels (Bq)	Curies (Ci)	2.7×10^{-11}
Becquerels (Bq)	Picocuries (pCi)	27
Becquerels (Bq)	Disintegrations per second (dps)	1
Disintegrations per second (dps)	Becquerels (Bq)	1

TABLE 9.4

Unit Conversions that Apply to Dose and Exposure

To Convert From	To	Multiply by
gray (absorbed dose of 1 J/kg)	rad (absorbed dose of 100 erg/g)	100
gray (absorbed dose of 1 J/kg)	Roentgen (R) (exposure dose; 1 R = radiation dose depositing 84 erg/g of air or 93 erg/g of water)	107
rad (absorbed dose of 100 erg/g)	gray (Gy) (absorbed dose of 1 J/kg)	0.01
rad (absorbed dose of 100 erg/g)	Roentgen (R) (exposure dose; 1 R = radiation dose depositing 84 erg/g of air or 93 erg/g of water)	1.07
rem (dose equivalent = rad × RBM factor)	sievert (Sv) (dose equivalent = gray × RBM factor)	0.01
sievert (Sv) (dose equivalent = gray × quality factor)	rem (total absorbed dose = rads × quality factor)	100
Roentgen (r, R) (exposure dose; 1 r = radiation dose depositing 84 erg per g of air or 93 erg per g of water)	rad (absorbed dose of 100 erg/g)	0.93
Roentgen (r, R) (exposure dose; 1 r = radiation dose depositing 84 erg/g of air or 93 erg/g of water)	gray (Gy) (absorbed dose of 1 J/kg)	0.0093

9.4.1 ACTIVITY

Activity is the number of disintegrations or emissions per second in a radioactive sample. Its International Standard (SI) unit is the becquerel (Bq) and the conventional unit more commonly used in the United States is the curie (Ci).

$$1 \text{ Bq} = 1 \text{ disintegration/s}$$

$$1 \text{ Ci} = 3.700 \times 10^{10} \text{ disintegrations/s}$$

The activity is a measure of the rate of nuclear disintegrations and, therefore, the half-life of the radionuclide; longer half-lives mean lower activity. The activity does not give any information about the kinds of particles emitted or their effects in the environment.

The definition of a curie originally was based on the disintegration rate of 1 g of radium, which is 3.700×10^{10} disintegrations/s. However, now it simply is defined as the above quantity and is independent of any experimentally determined value.

RULE OF THUMB

1 picoCurie (pCi) $= 10^{-12}$ Ci ≈ 2.2 disintegrations/min ≈ 1 disintegration every 27 s.

9.4.2 Absorbed Dose

Absorbed dose measures the amount of energy actually deposited per kilogram of a receiving body, regardless of the type of radiation. Its SI unit is the gray (Gy) and the conventional unit is the rad.

$$1 \text{ Gy} = \text{an absorbed dose of 1 J/kg}$$

$$1 \text{ rad} = \text{an absorbed dose of } 100 \text{ erg/g} = \text{absorbed dose of } 0.01 \text{ J/kg, } (1 \text{ erg} = 10^{-7} \text{ J})$$

Note that there is no time period specified. Every 100 ergs absorbed per gram of mass is a dose of 1 rad. Thus rads, which are a dose and not a rate, cannot be directly related to curies, which are a rate.

The number of rads per unit time that correspond to a specified activity depends on the nature of the particles emitted, their energy, and the absorbing, or stopping, power of the matter in which the particles deposit their energy.

The difference between the units of rad (or gray) and curie (or becquerel) are that rads indicate the amount of energy absorbed by matter, whereas curies indicate the number of nuclei disintegrating per second.

The roentgen, abbreviated r or R, is a dose unit related to photon emissions. It measures the ionizing ability of x-rays and γ-rays. One roentgen is the amount of photon activity that produces 2×10^9 ion pairs in 1 cm^3 of air. It represents an absorbed dose of photon energy where air is the absorbing matter instead of human tissue. In general, the exposure to radiation in roentgen units is numerically approximately equal to the absorbed dose in rads (see Table 9.4).

RULES OF THUMB

1. α Particles are heavier and more highly charged than β particles. Therefore, they lose their energy faster to their surrounding and penetrate shorter distances. β Particles lose their energy over longer paths.
2. α Particles produce approximately 20 times more tissue damage per unit length of travel than β particles. RBE for αs $= 20$, RBE for βs $= 1$.

9.4.3 Dose Equivalent

Dose equivalent measures the amount of energy that produces a certain biological effect. It is an empirical quantity that attempts to quantify the fact that the biological hazard from radiation depends on two factors: the amount of energy absorbed by tissues and the type of radiation. The dose equivalent is the absorbed dose multiplied by a quality factor, called the RBE factor, where RBE stands for relative biological effectiveness. Its SI unit is the sievert (Sv) and the conventional unit is the rem (roentgen equivalent man).

$$\text{dose equiv. (Sv)} = \text{absorbed dose in SI units} \times RBE = \text{grays} \times RBE$$

$$\text{dose equiv. (rem)} = \text{absorbed dose in conventional units} \times \text{quality factor} = \text{rads} \times RBE$$

When a dose of radiation is expressed in sieverts or rems, it does not matter what type of radiation it is since it has already been adjusted by an RBE factor. RBE factors, Table 9.5, are weighting factors for different kinds of radiation but do not consider different sensitivities of various organs. All internal organs or organ systems are treated the same as the whole body.

Another dose equivalent term that is sometimes used is the rem ede (effective dose equivalent), which is a dose adjusted for different radiation types and by an organ weighting factor to account for organ sensitivity to radiation. EPA continues to reevaluate the complicated issues involved in setting dose equivalent standards based on risks determined by rem ede models (EPA, 2000). At present, a 4 mrem/y maximum exposure to critical organs remains the drinking water standard. As Table 9.5 indicates, 1 rad of α particles absorbed causes about 20 times the biological damage as 1 rad of β particles, 20 rem versus 1 rem.

Sieverts and rems are the product of two quantities: the first is the energy absorbed, as given by grays or rads; the second is the quality factor, which depends on the type of radiation. The quality factor is an empirical quantity that relates a tissue dosage unit (sievert or rem) to an energy absorption unit (gray or rad) by

$$\text{Number of sieverts or rems} = \text{Number of grays or rads} \times RBE$$

TABLE 9.5

RBE (Relative Biological Effectiveness) Values for Several Types of Radiations, Based on Whole-Body Exposure

Radiation	RBE
x- and γ-Rays	1
β-Rays and electrons	1
Thermal neutrons	2
Fast neutrons	10
High-energy protons	10
α Particles	20
Fission fragments, heavy particles of unknown charge	20
Heavy ions	20

Note: RBE values are also called quality factors or weighting factors.

An allowable rem dose depends on the tissues irradiated and the amount of time over which the radiation is received. A given amount of radiation damage in a hand or foot is less significant than the same damage in the thyroid gland. Since many tissues can regenerate, a low rate of radiation energy deposition in tissues may be less significant than an equal total energy deposition occurring in a much shorter time.

Note again that there is no time factor and that sieverts and rems, like grays and rads, are dose units and not rate units. Rems per hour would indicate a dose rate.

9.4.4   Unit Conversion Tables

9.4.4.1   Converting between Units of Dose Equivalent and Units of Activity (Rems to Picocuries)

There is no direct conversion factor between rems and picocuries. Rems must be calculated from rads using equations that consider the particular radionuclides present and the type and magnitude of body exposure. Table 9.6 gives calculated concentrations (in picocurie per liter) of β, γ, and x-ray emitters in drinking water that result in an absorbed dose of 4 mrem/y to the total body or to any critical organ. (EPA, 2002) However, a provision is made for monitoring gross β activity without calculating a rem value. The gross β standard for drinking water is 4 mrem/y, but a β screening standard is set at 50 pCi/L. An analysis of the major radionuclides and a calculation of the rem value are required only if the sample gross β activity exceeds 50 pCi/L.

TABLE 9.6

Calculated Concentrations (pCi/L) of β and Photon Emitters in Drinking Water Yielding a Dose of 4 mrem/year to the Total Body or to Any Critical Organ as Defined in NBS Handbook 69

Nuclide	pCi/L	Nuclide	pCi/L	Nuclide	pCi/L	Nuclide	pCi/L
H-3	20,000	Sr-85 m	21,000	Sb-124	60	Er-169	300
Be-7	6,000	Sr-85	900	Sb-125	300	Er-171	300
C-14	2,000	Sr-89	20	Te-125m	600	Tm-170	100
F-18	2,000	Sr-90	8	Te-127	900	Tm-171	1,000
Na-22	400	Sr-91	200	Te-127m	200	Yb-175	300
Na-24	60	Sr-92	200	Te-129	2,000	Lu-177	300
Si-31	3,000	Y-90	60	Te-129m	90	Hf-181	200
P-32	30	Y-91	90	Te-131m	200	Ta-182	100
S-35 inorg	500	Y-91m	9,000	Te-132	90	W-181	1,000
Cl-36	700	Y-92	200	I-126	3	W-185	300
Cl-38	1,000	Y-93	90	I-129	1	W-187	200
K-42	900	Zr-93	2,000	I-131	3	Re-186	300

(Continued)

TABLE 9.6 (Continued)
Calculated Concentrations (pCi/L) of β and Photon Emitters in Drinking Water Yielding a Dose of 4 mrem/year to the Total Body or to Any Critical Organ as Defined in NBS Handbook 69

Nuclide	pCi/L	Nuclide	pCi/L	Nuclide	pCi/L	Nuclide	pCi/L
Ca-45	10	Zr-95	200	I-132	90	Re-187	9,000
Ca-47	80	Zr-97	60	I-133	10	Re-188	200
Sc-46	100	Nb-93m	1,000	I-134	100	Os-185	200
Sc-47	300	Nb-95	300	I-135	30	Os-191	600
Sc-48	80	Nb-97	3,000	Cs-131	20,000	Os-191m	9,000
V-48	90	Mo-99	600	Cs-134	80	Os-193	200
Cr-51	6,000	Tc-96	300	Cs-134m	20,000	Ir-190	600
Mn-52	90	Tc-96m	30,000	Cs-135	900	Ir-192	100
Mn-54	300	Tc-97	6,000	Cs-136	800	Ir-194	90
Mn-56	300	Tc-97m	1,000	Cs-137	200	Pt-191	300
Fe-55	2,000	Tc-99	900	Ba-131	600	Pt-193	3,000
Fe-59	200	Tc-99m	20,000	Ba-140	90	Pt-193m	3,000
Co-57	1,000	Ru-97	1,000	La-140	60	Pt-197	300
Co-58	300	Ru-103	200	Ce-141	300	Pt-197m	3,000
Co-58m	9,000	Ru-105	200	Ce-143	100	Au-196	600
Co-60	100	Ru-106	30	Ce-144	30	Au-198	100
Ni-59	300	Rh-103m	30,000	Pr-142	90	Au-199	600
Ni-63	50	Rh-105	300	Pr-143	100	Hg-197	900
Ni-65	300	Pd-103	900	Nd-147	200	Hg-197m	600
Cu-64	900	Pd-109	300	Nd-149	900	Hg-203	60
Zn-65	300	Ag-105	300	Pm-147	600	Tl-200	1,000
Zn-69	6,000	Ag-110m	90	Pm-149	100	Tl-201	900
Zn-69m	200	Ag-111	100	Sm-151	1,000	Tl-202	300
Ga-72	100	Cd-109	600	Sm-153	200	Tl-204	300
Ge-71	6,000	Cd-115	90	Eu-152	200	Pb-203	1,000
As-73	1,000	Cd-115m	90	Eu-154	60	Bi-206	100
As-74	100	In-113m	3,000	Eu-155	600	Bi-207	200
As-76	60	In-114m	60	Gd-153	600	Pa-230	600
As-77	200	In-115	300	Gd-159	200	Pa-233	300
Se-75	900	In-115m	1,000	Tb-160	100	Np-239	300
Br-82	100	Sn-113	300	Dy-165	1,000	Pu-241	300
Rb-86	600	Sn-125	60	Dy-166	100	Bk-249	2,000
Rb-87	300	Sb-122	90	Ho-166	90		

Source: From EPA, Implementation Guidance for Radionuclides, United States Environmental Protection Agency, Office of Ground Water and Drinking Water (4606M), EPA 816-F-00-002, March 2002, www.epa.gov/safewater.

EXAMPLE 10

The radioactive isotope ^{40}K (odd Z and odd N) is naturally present in biological tissues and subjects humans to a certain amount of unavoidable internal radiation. The body of an adult weighing 70 kg contains about 170 g of potassium, mostly in intracellular fluids. The relative natural abundance of ^{40}K is 0.0118%, its half-life is 1.28×10^9 years, and it emits β particles with an average kinetic energy of 1.40 MeV.

1. Calculate the total activity of ^{40}K, in millicurie, for a 70 kg adult.
2. Find the radiation dose rate in mrem/y.

Note: Potassium-40 is not included in Table 9.6 because human exposure to it is governed entirely by the body's metabolic processes and its natural abundance. Therefore, it cannot be controlled by regulation.

Answer:

1. First calculate the decay rate constant, k, with units of s^{-1}.

$$k = \frac{0.693}{t_{1/2}} = \frac{0.693}{(1.28 \times 10^9 \text{ y})(3.16 \times 10^7 \text{s y}^{-1})} = 1.71 \times 10^{-17} \text{ s}^{-1}$$

Let the number of ^{40}K atoms $= \mathcal{N}$

$$\mathcal{N} = \frac{170 \text{ g of K}}{40.0 \text{ g/mol}} \times \left(\frac{1.18 \times 10^{-4} \text{ mol }^{40}\text{K}}{\text{mol K}} \right)$$
$$\times \, 6.02 \times 10^{23} \text{ atoms/mol} = 3.02 \times 10^{20} \text{ atoms of } ^{40}\text{K}$$

Activity $= \mathcal{A} = k\mathcal{N} = (1.71 \times 10^{-17} \text{ s}^{-1})(3.02 \times 10^{20} \text{ atoms}) = 5.16 \times 10^3 \text{ atoms/s(dps)}$

Converting to curies gives

$$\frac{5.16 \times 10^3 \text{ s}^{-1}}{3.7 \times 10^{10} \text{ s}^{-1}/\text{Ci}} = 1.4 \times 10^{-7} \text{ Ci} = 0.14 \text{ μCi}$$

2. Number of curies per kilogram $= \dfrac{1.4 \times 10^{-7} \text{ Ci}}{70 \text{ kg}} = 2.0 \times 10^{-9} \text{ Ci/kg}$

Disintegrations per second per kilogram $= \dfrac{5.16 \times 10^3 \text{dps}}{70 \text{ kg}} = 73.7 \text{ s}^{-1} \text{ kg}^{-1}$

Joules per β particle $= (1.40 \times 10^6 \text{ eV})(1.60 \times 10^{-19} \text{ J/eV}) = 2.24 \times 10^{-13} \text{ J}$

Joules $\times$ dps per kilogram $= (2.24 \times 10^{-13} \text{ J})(73.7 \text{ s}^{-1} \text{ kg}^{-1}) = 1.65 \times 10^{-11} \text{ J/s/kg}$

Joules per kilogram per year $= (1.65 \times 10^{-11} \text{ J/s/kg})(3.16 \times 10^7 \text{ s/year})$
$$= 5.22 \times 10^{-4} \text{ J/kg/year}$$

The number of rems is the number of centijoules (10^{-2} J) of radiation energy absorbed per kilogram of tissue weight multiplied by the RBE factor. From Table 9.5, RBE $= 1$.

TABLE 9.7

Sources of Environmental Radiation in the United States and Estimated Annual Dose Contributions to Individuals. The Total Annual Dose Is about 399 mrem/year

Sources	Approximate Average Annual Dose to a Person in the United States (mrem/year)	Approximate Percent of Total Dose Received
Natural sources 84% of total (about 334 mrem/year)		
Inhaled radon	200	50
Ingested with food	40	10
Cosmic radiation	27	7
Terrestrial radiation	28	7
Internal body	39	10
Man-made sources 16% of total (about 65 mrem/year)		
Medical x-rays	39	10
Nuclear medicine	14	4
Consumer products	9	2
Occupational exposure	0.9	0.2
Weapons testing fallout	<0.9	<0.2
Nuclear fuel cycle	0.4	0.1
Miscellaneous	0.4	0.1

Since ^{40}K is distributed throughout all body tissues, we can assume that essentially all the β radiation energy is absorbed within the body.

$$\text{Number of rem per year} = \frac{5.22 \times 10^{-4} \text{ J/kg/year}}{10^{-2} \text{ J/kg/rem}}$$
$$= 5.22 \times 10^{-2} \text{ rem/year} = 52.2 \text{ mrem/year}$$

Table 9.7 shows that the annual dosage from all natural radiation sources received by a person is about 399 mrem, which shows that about 13% of our natural radiation exposure is from ^{40}K and is unavoidable because it is part of our body's chemical makeup.

As shown in Table 9.7, about 84% of the total average annual effective dose is from natural sources of radiation, and of that, most is from radon. Of the other 16%, the majority is from medical diagnosis and treatments.

9.5 NATURALLY OCCURRING RADIOISOTOPES IN THE ENVIRONMENT

Naturally occurring radionuclides become widely distributed in low concentrations in soil and water by weathering and erosion of rocks followed by transport dissolved in groundwater or as a gas. The geological distribution of the parent radionuclide determines where the first daughter elements are formed, but because many daughters are more mobile than the parent radionuclide, they can move long distances from their parent dissolved in groundwater or as a gas (radon). The radionuclides most commonly found in groundwater are radon-222, radium-226, uranium-238, and uranium-234 from the uranium-238 decay series, and radium-228 from the thorium-232 decay series. Other radionuclides of these two decay series, and all isotopes of the uranium-235 decay series, are usually not present in significant amounts in groundwater because they are present as highly insoluble compounds or have very short half-lives that preclude the buildup of large concentrations.

Naturally occurring radionuclides with $Z > 82$ are those in the three radioactive decay series of uranium-238, uranium-235, and thorium-232 (see Figures 9.3 through 9.5). Unless there have been releases from nearby facilities that make or use enriched uranium, the uranium-235 series is much less important than the others because of the low natural abundance of U-235 (Table 9.8).

Naturally occurring radionuclides with $Z < 82$, like hydrogen-3, carbon-14, potassium-40, rubidium-87, etc., either have long enough half-lives to have persisted since the earth's formation or are formed continuously by the action of radioisotope and cosmic radiation on stable atoms in the earth's atmosphere and geosphere.

Table 9.8 summarizes some important information for the major radionuclides found in surface and groundwaters.

9.5.1 CASE STUDY: RADIONUCLIDES IN PUBLIC WATER SUPPLIES

The widespread distribution of radionuclides in groundwater is a serious problem. Approximately 80% of public water systems in the United States get all or most of their water from underground sources. Many of these groundwater sources have concentrations of radionuclides above EPA maximum contaminant levels (MCLs) for drinking water (see Table 9.8). Ever since EPA first proposed drinking water standards for radionuclides in 1976, there continue to be many public water systems (PWSs) with exceedances of EPA maximum contaminant levels. In 1992, 415 PWSs (about 0.7% of U.S. total) serving nearly 2 million people reported exceedances of EPA radionuclide MCLs in their delivered water. The number of PWSs reporting exceedances dropped to 200 by 1998. (EPA, 1999) The main problem was that the treatment technologies most effective for removing dissolved radionuclides, ion exchange, and reverse osmosis, were not commonly used by PWSs, especially those serving small communities.

Even today, many PWSs continue to operate with radionuclide concentrations above EPA MCLs, even with present-day treatment technologies. A few examples of the widespread existence of systems operating above the radium MCL are listed below:

- The 2005 Illinois EPA Annual Compliance Report for Community Water Supplies reports 90 PWS facilities with 287 exceedances of the radium MCL of 5 pCi/L. At year end 101 PWSs had returned to compliance, with 186 still in violation (IEPA, 2006).
- The 2005 New Mexico EPA Annual Compliance Report for Community Water Supplies reports 15 PWS facilities (an increase of 4 over 2004) with

TABLE 9.8

Major Natural and Artificial Radionuclides Found in Surface and Groundwaters

Radionuclide	Half-life	Average Concentration in Natural Waters	Emissions (MeV)	Sources (Nuclear Reaction, Environmental Pathways, etc.)
Naturally occurring from cosmic radiations				
Carbon-14 (^{14}C)	5730 years	6 pCi/g-carbon	β	Thermal neutrons from cosmic radiation or nuclear weapons testing reacting with atmospheric ^{14}N, $^{14}_7N + ^1_0n \rightarrow ^{14}_6C + ^1_1H$
Potassium-40 (^{40}K)	1.28×10^9 years	<4 pCi/L	β, γ	0.0119% of natural K
Silicon-32 (^{32}Si)	104 years		β	Nuclear fission of atmospheric ^{40}Ar by protons from cosmic sources
H-3, Tritium (^3H)	12.3 years	10–25 pCi/L	β	Thermal neutrons from cosmic radiation or nuclear weapons testing reacting with atmospheric ^{14}N, $^{14}_7N + ^1_0n \rightarrow ^{12}_6C + ^3_1H$
Naturally occurring from ^{238}U and ^{232}Th decay series				
Uranium-238 (^{238}U)	4.5×10^9 years	0.1–10 μg/L ($^{238}U + ^{234}U$)	α	Parent of decay chain of ^{238}U; migration from point of generation
Uranium-234 (^{234}U)	2.5×10^5 years	0.1–10 μg/L ($^{238}U + ^{234}U$)	α	In decay chain of ^{238}U; migration from point of generation
Radium-223 (^{223}Ra)	11.4 days		α, γ	In decay chain of ^{235}U; migration from point of generation
Radium-224 (^{224}Ra)	3.7 days		α, γ	In decay chain of ^{232}Th; migration from point of generation
Radium-226 (^{226}Ra)	1620 years	0.01–0.1 pCi/L	α, γ	In decay chain of ^{238}U; migration from point of generation
Radium-228 (^{228}Ra)	6.7 years		β	In decay chain of ^{232}Th; migration from point of generation

Radionuclide	Half-life	Concentration	Decay	Comments
Radon-219 (^{219}Rn)	4.0 s		α	In decay chain of ^{238}U; migration from point of generation
Radon-220 (^{222}Rn)	56 s		α	In decay chain of ^{232}Th; migration from point of generation
Radon-222 (^{222}Rn)	3.8 days	1 pCi/L surface, >10^3 pCi/L groundwater	α	In decay chain of ^{238}U; migration from point of generation
Thorium-230 (^{230}Th)	7.54×10^4 years		α, γ	In decay chain of ^{238}U; migration from point of generation
Thorium-234 (^{234}Th)	24.1 days		β, γ	In decay chain of ^{238}U; migration from point of generation
Thorium-232 (^{232}Th)	1.4×10^{10} years	0.01–1 µg/L	α	In decay chain of ^{232}Th; migration from point of generation

From fission reactions in reactors and weapons

Radionuclide	Half-life	Concentration	Decay	Comments
Strontium-90 (^{90}Sr)	28.5 years	0.5 pCi/L	β	Fission products with highest yields and biological activity
Iodine-131 (^{131}I)	8.04 days		β, γ	
Cesium-137 (^{137}Cs)	30.0 years	0.1 pCi/L	β, γ	
Barium-140 (^{140}Ba)	12.75 days		β, γ	
Zirconium-95 (^{95}Zr)	64.02 days		β, γ	
Cerium-141 (^{141}Ce)	32.5 days		β, γ	Fission products of lesser yields; listed in order of decreasing yield
Strontium-89 (^{89}Sr)	50.55 days		β, γ	
Ruthenium-103 (^{103}Ru)	39.254 days		β, γ	
Krypton-85 (^{85}Kr)	10.72 years		β, γ	
Cobalt-60 (^{60}Co)	5.271 years		β, γ	From nonfission neutron reactions in reactors
Manganese-54 (^{54}Mn)	312.20 days		γ	From nonfission neutron reactions in reactors
Iron-55 (^{55}Fe)	2.73 years		x-ray	From high-energy neutrons acting on stable ^{56}Fe in weapon hardware

42 exceedances of the gross α and combined uranium MCLs of 15 pCi/L and 30 μg/L (micrograms per liter), respectively (NMED, 2006).

- In Colorado today (2007), there are approximately 90 to 135 water supply systems reported to be operating with radionuclide concentrations in excess of MCLs (Miller and Stanford, 2006).

It is important to keep in mind that most PWS exceedances of radionuclide MCLs are not considered to be acute problems requiring immediate correction. Radionuclide MCLs are determined by considering the potential health effects to adults resulting from a 70 year exposure, with a water consumption of 2 L per day. Typical exceedances occurring in a PWS do not pose immediate health risks, any more than do medical x-rays or high altitude airplane trips. EPA recognizes that, because of background radiation, risks from exposure to radiation cannot be eliminated; the goal is to minimize them.

In view of the 70 year, 2 L per day drinking water consumption basis for determining radionuclide MCLs, EPA regulations do not require treatment to the MCL limit within any specified time frame (Table 9.9). The action required for most radionuclide exceedances is not rejection of the water supply but continued acceptance, while corrective actions are undertaken. The exact manner of correction is decided by individual states. A common approach used by many states for most exceedances of naturally occurring radionuclides is to require that the water supplier investigates corrective measures, such as best available technology (BAT) treatment or alternative supplies, while continuing a stringent monitoring program. Unless the annual average radionuclide levels are deemed high enough to be of immediate concern, continued use of the water supply is considered acceptable (EPA, 1976).

9.5.2 URANIUM

Uranium (U, atomic number 92) has 18 isotopes with atomic masses ranging from 222 to 242. All are radioactive. Only ^{234}U, ^{235}U, and ^{235}U are found naturally. Pure uranium emits only α particles accompanied by a low level of γ radiation. The mass differences among the uranium isotopes are small and the isotopes do not normally

---

TABLE 9.9
EPA National Drinking Water Standards

Radionuclide	MCL	MCLG
Radium-226 and radium-228 combined	5 pCi/L	0
Gross α emitters	15 pCi/L	0
All β particles and photon emitters	4 millirem (mrem)/y; any organ or whole body	0
Uranium	0.030 mg/L	0
Radon	300 pCi/L (final rule pending 2007 or 2008)	—

Note: MCLs are based on a 70 year exposure, with a water consumption of 2 L per day.

TABLE 9.10

Properties of the Naturally Occurring Uranium Isotopes

Uranium Isotope	Half-life (years)	Specific Activity (disint./s/mg)	Decay Constant, k (y^{-1})	Abundance (atom %)	Abundance (weight %)
U-234	2.48×10^5	2.28×10^5	2.79×10^{-6}	0.0055	0.0054
U-235	7.13×10^8	79.0	9.72×10^{-10}	0.7204	0.7114
U-238	4.51×10^9	12.31	1.537×10^{-10}	99.2741	99.2830
Natural mix (pure U)	—	25.0	—	—	—

fractionate through natural physical or chemical processes. Uranium exists in U(III), U(IV), U(V), and U(VI) oxidation states, of which the U(IV) and U(VI) states are most commonly found in the environment. Properties of the three naturally occurring uranium isotopes are presented in Table 9.10.

Uranium is the only radionuclide for which the chemical toxicity has been found to be greater or equal to its radiotoxicity and for which its drinking water standard is expressed in terms of a reference dose (RfD).* Since α emission is the most common measurement of uranium, an activity-to-mass conversion is useful. EPA has assumed that the mix of uranium isotopes commonly found in public water systems will have a conversion factor of 0.9 pCi/μg, which means that a drinking water MCL of 30 μg/L will typically correspond to 27 pCi/L. EPA considers that 30 μg/L (27 pCi/L) is protective of both chemically caused kidney toxicity and radiation caused cancer.

9.5.2.1 Uranium Geology

Uranium occurs naturally in the earth's crust with an average concentration of around 2–3 mg/kg (Langmuir, 1997). Uranium is common in crustal rocks around the world and is found in trace amounts in nearly all rocks and soils. Ninety percent of the world's known uranium sources are contained in conglomerates and sandstone. The earth's crust contains 2–3 mg/kg uranium on average. It averages about 1.3 mg/kg in sedimentary rocks and ranges between 2.2 and 15 mg/kg in granites and between 20 and 120 mg/kg in phosphate rocks. Igneous rocks with high silicate content, such as granite, have a uranium content above average while the contents of sedimentary and basic rocks, such as basalts, are below average.

Because of its chemical reactivity, it is not present as free uranium in the environment. In nature, uranium is generally found as an oxide, such as in the olive-green-colored mineral pitchblende, which contains triuranium octaoxide (U_3O_8). Uranium dioxide (UO_2) is the chemical form most often used for nuclear

---

* The reference dose (RfD) is an estimate of a daily ingestion exposure to the population, including sensitive subgroups, that is likely to be without an appreciable risk of deleterious effects during a 70 year lifetime. Dissolved uranium is a kidney toxin and the risk of toxic kidney effects from uranium depends on the mass of uranium ingested. The EPA oral RfD for uranium is 0.6 μg/kg body weight/day.

reactor fuel. Both oxide forms are solids that have a low solubility in water and are relatively stable over a wide range of environmental conditions.

U_3O_8 is the most stable form of uranium and is the form found in nature. The most common form of U_3O_8 is "yellow cake," a solid named for its characteristic color that is produced during the uranium mining and milling process. UO_2 is a solid ceramic material and is the form in which uranium is most commonly used as a nuclear reactor fuel. At ambient temperatures, UO_2 will gradually convert to U_3O_8. Uranium oxides are extremely stable in the environment and are thus generally considered the preferred chemical form for storage or disposal.

Under reducing conditions, uranium is mainly found as the oxide UO_2, an insoluble compound found in minerals. With oxidizing conditions, it forms the oxide UO_3, a moderately soluble compound found in surface waters. Worldwide, uranium averages around 1.3 mg/kg in sedimentary rocks, between 2.2 and 15 mg/kg in granites, and between 20 and 120 mg/kg in phosphate rocks (Langmuir, 1997). Its concentration in most natural waters is between 0.1 and 10 μg/L, although concentrations as high as 1–15 mg/L may exist in water in contact with uranium ore deposits (Hem, 1992).

Uranium can exist in the oxidation states U(III), U(IV), U(V), and U(VI), but only the uranous [U(IV)] and uranyl [U(VI)] oxidation states are commonly found in the environment. Oxidized species [U(VI)], such as the uranyl cation UO_2^{2+} or anionic species formed at high pH, are the most soluble, favoring the wide distribution of uranium in the oxidized portion of the Earth's crust. The reduced species [U(IV)] are only slightly soluble (Hem, 1992). Uranium is a very mobile element in the environment because of the solubility of its oxidized forms.

Important U(IV) minerals include uraninite (UO_2 through $UO_{2.25}$, the main component of pitchblende) and coffinite ($USiO_4$). Aqueous U(IV) is inclined to form sparingly soluble precipitates, adsorb strongly to mineral surfaces, and partition into organic matter, thereby reducing its mobility in groundwater.

Important U(VI) minerals include carnotite [$(K_2(UO_2)_2(VO_4)_2)$], schoepite ($UO_3 \cdot 2H_2O$), rutherfordine (UO_2CO_3), tyuyamunite [$Ca(UO_2)_2(VO_4)_2$], autunite [$Ca(UO_2)2(PO_4)_2$], potassium autunite [$K_2(UO_2)_2(PO_4)_2$], and uranophane [$Ca(UO_2)_2(SiO_3OH)_2$]. Some of these are secondary phases, which may form when sufficient uranium is leached from contaminated wastes or a disposal system and migrates downstream. Uranium is also found in phosphate rock and lignite at concentrations that can be commercially recovered. In the presence of lignite and other sedimentary carbonaceous substances, uranium enrichment is believed to be the result of uranium reduction to form insoluble precipitates, such as uraninite.

9.5.2.2 Uranium in Water

Uranium in surface water is ovften from phosphate deposits and mine tailings, as well as from runoff of phosphate fertilizers (which can contain up to 150 ppm of uranium) from agricultural land. Greater than 99% of uranium transported by runoff from land to fresh water systems is in suspended particles and remains in the sediment. Levels in U.S. streams are usually between 0.1 and 10 mg/L (Hem, 1992), but can exceed 20 mg/L in irrigation return flows because of phosphates and evaporative concentration. In waters associated with uranium ore deposits, uranium concentrations may be greater than 1000 $\mu g/L$.

Uranium is widely dispersed in groundwater because of its presence in soluble minerals, its long half-life, and its relatively high abundance. The highest concentrations in groundwater are found in granite rock and granitic sediments. Uranyl cation (U^{6+}) forms strong carbonate complexes in most waters above pH 6. Complexation with carbonate anions greatly increases the solubility of uranium minerals, facilitating uranium mobility. Solubility of uranium is also enhanced by complexation with phosphate, sulfate, and fluoride ions, and with organic compounds, especially humic substances. Because it is readily dissolved and transported by oxidizing groundwaters, it can be transported to areas far from its original location.

Uranium is less mobile in reducing groundwater, where U(IV) forms solids of low solubility and the dominant uranous ion (U^{4+}) and its aqueous complexes tend to sorb very strongly to humic materials and mineral surfaces in the aquifer. U(IV) concentrations in reducing groundwater are usually less than 10 $\mu g/L$.

RULES OF THUMB

1. Uranium in surface runoff is mostly (>99%) in sediment form.
2. Uranium is the only radionuclide for which its chemical toxicity risk is greater or equal to its cancer risk from emissions.
3. The drinking water standard for uranium is 30 $\mu g/L$. This corresponds to 27 pCi/L in most public water systems. The conversion factor is $1\mu g/L = 0.9$ pCi/L.
4. Uranium, generally, is least mobile in reducing (anaerobic) environments free of complexing anions and most mobile in oxidizing (aerobic) environments with high concentrations of complexing anions.
5. High alkalinity (carbonate content) and high levels of sulfate, fluoride, and humic materials increase uranium solubility by complexation.
6. Reducing conditions in groundwater and lake sediments promote uranium sorption to solid surfaces and greatly lower concentrations and mobility.
7. Phosphate fertilizers can contain up to 150 ppm of uranium.

9.5.3 RADIUM

All the isotopes of radium between Ra-223 and Ra-230 have been observed. Only Ra-223, Ra-224, Ra-226, and Ra-228 occur naturally (see Table 9.8). Ra-226 and Ra-228 are usually the only radium isotopes of environmental interest because their half-lives (1620 and 6.7 years, respectively) are long enough to allow substantial environmental accumulation.

In addition to their own radiation emissions, three of these radium isotopes present additional environmental and health concerns due to the fact that they decay into radon. Radon is a gas at room temperature and is thus more mobile in the environment.

- Ra-223, in the uranium-235 decay series, decays to radon-219 by α emission.
- Ra-224, in the thorium-232 decay series, decays to radon-220 by α emission.
- Ra-226, in the uranium-238 decay series, decays to radon-222 by α emission.
- Ra-228, in the thorium-232 decay series, decays to actinium-228 by β emission.

Radium-226 is of greatest concern because it not only is the most abundant, but it also decays into the most abundant radon isotope, radon-222.

Radium belongs to periodic table group 2A, the alkaline earth metals, and its chemical properties are, therefore, most similar to those of barium, strontium, and calcium. In the body, radium behaves like calcium, and becomes incorporated into bone structure, where it poses a serious risk of bone cancer. In public water systems with radium problems, radium is often found precipitated as a carbonate or sulfate along with calcium and magnesium deposits in pipes of the distribution system.

Radium exists in only the Ra(II) oxidation state in solution and does not easily complex in water. It forms carbonate and sulfate salts of very low water solubility. Radium salts of chloride, nitrate, and bromide are soluble. Sorption can remove radium from solution by adsorption and coprecipitation by scavengers such as iron hydroxide and barium sulfate.

9.5.3.1   Radium in Soil

Radium-226 and radium-228 are found in soil throughout the world. In the United States, the mean concentration of Ra-226 in 356 surface soil samples collected from 33 states was 1.1 pCi/g (Myrick et al., 1981).

Typical concentrations by rock type are

- Sandstone, 0.71 pCi/g
- Limestone, 0.42 pCi/g
- Shale, 1.1 pCi/g

Coal burning and uranium mining/milling operations have produced elevated levels of radium in soil. The concentration of Ra-226 in soils contaminated by mining and milling activities can be 40,000 pCi/g or more. Using uranium concentrations as an indicator of radium levels, elevated radium levels in soil are expected in the western third of the continental United States, including large areas of California and Idaho, and in Wisconsin, Minnesota, the Appalachian Mountains, and Florida (ATSDR, 1990).

9.5.3.2   Radium in Water

Radium-226 and radium-228 are frequently found in concentrations that exceed EPA drinking water standards wherever uranium-238 and thorium-232 minerals are abundant. In general, shallow groundwater has less radium than deep aquifers, and treated water has less radium than raw groundwater. The radium content of surface water is usually very low, lower than most groundwater supplies.

The solubility of radium generally is low, but increases with decreasing pH. Dissolved radium occurs mainly as Ra^{2+} within the pH range 3–10. Radium tends to be most soluble in reducing waters high in iron and manganese and low in sulfate. Increased concentrations of total dissolved solids also increase the solubility of radium because fewer sorption sites are available for radium species.

Radium-226, a daughter product of ^{238}U, is the most common radium isotope in natural waters because of the abundance and mobility of its parent uranium-238 in groundwater. It is more abundant than ^{228}Ra, a daughter product of ^{232}Th, because uranium is generally more abundant and is more soluble than thorium.

Although ^{228}Ra is chemically similar to ^{226}Ra, its distribution in groundwater is very different for several reasons. The relatively short half-life of ^{228}Ra limits the potential for transport without the parent being present. Consequently, ^{228}Ra cannot migrate far from its source before it decays to another progeny. Thorium-232, the parent of ^{228}Ra, is extremely insoluble and is not subject to mobilization in most groundwater environments. The very low solubility of thorium (much lower than uranium) limits the distribution of ^{228}Ra in groundwater. Areas associated with the presence of ^{228}Ra include the east coastal plain and high plains aquifers.

9.5.4   RADON

Radon (Rn) is a naturally occurring chemically inert gas, in the same family as helium, neon, and argon. Radon is produced in the natural decay chains of ^{238}U, ^{235}U, and ^{232}Th and is the only naturally occurring gaseous element that is radio-active. It has no odor, color, or taste and is about 7.4 times more dense than air. All radon isotopes are radioactive and decay by α emission. Radon accounts for about 50% of the average radiation exposure of people in the United States (see Table 9.7).

There are three naturally occurring isotopes of radon:

- ^{219}Rn, half-life = 3.9 s; in decay chain of ^{235}U
- ^{220}Rn, half-life = 54.5 s; in decay chain of ^{232}Th
- ^{222}Rn, half-life = 3.8 days; in decay chain of ^{238}U

Because of its longer half-life, only ^{222}Rn is environmentally important. Wherever ^{238}U is in the soil, ^{222}Rn is formed continuously as a daughter product. Because radon is an inert gas, it moves freely through soil fissures and pore spaces without reacting to become a less mobile radionuclide.

Its half-life of 3.8 days is long enough for significant amounts of radon to travel far from their point of origin and dissolve in groundwater, diffuse to the surface atmosphere, and collect in underground voids such as caves, building basements, subways, water wells, sewers, etc. On the other hand, its half-life is short enough that after 40 days (about 10 half-lives) an isolated sample has decayed to a negligible activity. Radon is sufficiently soluble in water to reach concentrations as high as 300,000 pCi/L where uranium minerals in granites and uranium minerals in pegmatites associated with metamorphic rocks are in the area. However, radon concentrations around 10,000 pCi/L are more typical. As with most gases, its solubility in water varies inversely with water temperature; the colder the water, the greater is radon's solubility.

Because it is an inert gas and unreactive in the environment, it is the most mobile radioisotope and can travel long distances dissolved in groundwater and through voids and fissures in the vadose zone as a gas. Radon is found nearly everywhere in soil, air, and water; even outdoor air contains low levels of radon (typically about 0.4 pCi/L). Radon enters building structures through floor and wall cracks, drains, hollow concrete blocks, openings around pipes, windows, etc. Radon is denser than air (density of radon at STP = 9.73 g/L, density of air at STP = 1.29 g/L) and does not diffuse evenly into the atmosphere, but tends to collect in voids and low spots.

EPA's National Residential Radon Survey found that over 6% of all homes nationwide have average annual indoor radon levels above 4 pCi/L (Marcinowski et al., 1944). Most indoor radon comes from uranium and radium decay in rocks and soil around a home, although other, usually less significant, sources of indoor radon are water and some construction materials.

Radon also is drawn into the low pressure zone within a pumping well's cone of depression. The result is that radon is drawn from remote crevices and fractures into the well. For buildings whose water is supplied by a well, radon dissolved in groundwater can enter through the water distribution system, becoming volatilized into the indoor air space as the water is used, especially showers and washing machines. Public water supplies tend to have lower radon levels than wells because the longer residence times allow more complete radioactive decay. Also, where radon is a known problem, public supplies often treat their water by aeration, venting the radon gas to the atmosphere.

9.5.4.1 Health Issues

Radon is a known human lung carcinogen and is reported to be the second leading cause of lung cancer in the United States (National Academy of Sciences, 1999). It is estimated that radon is responsible for 15,000–20,000 lung cancer deaths per year in the United States (National Academy of Sciences, 1999). It is the largest source of radiation exposure to the public and is considered a serious health risk.

By far, the primary health risk for radon is exposure through inhalation, not ingestion. Drinking radon-rich water appears to result in only a very minor increase in the risk of stomach cancer (National Academy of Sciences, 1999). The primary concern of exposure to radon-rich water is its contribution to radon in indoor air as it volatilizes out of the water phase. This occurs during normal household water use, particularly during showers and during the washing of dishes and clothes.

Most radon that is inhaled is also exhaled. However, radon remaining in the lung decays in several steps to form daughter radioactive isotopes of polonium, lead, bismuth, and tellurium, all with short half-lives (see Figure 9.3). These radon decay products are chemically toxic metals that are readily retained in the respiratory system and, over time, will damage sensitive lung and bronchial tissues. Also, the short half-lives of the daughter products allow them to undergo further decay before the action of mucus in the bronchial tubes can clear them out. Thus, the lungs and bronchia are exposed to additional α and β emissions.

Even very small exposures to radon can, over time, result in lung cancer, and researchers have not yet been able to determine a safe lower threshold. Smokers exposed to elevated levels of radon are particularly susceptible to contracting lung cancer because of the synergistic relationship between radon, smoking, and lung cancer. Generally speaking, health risks associated with radon are due to long-term exposure, from about 5–25 years (National Academy of Sciences, 1999).

The EPA has proposed an MCL for radon in water of 300 pCi/L, and an alternate MCL of 4000 pCi/L for public water suppliers that have radon mitigation programs for their customer base (EPA, 1999). Radon dissolved in public water supply systems nationwide averages about 250 pCi/L.

EXERCISES

1. Complete the following nuclear equations: (In every case, the unknown is just one kind of particle.)

   a. $^{218}_{84}\text{Po} \rightarrow {}^{4}_{-2}\text{He} + \underline{\hspace{1cm}}$

   b. $^{248}_{98}\text{Cf} + {}^{18}_{8}\text{O} \rightarrow {}^{263}_{106}\text{Sg} + \underline{\hspace{1cm}}$

   c. $^{1}_{0}\text{n} + {}^{235}_{92}\text{U} \rightarrow {}^{90}_{38}\text{Sr} + 2{}^{1}_{0}\text{n} + \underline{\hspace{1cm}}$

   d. $^{2}_{1}\text{H} + {}^{1}_{1}\text{H} \rightarrow \underline{\hspace{1cm}} + {}^{0}_{0}\gamma$

2. The half-life for a certain radioisotope is 10 h. If you start with 1 g of the isotope, how long will it take for it to be reduced to 1 mg (1/1000 g)?

3. Briefly discuss radon gas. How is it formed and why is it hazardous?

4. After a uranium-238 atom emits one α particle, what isotope does it become? Use the periodic table inside the front cover, but do not refer to other figures in this chapter.

5. A scientist is studying isotopes of the element thorium (Th), which has the atomic number 90. He isolates two different samples, each of which contains a single pure isotope of thorium. All the atoms in one sample have an atomic weight of 230 mass units and all the atoms in the other sample have an atomic weight of 234 mass units.

   a. The lighter sample must have lost weight in a chemical reaction.

   b. Each atom of the heavier sample has four more protons in the nucleus than do the atoms in the lighter sample.

   c. Each atom of the heavier sample has four more neutrons in the nucleus than do the atoms in the lighter sample.

   d. He must have made some mistake because all atoms of the same element must be identical. The samples cannot both be thorium.

6. Using only the periodic table, complete the following table (Z = atomic number; A = isotope mass number):

Chemical Symbol	Z	A	Protons	Neutrons	Electrons
Mo	42			56	
	34	80			34
Ag		109			
		37	17		16

7. Use Figure 9.2 to write an equation for the probable mode of decay of each of the following radionuclides. Your equation should be in the form of the answers to question 1.

   a. $^{38}_{16}S$ b. $^{38}_{20}Ca$ c. $^{239}_{94}Pu$

8. Water containing radioactive waste material with a half-life of 0.5 years is stored in underground tanks. If the waste concentration is 500 pCi/L, how long will it take for it to decay to less than 0.1 pCi/L?

9. Radon-222 has a half-life of 3.8 days. The source water for a drinking water treatment plant contains 1200 pCi/L of radon. Because aeration is not an option at this plant, how long must they store the water to meet the EPA proposed MCL of 300 pCi/L?

10. Tritium ($^{3}_{1}H$) has a half-life of 12.3 years. It is formed in the atmosphere by reactions of thermal neutrons from cosmic rays reacting with nitrogen-14. Because of its short half-life, a steady state concentration of tritium is maintained in the atmosphere. Tritium becomes incorporated in atmospheric water molecules, replacing a stable hydrogen-1 atom, and falls to earth in precipitation. When the tritium level in groundwater from a spring was measured, it was 0.4 times that of surface water in the area of the spring.

    a. Estimate the subsurface transit time of the groundwater from the recharge point to the spring outflow.

    b. Write the nuclear reaction for the process that forms tritium in the atmosphere. Check your answer with Table 9.8.

REFERENCES

ATSDR, 1990, Toxicological Profile for Radium. Agency for Toxic Substances and Disease Registry, U.S. Public Health Service, December 1990.

EPA, 1976, Title 40, Part 141, National Interim Primary Drinking Water Regulations, *Federal Register* 41(133): 28402, July 9.

EPA, 1999, 25 Years of the Safe Drinking Water Act: History and Trends. EPA 816-R-99-007. Washington, DC, EPA Office of Water.

EPA, 2000, Radionuclides Notice of Data Availability, Technical Support Document, United States Environmental Protection Agency, Office of Ground Water and Drinking Water, March 2000.

EPA, 2002, Implementation Guidance for Radionuclides, United States Environmental Protection Agency, Office of Ground Water and Drinking Water (4606M), EPA 816-F-00-002, March 2002, www.epa.gov/safewater.

Hem, J.D., 1992, *Study and Interpretation of the Chemical Characteristics of Natural Water,* 3rd edition, U.S. Geological Survey Water-Supply Paper 2254, United States Government Printing Office, Washington DC.

IEPA, 2006, Illinois Environmental Protection Agency, 2005 Illinois EPA Annual Compliance Report for Community Water Supplies.

Langmuir, D., 1997, *Aqueous Environmental Chemistry,* Prentice Hall, Upper Saddle River, New Jersey, 600 pp.

de Marcillac, P., et al., 2003, Experimental detection of α-particles from the radioactive decay of natural bismuth, *Nature*, 422 (24): 876–878.

Marcinowski, F., Lucas, R.M., and Yeager, W.M., 1944, National and regional distributions of airborne radon concentrations in U.S. homes, *Health Physics*, 66 (6): 699–706.

Miller, J. and Stanford, D., 2006, Radium and uranium: Compliance strategies and emerging issues, Presented to Colorado Rural Water Association Annual Conference, Colorado Springs, Colorado, February 13–16.

Myrick, T.E., Berven, B.A., and Haywood, F.F., 1981, *State Background Radiation Levels: Results of Measurements Taken During 1975–1979.* Report to the U.S. Department of Energy, Oak Ridge National Laboratory, Oak Ridge, TN. ORNL/TM-7343.

National Academy of Sciences, 1999, Risk assessment of radon in drinking water, Committee on Risk Assessment of Exposure to Radon in Drinking Water, Board on Radiation Effects Research, Commission on Life Sciences, National Research Council, National Academy Press, Washington DC.

NMED, 2006, New Mexico Environment Department, Drinking Water bureau, 2006 Public Water System Compliance Report.

10 Selected Topics in Environmental Chemistry

10.1 AGRICULTURAL WATER QUALITY

Most water quality-related problems in irrigated agriculture fall into four general types:

1. *High salinity*: Dissolved salts (total dissolved salts [TDS]) in the water may reduce water availability to the plants to the extent that crop yield is affected. The effect is caused by a lowering of the osmotic pressure that the plants can exert across their root membranes for absorbing water. Salinity problems can often be mitigated by proper irrigation practices.
2. *Low water infiltration rate*: Relatively high sodium or low calcium and magnesium content of the water may reduce the water permeability of the soil to the extent where sufficient water cannot flow through the root zone at an adequate rate for optimal plant growth. The effect is caused when an excess of sodium ions adsorbed on clay particles causes the soils to swell, thereby reducing pore size and water permeability. An empirical indicator, the sodium absorption ratio (SAR), measures the excess of sodium over calcium and magnesium and provides a guide to potential soil permeability problems (see Section 10.2).
3. *Specific ion toxicity*: Certain ions can accumulate in the leaves of sensitive crops to concentrations high enough to cause crop damage and reduce yields. Ion toxicity arises mainly from sodium, chloride, and boron. Other trace elements are also toxic to plants in low concentrations but generally are present in groundwater in such low concentrations that they seldom are a problem. In general, concentrations of concern for specific ion toxicity are lower for sprinkler irrigation than for surface irrigation because toxic ions can be absorbed directly into the plant through leaves wetted by the sprinkler water. Direct leaf absorption speeds the rate of accumulation of toxic ions.
4. *Nutrient imbalance*: Nitrogen ion concentrations can be too high, resulting in excessive vegetative growth, weak supporting stalks, delayed plant maturity, and poor crop quality.

Measuring the following set of parameters will normally allow an adequate evaluation of agricultural water quality. The importance of these parameters are indicated in Tables 10.1 and 10.2.

TDS	Calcium	Chloride	Bicarbonate
SAR	Magnesium	Selenium	Nitrate + nitrite
Sodium	Boron	Copper	pH

TABLE 10.1
Suggested Maximum Parameter Levels in Groundwaters Used for Crop Irrigation

General Problem	Parameter	Units	Suggested Maximum Value
Salinity	Total dissolved solids (TDS)	mg/L	450
	Specific conductivity	mS/cm	700
Water infiltration	Sodium absorption ratio (SAR)		3–9[a]
Specific ion toxicity	Sodium (Na)	mg/L	70
	Chloride (Cl)	mg/L	100
	Boron (B)	mg/L	1–3[b]
	Trace elements[c]		
	Aluminum (Al)	mg/L	5
	Arsenic (As)	mg/L	0.1
	Beryllium (Be)	mg/L	0.1
	Cadmium (Cd)	mg/L	0.01
	Cobalt (Co)	mg/L	0.05
	Chromium (Cr)	mg/L	0.10
	Copper (Cu)	mg/L	0.20
	Fluoride (F)	mg/L	1.0
	Iron (Fe)	mg/L	5.0
	Lithium (Li)	mg/L	2.5
	Manganese (Mn)	mg/L	0.20
	Molybdenum (Mo)	mg/L	0.01
	Nickel (Ni)	mg/L	0.20
	Lead (Pb)	mg/L	5.0
	Selenium (Se)	mg/L	0.02
	Vanadium (V)	mg/L	0.10
	Zinc (Zn)	mg/L	2.0

Source: Based on data from "Water Quality for Agriculture," FAO Irrigation and Drainage Paper No. 29, Rev. 1, Food and Agriculture Organization of the United Nations, 1986, and Colorado water quality standards for agricultural uses.

[a] Also depends on specific conductivity (used as a measure of TDS). At given SAR, infiltration rate increases as water conductivity increases.

[b] Depends on sensitivity of the crop.

[c] Suggested maximum value is for a water application rate consistent with good agricultural practice (about 10,000 m^3/year). Toxicity and suggested maximum value depend strongly on the crop. Normally not monitored unless a problem is expected. Several trace elements are essential nutrients in low concentrations.

TABLE 10.2
Groundwater Parameter Levels of Concern for Crop Irrigation

Crop Growing Restrictions

Restriction Cause	Parameter Value	Degree of Restriction
Chloride toxicity (surface irrigation)[a]	Less than 142 mg/L	None
	Between 142 and 355 mg/L	Moderate
	Greater than 355 mg/L	Severe
Chloride toxicity (sprinkler irrigation)[b]	Less than 107 mg/L	None
	Greater than 107 mg/L	Moderate
Sodium toxicity (surface irrigation)[a]	Less than 69 mg/L	None
	Between 69 and 207 mg/L	Moderate
	Greater than 207 mg/L	Severe
Sodium toxicity (sprinkler irrigation)[b]	Less than 69 mg/L	None
	Greater than 69 mg/L	Moderate
Sodium adsorption ratio[c]	SAR less than 3	None
	SAR between 3 and 9	Moderate
	SAR greater than 9	Severe
Nitrate[d]	Less than 5 mg/L	None
	Between 5 and 12 mg/L	Slight
	Between 12 and 30 mg/L	Moderate
	Greater than 30 mg/L	Severe

Source: Based on data from "Water Quality for Agriculture," FAO Irrigation and Drainage Paper No. 29, Rev. 1, Food and Agriculture Organization of the United Nations, 1986, and Colorado water quality standards for agricultural uses.

[a] With surface irrigation, sodium and chloride ions are absorbed along with water through plant roots, move with the transpiration stream, and accumulate in the leaves where leaf burn and drying may result. Most tree crops and woody plants are sensitive to sodium and chloride toxicity. Most annual plants are not sensitive.

[b] With sprinkler irrigation, toxic sodium and chloride ions can be absorbed directly into the plant through leaves wetted by the sprinkler water. Direct leaf absorption speeds the rate of accumulation of toxic ions.

[c] SAR values greater than 3 may reduce soil permeability and restrict the availability of water to plant roots.

[d] NO_3 levels greater than 5 mg/L may cause excessive growth, weakening grain stalks, and affecting production of sensitive crops (e.g., sugar beets, grapes, apricots, citrus, avocados, etc.). Grazing animals may be harmed by pasturing where NO_3 levels are high.

Tables 10.1 and 10.2 list groundwater quality parameters of interest in groundwaters that will be used for agricultural irrigation purposes. Most of the parameters listed as trace elements need only be monitored for certain sensitive crops.

Table 10.1 lists parameters and maximum levels that will normally cause no crop growing restrictions for sensitive plants. Table 10.2 gives additional information concerning degrees of restriction for different parameter levels and the influence of the form of irrigation (sprinkler or surface watering). Much of the information has been adapted from FAO (1986).

10.2 SODIUM ADSORPTION RATIO

The sodium adsorption ratio indicates the amount of sodium present in soils, relative to calcium and magnesium. If the sodium fraction is too large, soil permeability may be low and the movement of water through the soil is restricted. SAR is important for plant growth because its magnitude is an indication of the availability of soil pore water to plant roots.

A sodium ion has a large radius of hydration. This means that its hydration sphere of bound water molecules is larger than most cations. When dissolved sodium is adsorbed into clay-bearing soils, it causes them to disperse and swell. This reduces the soil pore size and causes the permeability to water to be greatly reduced. If the sodium ion is replaced by ion exchange with cations having a smaller radius of hydration, the soil dispersion is reduced.

In most natural waters, Ca^{2+}, Mg^{2+}, and Na^+ are by far the most abundant ions, so other ions can usually be neglected in their effect on soil dispersion. Also, Ca^{2+} and Mg^{2+}, being doubly charged, are more tightly adsorbed to clay surfaces than sodium and, so, are preferentially adsorbed. They also have smaller radii of hydration and cause less soil dispersion. The relative amounts of sodium, calcium, and magnesium adsorbed to soil are proportional to the amounts dissolved in groundwater. By measuring the concentrations of sodium, calcium, and magnesium in water used for irrigation, the SAR can be calculated using Equation 10.1 and the risk of low soil permeability is evaluated.

$$SAR = \frac{[Na^+]}{\sqrt{\dfrac{[Ca^{2+}] + [Mg^{2+}]}{2}}} \tag{10.1}$$

where
$\quad$ $[Na^+]$ is the water concentration of sodium in eq/L
$\quad$ $[Ca^{2+}]$ is the water concentration of calcium in eq/L
$\quad$ $[Mg^{2+}]$ is the water concentration of magnesium in eq/L

10.2.1 WHAT SAR VALUES ARE ACCEPTABLE?

Acceptable SAR values for irrigation water depend on the particular water and soil characteristics and are often considered in conjunction with the specific conductivity of irrigation water. Since high conductivity means that the water contains high levels of dissolved solids, it generally will contain high levels of calcium and magnesium. Even if the concentrations of calcium and magnesium are low, high conductivity means that the water contains many other cations that will compete with sodium for soil adsorption sites. Thus, high conductivity limits the deleterious effects of sodium (Figure 10.1). It should be noted that water with specific conductivity greater than about 3000 $\mu S/cm$ is of poor quality for irrigation regardless of the SAR value. When plant roots encounter water with too high a concentration of TDS, the osmotic pressure balance across the outer root membrane drives water from within the plant out into the soil, desiccating the plant.

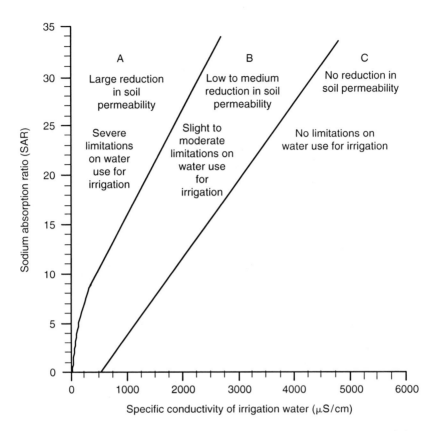

FIGURE 10.1 Effects of sodium absorption ratio (SAR) and specific conductivity on soil permeability. Zone A exhibits large reductions in soil permeability to water and will severely restrict the amount of water available to plant roots. Zone B has low to moderate reductions in soil permeability and may require care in irrigation practices and choice of crops. Zone C water has little or no reduction in soil permeability. Note, however, that water with specific conductivity greater than 3000 μS is of poor quality for agriculture because of osmotic stress, regardless of soil permeability.

RULES OF THUMB

1. Water having SAR values less than 3 will not diminish soil permeability when specific conductivity is greater than about 200 μS/cm.
2. Water having SAR values between 3 and 6 and specific conductivity between 500 and 2000 μS/cm may require care with irrigation methods. Soil permeability may be diminished.
3. Water having SAR values greater than 10 and specific conductivity less than 500 μS/cm will require considerable care with irrigation methods. Soil permeability may be too low to allow sufficient water to reach plant roots.

EXAMPLE 1

DETERMINATION OF SODIUM ADSORPTION RATIO

Determine the SAR for two different water supplies, A and B. Measured concentrations were

Supply A	$Ca^{2+} = 4.6$ mg/L	$Mg^{2+} = 1.1$ mg/L	$Na^+ = 144$ mg/L	TDS $= 255$ mg/L
Supply B	$Ca^{2+} = 108$ mg/L	$Mg^{2+} = 33$ mg/L	$Na^+ = 63$ mg/L	TDS $= 1050$ mg/L

Answer:

1. Determine the Ca^{2+}, Mg^{2+}, and Na^+ concentrations in meq/L (see discussion of equivalent weight in Chapter 1).

	Equivalent Weight (g/eq or mg/meq)	Concentration			
		Water Supply A		Water Supply B	
Constituent		(mg/L)	(meq/L)	(mg/L)	(meq/L)
Ca^{2+}	20.0	4.6	0.23	108	5.40
Mg^{2+}	12.15	1.1	0.090	33	2.72
Na^+	23.0	255	11.1	63	2.74

2. Compute the SAR of the two waters.
   a. For Supply A

$$SAR = \frac{[Na^+]}{\sqrt{\dfrac{[Ca^{2+}] + [Mg^{2+}]}{2}}} = \frac{11.1}{\sqrt{\dfrac{0.23 + 0.090}{2}}} = 27.8$$

This is a very high SAR value, and the water would be generally unacceptable for irrigation. Note that the TDS value for water supply A is moderate (255 mg/L) and that Supply A water is generally excellent for domestic use.
   b. For Supply B

$$SAR = \frac{2.74}{\sqrt{\dfrac{5.40 + 2.72}{2}}} = 1.36$$

Water supply B has a low SAR value and a TDS value (1050 mg/L, equivalent to about 1750 μS/cm, see Chapter 3, Section 3.12.3) that is beneficial for agriculture. It would be an excellent source of irrigation water. The TDS value is high for potable water and water supply B is an inferior source of domestic water compared with Supply A. Supply B will require more treatment to produce acceptable finished drinking water.

10.3 DEICING AND SANDING OF ROADS: CONTROLLING ENVIRONMENTAL EFFECTS

Road sanding and deicing, undertaken to enhance winter highway safety, have the potential to contribute significant amounts of sediment and chemicals to the receiving waters of surface runoff. To minimize impacts on surface waters, it is often necessary to incorporate physical and operational controls designed to

- Reduce the application of sand and deicing chemicals as far as practicable.
- Manage surface flow from treated roads and stockpiled materials in a manner that retains sediment and infiltrates dissolved chemicals.

10.3.1 METHODS FOR MAINTAINING WINTER HIGHWAY SAFETY

Snow and ice on roads reduce wheel traction and cause drivers to have less control of their vehicles. Highway departments currently use a site- and event-specific combination of three approaches for mitigating the effects of highway snow and ice:

- Apply antiskid materials, such as sand or other gritty solids, to road surfaces to improve traction.
- Apply deicing chemicals that melt snow and ice by lowering the freezing point of water.
- Plow roads to remove the snow and ice.

Although highway safety is the first concern in the use of snow control measures, environmental impacts are also important. Many highway departments are evaluating the effectiveness of alternative chemicals and operating procedures for minimizing the environmental impacts of sanding, deicing, and snow removal, while not compromising road safety.

10.3.2 ANTISKID MATERIALS

The most commonly used antiskid material is sand, usually derived either from rivers or crushed aggregate. Other abrasives such as volcanic cinders, coal ash, mine tailings, etc. are sometimes used, based on their local availability and cost. River sand is rounded and smooth and is somewhat less effective than crushed aggregate, which is rough and angular. However, river sand is generally cleaner and less contaminated than crushed aggregate. Between 3% and 30% by volume of deicing chemicals are often mixed with sand for increased effectiveness. The amount of sand required is very site- and event-specific. For example, in the Denver, Colorado Metro Area, the average amount of sand applied per snow event is 800–1200 lb per lane-mile of treated road, more sand generally being required in the western part of the city than in the eastern part (Chang et al., 1994). In Glenwood Canyon, Colorado, where post-event sand removal is especially difficult, highway maintenance personnel have reduced the use of sand recently from 280 to 60 lb per lane-mile by increasing the use of chemical deicers (Rocky Mountain Construction, 1995).

10.3.2.1   Environmental Concerns of Antiskid Materials

Air and water contamination are potential concerns with the use of sand and other antiskid grits. In Denver, fine particulates generated by traffic abrasion of road sand have been found to contribute around 45% of the atmospheric PM_{10} (airborne particulate matter less than 10 μm in diameter) load during winter. In 1997, Environmental Protection Agency (EPA) standards for PM_{10} were 50 mg/m^3, annual arithmetic mean, and 150 mg/m^3, 24 h arithmetic mean. Efforts to attain compliance with these standards have compelled communities to increasingly use chemical deicers in place of antiskid grits (Chang et al., 1994).

Although airborne particulates from road sand are significant as atmospheric pollution, they represent an insignificant fraction of the total mass of sand applied to the roads. Essentially all the sand applied for traction control becomes potential wash-load that eventually is either flushed to receiving waters (including sewers, streams, and lakes), trapped in sediment control structures, or swept up and deposited in landfills.

10.3.3   Chemical Deicers

A variety of water-soluble inorganic salts and organic compounds are used to melt snow and ice from roads. The most commonly used road deicer is sodium chloride (NaCl), because of its relatively low cost and high effectiveness. Other accepted road deicing agents are potassium chloride, calcium chloride, magnesium chloride, calcium magnesium acetate (CMA), potassium acetate, and sodium acetate.* These chemicals may be used in solid or liquid forms and are frequently combined with one another in various ratios. Different deicer formulations have been rated for overall value based on performance (melting, penetration, and disbonding snow from the road surface), corrosivity, spalling of road surface, environmental impacts, and cost (Chang et al., 1994). Commercial formulations that use chloride salts usually include corrosion inhibitors, generally regarded as effective and worth the additional cost.

10.3.3.1   Chemical Principles of Deicing

Water containing dissolved substances always has a lower freezing point than pure water. Any soluble substance will have some deicing properties. How far the freezing point of water is lowered depends only on the concentration of dissolved particles and not on their nature. Given the same concentration of dissolved particles, the freezing point lowering of water will be the same for sodium chloride, calcium chloride, ethylene glycol, or any other solute. The solubility of each deicing substance at the final solution temperature determines how many particles can go into solution, and this is the ultimate limit on the lowest freezing point attainable; ice will melt as long as the outdoor temperature is above the lowest freezing point of the solute–water mixture. Pure NaCl theoretically can melt ice at temperatures as low as $-6°F$, but no lower. Calcium chloride is effective down to $-67°F$.

---

* Several deicers, ethylene glycol, methanol, and urea, are useful mainly for specialized purposes such as airplane and runway deicing, but seldom find highway use for reasons of poor performance, high costs, toxicity, and difficulty in application.

When a salt dissolves to form positive and negative ions, each ion counts as a dissolved particle. Ionic compounds such as sodium chloride (NaCl) and calcium chloride ($CaCl_2$) are efficient deicers because they always dissociate into positive and negative ions upon dissolving, forming more dissolved particles per mole than nonionizing solutes. One NaCl molecule dissolves to form two particles, Na^+ and Cl^-; one $CaCl_2$ molecule forms three particles, one Ca^{2+} and two Cl^-; whereas the organic molecule ethylene glycol ($C_2H_6O_2$) does not dissociate and dissolves as one particle. Three molecules of dissolved ethylene glycol are needed to lower the freezing point the same amount as one molecule of calcium chloride. Another advantage of calcium and magnesium chlorides is that they dissolve exothermically, releasing a significant amount of heat that further helps to melt snow and ice. Conversely, NaCl does not release heat upon dissolving; in fact, the dissolution of NaCl is slightly endothermic and has a small cooling effect.

The differences in effectiveness for different deicing chemicals are related primarily to their different solubilities at environmental temperatures, number of dissolved particles formed per pound of material, and exothermicity of dissolution. Organic deicers (e.g., calcium magnesium acetate (CMA) and ethylene glycol) are said to be more effective than salts at breaking the bond between pavement and snow, allowing for easier plowing and snow removal. Organic deicers are also believed to be stored in surface pores of pavement, aiding in disbonding the snow and possibly prolonging the period of effective deicing.

10.3.3.2 Corrosivity

The main advantage of organic deicers, such as CMA, over inorganic chloride salts, such as NaCl, is their lower corrosivity. Corrosivity results from chemical and electrolytic reactions with solid materials. The chemical corrosivity of chloride salts arises mostly from the chemical reactivity of chloride ions and does not depend strongly on which salt is the source of the chloride ions. Electrolytic corrosivity affects metals, mainly iron alloys, and occurs when dissolved salt ions transfer electrons between zones of the metal surface having slightly different composition, allowing atmospheric or dissolved oxygen (DO) to chemically react with the metal. Electrolytic corrosivity depends, in a complicated fashion, on the nature of the metal surface and the nature of the dissolved ions. However, electrolytic corrosion for any surface will always increase as the total ion concentration (often measured as TDS or specific conductivity) increases.

When chloride is present, chemical and electrolytic corrosion acts synergistically to accelerate the overall corrosion rate. The addition of corrosion inhibitors to commercially formulated salt deicers is reported to reduce salt corrosivity. The main reason for using chloride salts rather than nitrates, fluorides, bromides, etc. is the relatively low toxicity of chlorides to plants and aquatic life.

10.3.4 Environmental Concerns of Chemical Deicers

Corrosivity, not adverse environmental impact, has been the main problem associated with the use of chemical deicers. While each of the common deicers has potential environmental effects, studies show that none pose strong adverse threats (Michigan Department of Transportation, 1993; Watershed Research, 1995). Most deicing residues are highly soluble and have low toxicity; they flush quickly through soils and waterways and, in general, rapidly become diluted to levels that pose no environmental problems. Even under a worst-case scenario, undesirable effects are likely to be observed only close to the points of application, where concentrations are the highest. Studies by the Michigan Department of Transportation show that the greatest impacts have been to sensitive vegetation adjoining treated roadways; stream and lake concentrations of chloride and other deicing chemicals seldom reach levels that are detrimental to aquatic life. The State of Michigan has found that, although little surface water and groundwater contamination has been observed that is directly attributable to deicing practices, much of it has resulted from spillage and poor storage practices (Michigan Department of Transportation, 1993).

10.3.5 Deicer Components and Their Potential Environmental Effects

10.3.5.1 Chloride Ion

There is no stream standard for chloride ion; it is generally regarded as a non-detrimental chemical component of state waters. Tests on fish showed no effect for concentrations of NaCl between 5,000 and 30,000 mg/L, depending on the species, exposure time, and water quality. Concentrations of chloride ion required to immobilize Daphnia in natural waters ranged between 2100 and 6143 mg/L (McKee and Wolf, 1963). It has been recommended that concentrations above 3000 mg/L can be considered deleterious to both fish-food organisms and fish fry and that a permissible limit of 2000 mg/L can be established for fresh waters. However, these recommendations have not been acted upon at either the Federal level or in Colorado. The United States Environmental Protection Agency (USEPA) secondary drinking water standard of 250 mg/L for chloride, based on average taste thresholds, was found to be seldom exceeded in a Michigan study of road deicing impacts (Michigan Department of Transportation, 1993).

Chloride in road splash can "burn" sensitive vegetation adjacent to treated roads; plant uptake may cause osmotic stress. Theoretically, chloride can form complexes that increase mobility of metals in soils, but very few such cases have been confirmed (Michigan Department of Transportation, 1993). Spring thaw surges may temporarily create surface water chloride levels detrimental to aquatic biota. However, dilution quickly occurs and flowing streams have not been significantly impacted (Michigan Department of Transportation, 1993). Spring thaw surges may temporarily cause high chloride levels in groundwater, although reports of excessive levels (>250 mg/L) are rare (Michigan Department of Transportation, 1993).

10.3.5.2 Sodium Ion

Sodium is even less toxic to aquatic biota than chloride. There are no water quality standards for sodium ion. The main problems associated with sodium ion are its effects on agricultural soil permeability (see Section 10.2) and the necessity for restricted sodium intake by hypertensive people. Na$^+$ levels in groundwater can increase temporarily during spring thaws, posing a health threat to persons requiring low sodium intake.

10.3.5.3   Calcium, Magnesium, and Potassium Ions

All are plant, animal, and human nutrients and there are no stream or drinking water standards for these substances. Calcium and magnesium improve soil aeration and permeability by decreasing the SAR. Calcium and magnesium also increase water hardness beneficially, reducing the toxic effects of dissolved heavy metals on aquatic life. Theoretically, these cations could increase heavy metal mobility in soils by exchange processes, but there is little documentation of such behavior.

10.3.5.4   Acetate

Acetate has no drinking water standards and has lower toxicity than NaCl. It biodegrades rapidly and does not accumulate in the environment. The only reported potential environmental problem with acetate is that, in large concentrations, it can deplete oxygen levels in surface waters by increasing biological oxygen demand (BOD) during biodegradation.

10.3.5.5   Impurities Present in Deicing Materials

Deicers contain trace amounts of heavy metals and sometimes phosphorus and nitrogen. These can be released with snowmelt, especially during spring thaw. Because the heavy metal impurities become mostly associated with solids, they are best controlled by sediment containment. Phosphorus and nitrogen will be controlled by infiltration of snowmelt into pervious areas, where they encourage vegetative growth.

10.4   DRINKING WATER TREATMENT

Clean drinking water is the most important public health factor. But over 2 billion people worldwide do not have adequate supplies of safe drinking water. Worldwide, between 15 and 20 million babies die every year due to waterborne diarrheal diseases such as typhoid fever, dysentery, and cholera. Contaminated water supplies and poor sanitation cause 80% of the diseases that afflict people in the poorest countries. The development of municipal water purification in the last century has allowed cities in the developed countries to be essentially free of waterborne diseases. Since the introduction of filtration and disinfection of drinking water in the United States, waterborne diseases such as cholera and typhoid have been virtually eliminated.

However, in 1974, it was discovered that water disinfectants themselves react with organic compounds naturally occurring in water to form unintended disinfection by-products (DBPs) that may pose health risks (Bellar et al., 1974; Rook, 1974; Bull and Kopfler, 1991). Trihalomethane DBPs were regulated by EPA in 1979 (USEPA, 1979).

Since then, several DBPs (bromodichloromethane, bromoform, chloroform, dichloroacetic acid, and bromate) have been shown to be carcinogenic in laboratory animals at high doses. Some DBPs (bromodichloromethane, chlorite, and certain haloacetic acids) can also cause adverse reproductive or developmental effects in laboratory animals. Believing that DBPs present a potential public health risk, EPA published guidelines for minimizing their formation (USEPA, 1997, 1999) and established standards in 1998 for drinking water concentrations of DBPs and disinfectant residuals. The goal of EPA disinfectant and disinfection by-product (DBP) regulations is to balance the health risks of pathogen contamination against DPB formation.

10.4.1   WATER SOURCES

Drinking water supplies are either from surface waters or groundwater. In the United States, groundwater sources (wells) supply about 53% of all drinking water and surface water sources (reservoirs, rivers, and lakes) supply the remaining 47%. Groundwater comes from underground aquifers, where wells are bored to recover the water. Wells may be from tens to hundreds of meters deep. Generally, water in deep aquifers is replaced by percolation from the surface very slowly, hundreds to thousands of years. Water in the deep Ogallala aquifer in the Great Plains region of the United States is estimated to be thousands of years old and is called "fossil water." Replenishment of such aquifers occurs over thousands of years and it is easy to withdraw water from them at a rate that greatly exceeds replacement. Such aquifers are essentially nonrenewable resources in our lifetimes. The Ogallala aquifer has been depleted significantly over the past several decades, principally by agricultural irrigation.

Groundwater tends to be less contaminated than surface water. It is normally more protected from surface contamination and, because it moves more slowly, organic matter has time to be decomposed by soil bacteria. The soil itself acts as a filter so that less suspended matter is present.

Surface water comes from lakes, rivers, and reservoirs. It usually has more suspended materials than groundwater, and so requires more processing to make it safe to drink. Surface waters are used for other purposes besides drinking and often become polluted by sewage, industrial, and recreational activities. On most rivers, the fraction of "new" water diminishes with distance from the headwaters, as the water becomes more and more used and reused. On the Rhine River in Europe, for example, communities near the mouth of the river receive as little as 40% "new" water in the river. All the other water has been previously discharged by an upstream city or originates as nonpoint source return flow from agricultural activities. Water treatment must make this quality of river water fit to drink. Filtration through sand was the first successful method of municipal water treatment, used in London in the middle 1800s. It led to an immediate decline in the amount of waterborne disease.

10.4.2   WATER TREATMENT

Major changes are occurring in the water treatment field, driven by increasingly tighter water quality standards, a steady increase in the number of regulated drinking water contaminants (from about 5 in 1940 to around 100 in 1999), and new regulations affecting disinfection and DBPs. Municipalities constantly seek to refine their water treatment and provide high-quality water by more economical means. A recent development in water treatment is the application of membrane filtration to drinking water treatment. Membrane filters have been refined to the point where, in certain cases, they are suitable as stand-alone treatment for small systems. More often, they are used in conjunction with other treatment methods to economically improve the overall quality of finished drinking water.

10.4.3 BASIC DRINKING WATER TREATMENT

The purpose of water treatment is, first, to make water safe to drink by ensuring that it is free of pathogens and toxic substances. The second goal is to make it a desirable drink by removing offensive turbidity, tastes, colors, and odors.

Conventional drinking water treatment addresses both of these goals. It consists of four steps:

1. Primary settling
2. Aeration
3. Coagulation
4. Disinfection

Not all four of the basic steps are needed in every treatment plant. Groundwaters, in particular, usually need much less treatment than surface waters. Groundwaters may need no settling, aeration, or coagulation. For clean groundwaters, only a little chlorine (≈ 0.16 ppm) is added to protect the water while in the distribution system. The relatively new treatment technology of membrane filtration is increasingly being used in conjunction with the more traditional treatments and, also, as a stand-alone treatment.

10.4.3.1 Primary Settling

Water, which has been coarsely screened to remove large particulate matter, is brought into a large holding basin to allow finer particulates to settle. Chemical coagulants may be added to form floc. Lime may be added at this point to help clarification if pH < 6.5. The floc settles by gravity, removing solids larger than about 25 μm.

10.4.3.2 Aeration

The clarified water is agitated with air. This promotes oxidation of any easily oxidizable substances, i.e., those which are strong reducing agents. Chlorine will be added later. If chlorine was added at this point and reducing agents were still in the water, they would reduce the chlorine and make it ineffective as a disinfectant.

Ferrous iron (Fe^{2+}) is a particularly troublesome reducing agent. It may arise from the water passing through iron pyrite (FeS_2) or iron carbonate ($FeCO_3$) minerals. If DO is present, Fe^{2+} is oxidized to Fe^{3+}, which precipitates as ferric hydroxide, $Fe(OH)_3$, at pH greater than 3.5. $Fe(OH)_3$ gives a metallic taste to the water and causes the ugly red-brown stain commonly found in sinks and toilets in iron-rich regions. The stain is easily removed with weak acid solutions, such as vinegar.

10.4.3.3   Coagulation

The finest sediments, such as pollen, spores, bacteria, and colloidal minerals, do not settle out in the primary settling step. For the finished water to look clear and sparkling, these fine sediments must be removed. Hydrated aluminum sulfate, $Al_2(SO_4)_3 \cdot 18H_2O$, sometimes called alum or filter alum, applied with lime, $Ca(OH)_2$, is the most common filtering agent used for secondary settling:

$$Al_2(SO_4)_3 + Ca(OH)_2 \rightarrow Al(OH)_3(s) + CaSO_4 \qquad (10.2)$$

At pH 6–8, $Al(OH)_3(s)$ is formed as a light, fluffy, gelatinous flocculant having an extremely large surface area that attracts and traps small suspended particles, carrying them to the bottom of the tank as the precipitate slowly settles. In this pH range, $Al(OH)_3$ is near its minimum solubility and very little Al^{3+} is left in solution.

10.4.3.4   Disinfection

Killing bacteria and viruses is the most important part of water treatment. Proper disinfection provides a residual disinfectant level that persists throughout the distribution system. This not only kills organisms that pass through filtration and coagulation at the treatment plant, but it also prevents reinfection during the time the water is in the distribution system. In a large city, water may remain in the system for 5 days or more before it is used. Five days is long enough for any missed microorganisms to multiply. Leaks and breaks in the water mains can permit recontamination, especially at the extremities of the system where the pressure is low. High pressure always causes the flow at leaks to be from inside to outside. But at low pressure, bacteria can seep in.

As a result of concerns about DBPs, EPA and the water treatment industry are placing more emphasis on the use of disinfectants other than chlorine, which at present is the most commonly used water disinfectant. Another approach to reducing the probability of DBP formation is by removing DBP precursors (naturally occurring organic matter [NOM]) from water before disinfection. However, use of alternative disinfectants has also been found to produce DBPs and current regulations try to balance the risks between microbial pathogens and DBPs. DBPs include the following, not all of which pose health risks:

- Halogenated organic compounds, such as trihalomethanes (THMs), haloacetic acids, haloketones, and other halogenated compounds that are formed primarily when chlorine or ozone (in the presence of bromide ion) are used for disinfection.
- Organic oxidation by-products, such as aldehydes, ketones, assimilable organic carbon (AOC), and biodegradable organic carbon (BDOC). The latter two DBPs result from large organic molecules being oxidized to

smaller molecules, which are more available to microbes, plant, and aquatic life as a nutrient source. Oxidized organics are formed when strong oxidizing agents (ozone, permanganate, chlorine dioxide, or hydroxyl radical) are used.
• Inorganic compounds, such as chlorate, chlorite, and bromate ions, formed when chlorine dioxide and ozone disinfectants are used.

10.4.3.5 Disinfection Procedures

Most disinfectants are strong oxidizing agents that react with organic and inorganic oxidizable compounds in water. In some cases, the oxidant is produced as a reaction by-product; hydroxyl radical is formed in this way. In addition to destroying pathogens, disinfectants are also used for removal of disagreeable tastes, odors, and color. They also can assist in the oxidation of dissolved iron and manganese, prevention of algal growth, improvement of coagulation and filtration efficiency, and control of nuisance water organisms such as Asiatic clams and zebra mussels.

The most commonly used water treatment disinfectant is chlorine. It was first used on a regular basis in Belgium in the early 1900s. Other disinfectants sometimes used are ozone, chlorine dioxide, and ultraviolet (UV) radiation. Of these, only chlorine and chlorine dioxide have residual disinfectant capability. With chlorine or chlorine dioxide, addition of a small excess of disinfectant maintains protection of the drinking water throughout the distribution system. Normally, residual chlorine or chlorine dioxide concentration of about 0.2–0.5 mg/L is sought. Disinfectants that do not have residual protection are normally followed by a low dose of chlorine in order to preserve the disinfection capability throughout the distribution system.

Part of the disinfection procedure involves removing DBP precursors, mainly total organic carbon (TOC), by coagulation, water softening, or filtration. A high TOC concentration (greater than 2.0 mg/L) indicates a high potential for DBP formation. Typical required reduction percentages of TOC for conventional treatment plants are given in Table 10.3.

10.4.4 Disinfection By-Products and Disinfection Residuals

The principal precursor of organic DBPs is naturally occurring organic matter. NOM is usually measured as TOC or dissolved organic carbon (DOC). Typically, about 90% of TOC is in the form of DOC. DOC is defined as the part of TOC that passes through a 0.45 μm filter. Halogenated organic by-products are formed in water when NOM reacts with free chlorine (Cl_2) or free bromine (Br_2). Free chlorine may be introduced when chlorine gas, chlorine dioxide, or chloramines are added for disinfection. Free bromine is a product of the oxidation by disinfectants of bromide ion already present in the source water.

Reactions of strong oxidants with NOM also form nonhalogenated DBPs, particularly when nonchlorine oxidants such as ozone and peroxone are used. Common nonhalogenated DBPs include aldehydes, ketones, organic acids, ammonia, and hydrogen peroxide.

Bromide ion (Br^-) may be present, especially where geothermal waters impact surface and groundwaters, and in coastal areas where saltwater incursion is occurring.

TABLE 10.3

Required Percentage Removal of Total Organic Carbon by Enhanced Coagulation for Conventional Water Treatment Systems[a]

Source Water TOC (mg/L)	Source Water Alkalinity (mg/L as $CaCO_3$)		
	0–60	>60–120	>120
>2.0–4.0	35.0%	25.0%	15.0%
>4.0–8.0	45.0%	35%	25.0%
>8.0	50.0%	40.0%	30.0%

Note: Enhanced coagulation is determined, in part, as the coagulant dose where an incremental addition of 10 mg/L of alum (or an equivalent amount of ferric salt) results in a TOC removal to below 0.3 mg/L.

[a] Conventional water treatment systems here apply to utilities using surface water and groundwater impacted by surface water.

Ozone or free chlorine oxidizes Br^- to form brominated DBPs, such as bromate ion, bromoform, cyanogen bromide, bromopicrin, and brominated acetic acid.

10.4.5   STRATEGIES FOR CONTROLLING DISINFECTION BY-PRODUCTS

Once formed, DBPs are difficult to remove from a water supply. Therefore, DBP control is focused on preventing their formation. The chief control measures for DBPs are:

- Lowering NOM concentrations in source water by coagulation and settling, filtering, and oxidation.
- Using sorption on granulated activated carbon (GAC) to remove DOC.
- Moving the disinfection step later in the treatment train, so that it comes after all processes that decrease NOM.
- Limiting chlorine to providing residual disinfection, following primary disinfection with ozone, chlorine dioxide, chloramines, or UV radiation.
- Protection of source water from bromide ion.

Table 10.4 is a list of the cancer classifications assigned by EPA for disinfectants and DBPs as of January 1999.

10.4.6   CHLORINE DISINFECTION TREATMENT

Chlorine is a corrosive and toxic yellow-green gas at room temperature, with a strong irritating odor. It is stored and shipped as liquefied gas. Chlorine is the most widely used water treatment disinfectant because of its many attractive features:

TABLE 10.4
EPA Cancer Classifications for Disinfectants and DBPs

Compound	Cancer classification[a]
Chloroform	B2
Bromodichloromethane	B2
Dibromochloromethane	C
Bromoform	B2
Monochloroacetic acid	
Dichloroacetic acid	B2
Trichloroacetic acid	C
Dichloroacetonitrile	C
Bromochloroacetonitrile	
Dibromoacetonitrile	C
Trichloroacetonitrile	
1,1-Dichloropropanone	
1,1,1-Trichloropropanone	
2-Chlorophenol	D
2,4-Dichlorophenol	D
2,4,6-Trichlorophenol	B2
Chloropicrin	
Chloral hydrate	C
Cyanogen chloride	
Formaldehyde	B1[b]
Chlorate	
Chlorite	D
Bromate	B2
Ammonia	D
Hypochlorous acid	
Hypochlorite	
Monochloramine	
Chlorine dioxide	D

Source: USEPA, 1999, *Alternative Disinfectants and Oxidants Guidance Manual*, EPA 815-R-99–014, April.

[a] The EPA classifications for carcinogenic potential of chemicals are (USEPA, 1996)

A: Human carcinogen	Sufficient evidence in epidemiologic studies to support causal association between exposure and cancer.
B: Probable human carcinogen	Limited evidence in epidemiologic studies (B1) and/or sufficient evidence from animal studies (B2).
C: Possible human carcinogen	Limited evidence from animal studies and inadequate or no data in humans.
D: Not classifiable	Inadequate or no animal and human evidence of carcinogenicity.
E: No evidence of carcinogenicity for humans	No evidence of carcinogenicity in at least two adequate animal tests or in adequate epidemiologic and animal studies.

- It is effective against a wide range of pathogens commonly found in water, particularly bacteria and viruses.
- It leaves a residual that stabilizes water in distribution systems against reinfection.
- It is economical and easily measured and controlled.
- It has been used for a long time and represents a well-understood treatment technology. It maintains an excellent safety record despite the hazards of handling chlorine gas.
- Chlorine disinfection is available from sodium and calcium hypochlorite salts, as well as from chlorine gas. Hypochlorite solutions may be more economical and convenient than chlorine gas for small treatment systems.

In addition to disinfection, chlorination is used for

- Taste and odor control, including destruction of hydrogen sulfide
- Color bleaching
- Controlling algal growth
- Precipitation of soluble iron and manganese
- Sterilizing and maintaining wells, water mains, distribution pipelines, and filter systems
- Improving some coagulation processes

Problems with chlorine usage include:

- Not effective against *Cryptosporidium* and limited effectiveness against *Giardia lamblia* protozoa.
- Reactions with NOM can result in the formation of undesirable DBPs.
- The hazards of handling chlorine gas require special equipment and safety programs.
- If site conditions require high chlorine doses, taste and odor problems may arise.

Chlorine dissolves in water by the following equilibrium reactions:

$$Cl_2(g) \leftrightarrow Cl_2(aq) \tag{10.3}$$

$$Cl_2(aq) + H_2O \leftrightarrow H^+(aq) + Cl^-(aq) + HOCl(aq) \tag{10.4}$$

$$HOCl(aq) \leftrightarrow H^+(aq) + OCl^-(aq) \tag{10.5}$$

At pH values below 7.5, hypochlorous acid (HOCl) is the dominant dissolved chlorine species (see Figure 10.2). Above pH 7.5, chlorite anion (OCl^-) is dominant. The formation of H^+ indicates that chlorination reduces total alkalinity.

The active disinfection species, Cl_2, HOCl, and OCl^-, are called the total free available chlorine, although Cl_2 is insignificant above pH 2. All these species are oxidizing agents, but chloride ion (Cl^-) is not. HOCl is about 100 times more effective as a disinfectant than OCl^-. Thus, the amount of chlorine required for a given level of disinfection depends on the pH. Higher doses are needed at higher pH.

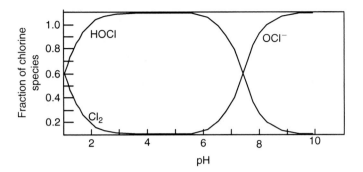

FIGURE 10.2 Distribution diagram for dissolved chlorine species. Free chlorine molecules, Cl_2, exist only below about pH 1. At pH 7.5, [HOCl] = [OCl$^-$].

At pH 8.5, 7.6 times as much chlorine must be used as at pH 7.0 for the same amount of disinfection. HOCl is more effective than OCl$^-$ because, as a neutral molecule, it can penetrate into cell membranes of microorganisms more easily than can OCl$^-$.

When chlorine gas is added to a water system, it dissolves according to Equations 10.3 through 10.5. All substances present in the water that are oxidizable by chlorine constitute the chlorine demand, see Figure 10.3. Until oxidation of these substances is complete, all the added chlorine is consumed; the net dissolved chlorine concentration remains zero as chlorine is added. When no chlorine-oxidizable matter is left, i.e., the chlorine demand has been met, the dissolved chlorine concentration (chlorine residual) increases in direct proportion to the additional dose.

If chlorine demand is zero, residual always equals the dose, and the plot is a straight line of slope = 1, passing through the origin of Figure 10.3. Because the boiling point of molecular chlorine is −35°C at 1 atm pressure, chlorine is supplied

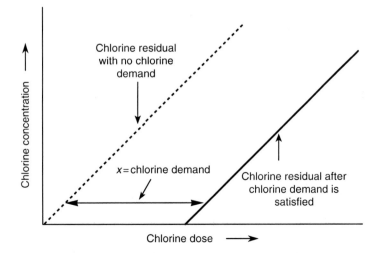

FIGURE 10.3 Relations among chlorine dose, chlorine demand, and chlorine residual.

and stored as the bulk liquid under pressure. The total time of water in the chlorine disinfection tank is generally about 20–60 min. A typical concentration of residual chlorine in the finished water is 1 ppm or less.

10.4.6.1   Hypochlorite

In addition to chlorine gas, the active disinfecting species HOCl and OCl^- can be obtained from hypochlorite salts, chiefly sodium hypochlorite (NaOCl) and calcium hypochlorite ($Ca(OCl)_2$). The salts react in water according to Equations 10.6 and 10.7.

$$NaOCl + H_2O \rightarrow HOCl + Na^+ + OH^- \qquad (10.6)$$

$$Ca(OCl)_2 + 2H_2O \rightarrow 2HOCl + Ca^{2+} + 2OH^- \qquad (10.7)$$

Note that adding chlorine gas to water lowers the pH, Equations 10.3 through 10.5, while hypochlorite salts raise the pH.

Sodium hypochlorite salts are available as the dry salt or in aqueous solution. The solution is corrosive with a pH of about 12. One gallon of 12.5% sodium hypochlorite solution is equivalent of about 1 lb of chlorine gas as a disinfectant. Unfortunately, sodium hypochlorite presents storage problems; after 1 month of storage under the best of conditions (low temperature, dark, and no metal contact), a 12.5% solution will have degraded to about 10%. On-site generation of sodium hypochlorite is accomplished by passing low voltage electrical current through a NaCl solution. On-site generation allows smaller quantities to be stored and makes the use of more stable dilute solutions (0.8%) feasible.

Calcium hypochlorite is commonly available as dry salt, which contains about 65% available chlorine. Almost 1.5 lb of calcium hypochlorite are equivalent to about 1 lb of chlorine gas as a disinfectant. Storage is less of a problem with calcium hypochlorite; normal storage conditions result in a 3%–5% loss of its available chlorine per year.

10.4.6.2   Definitions

Chlorine dose: The amount of chlorine originally used.

Chlorine residual: The amount remaining at time of analysis.

Chlorine demand: The amount used up in oxidizing organic substances and pathogens in the water, i.e., the difference between the chlorine dose and the chlorine residual.

Free available chlorine: The total amount of HOCl and ClO^- in solution. (Cl_2 is not present above pH 2.)

10.4.7   Drawbacks to Use of Chlorine: Disinfection By-Products

10.4.7.1   Trihalomethanes

The problem of greatest concern with the use of chlorine is the formation of chlorination by-products, particularly THMs ($CHCl_3$, $CHBrCl_2$, $CHBr_2Cl$, $CHBr_3$,

$CHCl_2I$, $CHBrClI$) and carbon tetrachloride (CCl_4) as possible carcinogens. It was once thought that THMs were formed by chlorination of dissolved methane. It is now known that they come from the reaction of $HOCl$, with acetyl groups in NOM, chiefly humic acids. Humic acids are breakdown products of plant materials like lignin. There is no evidence that chlorine itself is carcinogenic.

10.4.7.2 Chlorinated Phenols

If phenol or its derivatives from industrial activities are in the water, taste and color can be a problem. Phenols are easily chlorinated, forming compounds with extremely penetrating antiseptic odors. The most common chlorinated phenols arising from chlorine disinfection are shown in Table 10.5, with their odor thresholds, which are in the ppb ($\mu g/L$) range. At the ppm level, chlorinated phenols make water completely unfit for drinking or cooking. If phenol is present in the intake water, treatment choices are to employ additional nonchlorine oxidation for removing phenol, remove phenol with activated charcoal, or use a different disinfectant. The activated charcoal treatment is expensive and few communities use it.

In addition to the general strategies for controlling DBPs listed earlier, another option is available with chlorine use. Addition of ammonia with chlorination forms chloramines (see Section 3.9.7.3). Chloramines are weaker oxidants than chlorine and are useful for providing a residual disinfectant capability with a lower potential for forming DBPs.

EXAMPLE 2

Water has begun to seep into the basement of a home. The home's foundation is well above the water table and this problem had not been experienced before. The house is located about 50 ft downgradient from a main water line and one possibility is that a leak has occurred in the pipeline. The water utility company tested water entering the basement for the presence of chlorine, reasoning that if the water source was the pipeline, the chlorine residual should be detected. When no chlorine was found, the utility concluded that they were not responsible for the seep. Was this conclusion justified? Note that the maximum residual disinfection level (MRDL) mandated by EPA is 4.0 mg/L.

Answer:

No. Water would have to travel at least 50 ft through soil from the pipeline to the house. The chlorine residual should not exceed 4 mg/L and would almost certainly contact enough oxidizable organic and inorganic matter in the soil to be depleted below detection. A better water source marker would be fluoride, assuming the water supply is fluoridated. Although fluoride might react with calcium and magnesium in the soil to form solid precipitates, it is more likely to be detectable at the house than chlorine. However, neither test is conclusive. The simplest and best test would be to turn off the water in the pipeline for long enough to observe any change in water flow into the house. This, however, might not be possible. Another approach would be to examine the water line for leaks, using a video camera probe or soil conductivity measuring equipment.

TABLE 10.5

Odor Thresholds of Phenol and Chlorinated Derivatives from Drinking Water Disinfection with Chlorine

Phenol Compound	Chemical Structure	Odor Threshold in Water (ppb)
Phenol		>1000
2-Chlorophenol		2
4-Chlorophenol		250
2,4-Chlorophenol		2
2,6-Chlorophenol		3
2,4,6-Chlorophenol		>1000

10.4.8 CHLORAMINES

Many utilities use chlorine for disinfection and chloramines for residual mainten-ance. Chloramines are formed in the reaction of ammonia with HOCl from chlorine, a process that is inexpensive and easy to control. The reactions are described in the discussion of breakpoint chlorination in Chapter 3, Section 3.9.7.3. Although the reaction of chlorine with ammonia can be used for the purpose of destroying ammonia, it also serves to generate chloramines, which are useful disinfectants that are more stable and longlasting in a water distribution system than free chlorine.

Thus, chloramines are effective for controlling bacterial regrowth in water systems, although they are not very effective against viruses and protozoa. The primary role of chloramine use in the United States is as a secondary disinfectant to provide residual treatment. Being weaker oxidizers than chlorine, chloramines also form far fewer DBPs. However, they are not useful for oxidizing iron and manganese. When chloramine disinfection is the goal, ammonia is added in the final chlorination step. Chloramines are always generated on-site.

Optimal chloramine disinfection occurs when the chlorine:ammonia–nitrogen (Cl_2:N) ratio by weight is around 4, before the chlorination breakpoint occurs. Under these conditions, monochloramine (NH_2Cl) and dichloramine ($NHCl_2$) are the main reaction products and these are the effective disinfectant species. The normal dose of chloramines is between 1 and 4 mg/L. Residual concentrations are usually maintained between 0.5 and 1 mg/L. The MRDL mandated by EPA is 4.0 mg/L.

10.4.9 CHLORINE DIOXIDE DISINFECTION TREATMENT

Chlorine dioxide (ClO_2) is a gas above 12°C, with high water solubility. Unlike chlorine, it reacts quite slowly with water, remaining mostly dissolved as a neutral molecule. It is a very good disinfectant, about twice as effective as HOCl from Cl_2, but also about twice as expensive. ClO_2 was first used as a municipal water disinfectant in Niagara Falls, New York, in 1944. In 1977, about 100 municipalities in the United States were using it and thousands in Europe. The main drawback to its use is that it is unstable and cannot be stored. It must be made and used on-site, whereas chlorine can be delivered in tank cars.

Much of its reactivity is because it is a free radical. ClO_2 cannot be compressed for storage because it is explosive when pressurized or at concentrations above 10% by volume in air. It decomposes in storage and can decompose explosively in sunlight, when heated or agitated suddenly. So it is never shipped and is always prepared on-site and used immediately. Typical dose rates are 0.1–1.0 ppm.

Sodium chlorite is used to make ClO_2 by one of the three methods:

$$5NaClO_2 + 4HCl \leftrightarrow 4ClO_2(g) + 5NaCl + 2H_2O \qquad (10.8)$$

$$2NaClO_2 + Cl_2(g) \rightarrow 2ClO_2(g) + NaCl \qquad (10.9)$$

$$2NaClO_2 + HOCl \rightarrow 2ClO_2(g) + NaCl + NaOH \qquad (10.10)$$

Sodium chlorite is extremely reactive, especially in the dry form, and it must be handled with care to prevent potentially explosive conditions. If chlorine dioxide generator conditions are not carefully controlled (pH, feedstock ratios, low feedstock concentrations, etc.), the undesirable by-products chlorite (ClO_2^-) and chlorate (ClO_3^-) may be formed.

Chlorine dioxide solutions below about 10 g/L will not have sufficiently high vapor pressures to create an explosive hazard under normal environmental conditions of temperature and pressure. For drinking water treatment, ClO_2 solutions are generally less than 4 g/L and treatment levels are generally between 0.07 and 2.0 mg/L.

As ClO_2 is an oxidizer and not a chlorinating agent, it does not form THMs or chlorinated phenols. Hence, it does not have taste or odor problems. Common applications for ClO_2 have been to control taste and odor problems associated with algae and decaying vegetation, reducing the concentrations of phenolic compounds, and oxidizing iron and manganese to insoluble forms. Chlorine dioxide can maintain a residual disinfection concentration in distribution systems. The toxicity of ClO_2 restricts the maximum dose. At 50 ppm, ClO_2 can lead to breakdown of red corpuscles with the release of hemoglobin. Therefore, the dose of ClO_2 is limited to 1 ppm.

10.4.10 OZONE DISINFECTION TREATMENT

Ozone (O_3) is a colorless, highly corrosive gas at room temperature, with a pungent odor, easily detectable at concentrations as low as 0.02 ppmv, well below a hazardous level. It is one of the strongest chemical oxidizing agents available, second only to hydroxyl free radical (HO•)*, among disinfectants commonly used in water treatment. Ozone use for water disinfection started in 1893 in the Netherlands and in 1901 in Germany. Significant use in the United States did not occur until the 1980s. Ozone is one of the most potent disinfectants used in water treatment today. Ozone disinfection is effective against bacteria, viruses, and protozoan cysts, including *Cryptosporidium* and *G. lamblia*.

Ozone is made by passing a high voltage electric discharge of about 20,000 V through dry, pressurized air.

$$3O_2(g) + energy \rightarrow 2O_3(g) \tag{10.11}$$

Equation 10.11 is endothermic and requires a large input of electrical energy. The ozone gas is transferred to water through bubble diffusers, injectors, or turbine mixers. Once dissolved in water, ozone reacts with pathogens and oxidizable organic and inorganic compounds. Gas not dissolved is released to the surroundings as off-gas and must be collected and destroyed by conversion back to oxygen before release to the atmosphere. Ozonator off-gas may contain as much as 3000 ppmv of ozone, well above a fatal level. Ozone is readily converted to oxygen by heating to above 350°C or by passing it through a catalyst held above 100°C. Occupational Safety and Health Administration (OSHA) currently requires released gases to contain no more than 0.1 ppmv ozone for worker exposure. Typical dissolved ozone concentrations in water near the ozonator are around 1 mg/L.

Dissolved ozone gas decomposes spontaneously in water by a complex mechanism that includes the formation of hydroxyl free radical, which is the strongest oxidizing agent available for water treatment. Hydroxyl radical essentially reacts at every molecular collision with many organic compounds. The very high reaction rate of hydroxyl radicals limits their half-life in water to the order of microseconds and their concentration to less than about 10^{-12} mol/L. Both ozone molecules and hydroxyl free radicals play prominent oxidant roles in water treatment by ozonation.

---

* The dot following OH represents an unpaired orbital electron, which gives the radical, its very high degree of reactivity.

Because ozone is unstable and highly reactive, it cannot be stored and must be generated as needed. Ozone concentrations of about 4%–6% are achieved in municipal and industrial ozonators. Ozone reacts quickly and completely in water, leaving no active residual concentration. Decomposition of ozone in water produces DO in addition to hydroxyl radical (a very reactive short-lived oxidant) both of which further aid in disinfection and reducing BOD, chemical oxygen demand (COD), color, and odor problems. The air–ozone mixture is typically bubbled through water for a 10–15 min contact time. The main drawbacks to ozone use have been its high capital and operating costs and the fact that it leaves no residual disinfection concentration. Since it offers no residual protection, ozone can only be used as a primary disinfectant. It must be followed by a light dose of secondary disinfectant, such as chlorine, chloramine, or chlorine dioxide, for a complete disinfection system.

Several ways to assist ozonation are by adding hydrogen peroxide (H_2O_2), using UV radiation, and raising the pH to around 10–11. Hydrogen peroxide decomposes to form the very reactive hydroxyl radical, greatly increasing the hydroxyl radical concentration above that generated by simple ozone reaction with water. Reactions of hydroxyl radicals with organic matter cause structural changes that make organic matter still more susceptible to ozone attack. Adding hydrogen peroxide to ozonation is known as the advanced oxidation process (AOP) or PEROXONE process. UV radiation dissociates peroxide, forming hydroxyl radicals at a more rapid rate. Raising the pH allows ozone to react with hydroxyl ions (OH^-, not the radical $HO\bullet$) to form additional hydrogen peroxide. In addition to increasing the effectiveness of ozone oxidation, peroxide and UV radiation are effective as disinfectants in themselves. The use of these ozonation enhancers is known as the AOP.

The equipment for ozonation is expensive, but the cost per gallon decreases with large-scale operations. Generally, only large cities use ozone. ClO_2 is not as problem-free as ozone, but it is cheaper to use for small systems.

In addition to disinfection, ozone is used for

- DBP precursor control
- Protection against *Cryptosporidium* and *Giardi*
- Taste and odor control, including destruction of hydrogen sulfide
- Color bleaching
- Precipitation of soluble iron and manganese
- Sterilizing and maintaining wells, water mains, distribution pipelines, and filter systems
- Improving some coagulation processes

10.4.10.1 Ozone DBPs

Although it does not form the chlorinated DBPs that are of concern with chlorine use, ozone can react to form its own set of oxidation by-products. When bromide ion (Br^-) is present, as where geothermal waters impact surface and groundwaters or in coastal areas where saltwater incursion is occurring, ozonation can produce bromate ion (BrO_3^-), a suspected carcinogen, as well as brominated THMs and other brominated DBPs. Controlling the formation of unwanted ozonation by-products is accomplished by pretreatment to remove organic matter (activated carbon filters and membrane filtration) and scavenge BrO_3^- (pH lowering and hydrogen peroxide addition).

When bromide is present, addition of ammonia with ozone forms bromamines, by reactions analogous to the formation of chloramines with ammonia and chlorine, and lessens the formation of bromate ion and organic DBPs.

10.4.11 POTASSIUM PERMANGANATE

Potassium permanganate salt ($KMnO_4$) dissolves to form the permanganate anion (MnO_4), a strong oxidant effective at oxidizing a wide variety of organic and inorganic substances. In the process, manganese is reduced to manganese dioxide (MnO_2), an insoluble solid that precipitates from solution. Permanganate imparts a pink to purple color to water and is, therefore, unsuitable as a residual disinfectant.

Although easy to transport, store, and apply, permanganate is generally too expensive for use as a primary or secondary disinfectant. It is used in drinking water treatment primarily as an alternative to chlorine for taste and odor control, iron and manganese oxidation, oxidation of DPB precursors, control of algae, and control of nuisance organisms such as zebra mussels and the Asiatic clam. It contains no chlorine and does not contribute to the formation of THMs. When used to oxidize NOM early in a water treatment train that includes posttreatment chlorination, permanganate can reduce the formation of THMs.

10.4.11.1 Peroxone (Ozone Plus Hydrogen Peroxide)

The peroxone process is among those called AOPs, which employ highly reactive hydroxyl radicals (OH•) as a major oxidizing species. Hydroxyl radicals are produced when ozone decomposes spontaneously. Accelerating ozone decomposition using, for example, UV radiation or addition of hydrogen peroxide, elevates the hydroxyl radical concentration and increases the rate of contaminant oxidation. When hydrogen peroxide is used, the process is called peroxone.

Like ozonation, the peroxone process does not provide a lasting disinfectant residual. Oxidation is more complete and much faster with peroxone than with ozone. Peroxone is the treatment of choice for oxidizing many chlorinated hydrocarbons, difficult to treat by any other oxidant. It is also used for inactivating pathogens and destroying pesticides, herbicides, and volatile organic compounds (VOCs). It can be more effective than ozone for removing taste and odor causing compounds such as geosmin and 2-methyliosborneol (MIB). However, it is less effective than ozone for oxidizing iron and manganese. Because hydroxyl radicals react readily with carbonate, it may be necessary to lower the alkalinity in water with a high carbonate level to maintain a useful level of radicals. Peroxone treatment produces similar DBPs as does ozonation. In general, it forms more bromate than does ozone under similar water conditions and bromine concentrations.

10.4.11.2 Ultraviolet Disinfection Treatment

Ultraviolet radiation at wavelengths below 300 nm is very damaging to life-forms, including microorganisms. Low-pressure mercury lamps, known as germicidal lamps, have their maximum energy output at 254 nm. They are very efficient with about 40% of their electrical input being converted to 254 nm radiation. Protein and DNA in microorganisms absorb radiation at 254 nm, leading to photochemical reactions that destroy the ability to reproduce. UV doses required to inactivate bacteria and viruses are relatively low, of the order of 20–40 $mW \cdot s/cm^2$. Much higher doses, 200 $mW \cdot s/cm^2$ or higher, are needed to inactivate *Cryptosporidium* and *G. lamblia*.

Color or high levels of suspended solids can interfere with transmission of UV through the treatment cell and UV absorption by iron species diminishes the UV energy absorbed by microorganisms. Such problems may necessitate higher UV dose rates or pretreatment filtration. UV reaction cells are designed to induce turbulent flow, have long water flow paths and short light paths (around 3 in.), and provide for cleaning of residues from the lamp housings to minimize these problems. Where used, usually in small water treatment systems, UV irradiation is generally the last step in the water treatment process, just after final filtration and before entering the distribution system. UV systems are normally easy to operate and maintain, although severe site conditions, such as high levels of dissolved iron or hardness, may require pretreatment.

UV does not introduce any chemicals into the water and causes little, if any, chemical change in water. Therefore, overdosing does not cause water quality problems. UV is used mostly for inactivating pathogens to regulated levels. Since it leaves no residual, it can serve only as a primary disinfectant and must by followed by some form of chemical secondary disinfection, generally chlorine or chloramine. UV water treatment is used more in Europe than in the United States. Small-scale units are available for individuals who have wells with high microbial levels.

10.4.11.2.1 Characteristics of UV treatment
- Short contact time of 1–10 s. Ozone and chlorine require 10–50 min, necessitating large reaction tanks. Ozonation can be run on a flow-through basis.
- Destroys most viruses and bacteria without chemical additives. The destruction of *G. lamblia*, however, requires prefiltration. Leaves no residual disinfection potential in the water so that, for water entering a distribution system, light chlorination is still needed to provide prolonged disinfection.
- Low overall installation costs. Ozone generators are expensive. Chlorine metering systems are not especially expensive but large reaction tanks and safety systems are high cost items.
- Not influenced by pH or temperature. Chlorination and ozonation work best at lower pH (chlorine because it is in the HOCl form; ozone because it decomposes more rapidly at higher pH). Chlorination and ozonation both require longer contact time at lower temperatures.
- No toxic residues. It adds nothing to the water unless some organics are present that photoreact to form toxic compounds. The formation of THMs or other DBPs is minimal.

10.4.11.3 Membrane Filtration Water Treatment

Membrane filters are being used to treat groundwater, surface water, and reclaimed wastewater. Membrane filtration is a physical separation process and removes

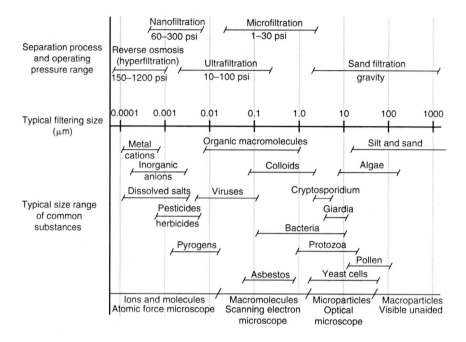

FIGURE 10.4 Comparison of filter processes and size ranges.

unwanted substances from water without utilizing chemical reactions that can lead to undesirable by-products. The range of membrane filters available is shown in Figure 10.4, along with common substances that can be removed by filtering. Although membranes sometimes serve as a stand-alone treatment, they are more often combined with other treatment technologies. For example, currently available microfiltration (MF) and ultrafiltration (UF) membranes are not very effective in removing DOC, some synthetic organic compounds, or THM precursors. Their performance in these respects is improved by adding powdered adsorbent material to the wastewater flow. Contaminants that might pass through the filters are adsorbed to the larger adsorbent particles and rejected by the filters.

Organic membrane filters are made from several different organic polymer films, normally formed as a thin film on a supporting woven or nonwoven fabric. Inorganic membrane filters are made from ceramics, glass, or carbon. They generally consist of a more porous supporting layer on which a thin microporous layer is chemically deposited. Inorganic membranes resist higher pressures, a wide pH range, and more extreme temperatures than do organic membranes. Their main disadvantages are greater weight and expense.

Filters can be fabricated to remove substances as small as dissolved ions, so they are useful for removing TDS, nitrate sulfate, radium, iron, manganese, DBP precursors, bacteria, viruses, and other pathogens from water without adding chemicals. It must also be recognized that there will be imperfections in manufactured membrane filters through which contaminants may pass. This is of particular concern with pathogens. Therefore, filters must never be regarded as having 100% rejection for any size range. Furthermore, filters do not protect water from reinfection after it has entered a distribution system, so it is common to add chlorine or another residual disinfectant at the end of the treatment chain for this purpose. As most organic matter has already been removed, end-of-treatment chlorination does not generate significant DBPs.

Unlike coarser filters operating in a "normal" mode, where all of the water passes through the filter surface, membrane filters operate in a *crossflow* mode. In crossflow filtration, the feed, or influent, stream is separated into two separate effluent streams. The pressurized feed water flows parallel to the membrane filter surface and some of the water diffuses through the filter. The remaining feed stream continues parallel to the membrane to exit the system without passing through the membrane surface. Filtered contaminants remain in the feed stream water, increasing in concentration until the feed stream exits the filter unit. The filtered water is called the permeate effluent, and the exiting feed stream water is called the concentrate effluent. Crossflow filtration provides a self-cleaning effect that allows continuous flushing away of contaminants, which, in "normal" filtration, would plug filters of small pore size very quickly.

Depending on the nominal size of the pores engineered into the membrane, crossflow filters are used in filtering applications classified as reverse osmosis (RO), nanofiltration (NF), ultrafiltration (UF), and microfiltration (MF), listed in order of increasing pore size range. The pore sizes in these membranes are so small that the significant pressure is required to force water through them; the smaller the pore size, the higher the required pressure. Figure 10.4 illustrates some of the uses for the different membrane filter types.

10.4.11.3.1 Reverse Osmosis
Reverse osmosis, sometimes called hyperfiltration, was the first crossflow membrane separation process to be widely used for water treatment. It requires operating pressures of 150–1200 pound per square inch (psi). It removes up to 99% of ions and most dissolved organic compounds. RO can meet most drinking water standards with a single-pass system. Although it might not be the most economical approach, using RO in multiple-pass systems allows the most stringent drinking water standards to be met. For example, rejection of 99.9% of viruses, bacteria, and pyrogens is achievable with a double-pass system.

10.4.11.3.2 Nanofiltration
Nanofiltration membranes can separate organic compounds with molecular weights as small as 250 Da.* It will also separate most divalent ions, and is effective for softening water (removing Ca^{2+} and Mg^{2+}). It allows greater water flow through at lower operating pressure (60–300 psi) than RO.

10.4.11.3.3 Ultrafiltration
Ultrafiltration membranes do not remove ions. They reject compounds greater than about 700 Da. The larger pore size permits lower operating pressures, in the range of

---

* 1 Da = 1 mol. wt. For example, the molecular weight of chloroform ($CHCl_3$) is 119 g/mol. The mass of one chloroform molecule is 119 Da. Chloroform will not be separated by nanofiltration, which does not reject molecules smaller than 250 Da.

10–100 psi. UF is useful for separating larger organic compounds, colloids, bacteria, pyrogens, *G. lamblia*, and *Cryptosporidium*. Most ions and smaller molecules such as chloroform and sucrose will pass through UF membranes.

10.4.11.3.4 Microfiltration

Microfiltration membranes reject contaminants in the 0.05–3.0 μm range and operate at pressures of 1–30 psi. They are available in polymer, metal, and ceramic materials. MF is sometimes used as a pretreatment for RO, increasing the efficiency and duty cycle of RO membranes significantly. MF will remove *G. lamblia* and *Cryptosporidium*, whose spores range between 3 and 18 μm in diameter, but not viruses and most bacteria. MF–RO treatment trains are reported to be economical, easy to operate, and very reliable.

10.5 ION EXCHANGE

Ion exchange is the reversible interchange of ions between a solid and a liquid. Hydrated ions on a solid are exchanged, equivalent for equivalent, for hydrated ions in solution. Cation exchange involves the interchange of positive ions. Anion exchange involves the interchange of negative ions.

In the natural environment, solid particles generally carry a surface charge, either positive or negative. This is true for both organic and inorganic solids. As water passes through soils, dissolved ions can leave the water to become attached to oppositely charged sites on soil surfaces. This displaces ions of the same charge sign previously attached to the surface, so that they become dissolved and mobile in the water. In general, ions of higher charge and smaller hydrated diameter will displace ions of lower charge and larger hydrated diameter. Larger hydrated diameter correlates with smaller ionic diameter and smaller ionic charge. Thus, smaller ions of the same charge have the largest hydrated diameters, as do ions of approximately the same ionic diameter but with smaller charge. Such ions (small ionic diameter and small charge) coordinate with more water molecules in their hydration sphere, resulting in a larger hydrated diameter. This is why nonhydrated sodium cation, Na^+, which is smaller than nonhydrated K^+, and has a smaller ionic charge than both Mg^{2+} and Ca^{2+}, causes greater swelling and loss of permeability of clayey soils than K^+, Mg^{2+}, or Ca^{2+} (see discussion of Section 10.2).

RULES OF THUMB

1. Dissolved ions with higher binding strength tend to displace surface-bound ions of lower binding strength.
2. Binding strength increases with larger nonhydrated ionic diameter (smaller hydrated diameter) of the ions.
3. Binding strength increases with the charge on the ions.
4. There is also a concentration effect. Continual high concentrations of any ion eventually displace most of the other ions having the same charge sign.

The order of cation binding strengths to a negatively charged surface is: (strongest) $Cr^{3+} > Al^{3+} \gg Ba^{2+} > Sr^{2+} > Ca^{2+} > Mg^{2+} \gg Cs^+ > NH_4^+ > K^+ > Na^+ > H_3O^+ > Li^+$ (weakest).

For example, Cr^{3+} will displace Al^{3+} on a surface; Ca^{2+} will displace K^+; H_3O^+ will displace Li^+. There is no permanent change in the structure of the solid that serves as the ion-exchange material.

10.5.1 WHY DO SOLIDS IN NATURE CARRY A SURFACE CHARGE?

Solid particle surfaces can acquire an electric charge in four ways. All four surface charge mechanisms can exist at the same time on mineral surfaces, and the latter two can exist at the same time on nonmineral (organic) surfaces also.

1. *Lattice imperfections*: During the crystal growth of silica minerals, an Al^{3+} cation may enter a lattice location intended for Si^{4+}, or a Mg^{2+} may substitute for Al^{3+} (all are isoelectronic third period cations and, thus, are of similar sizes), resulting in a net negative charge on the crystal.
2. *Differential solubilities*: Ions at different locations on the surface of slightly soluble salt crystals may have different tendencies to dissolve into water, resulting in either a negative or positive charge imbalance.
3. *pH-dependent chemical reactions at the particle surface*: Many solid surfaces (oxides, hydroxides, organics) contain ionizable functional groups, such as $-OH$, $-COOH$, or $-SH$. At high pH, these groups lose an H^+ (by: $H^+ + OH^- \rightarrow H_2O$), becoming charged as $-O^-$, $-COO^-$, or $-S^-$. At low pH, these groups gain an H^+, becoming $-OH_2^+$, $-COOH_2^+$, or $-SH_2^+$.
4. *Adsorption of hydrophobic (low solubility) or surfactant ions*: This can result in either positive or negative surfaces, and is not pH dependent.

RULES OF THUMB FOR PERMANENT SURFACE CHARGE

1. Surface charge caused by lattice imperfections is permanent and is not pH dependent.
2. Permanent surface charge occurs on clays and most minerals.
3. Permanent surface charge on minerals and clays is generally negative.

RULES OF THUMB FOR pH-DEPENDENT SURFACE CHARGE

1. At high pH, a negatively charged surface prevails.
2. At low pH, a positively charged surface prevails.
3. At some intermediate pH, the pH-dependent surface charge is zero. This pH is called the point of zero charge (pzc).

10.5.2 CATION- AND ANION-EXCHANGE CAPACITY (CEC)

Cations are attracted to negative sites on a solid surface. Cation-exchange capacity (CEC) is defined as the total number of negatively charged sites in a material at which reversible cation adsorption and desorption can occur. Operationally, it is measured by determining the total concentration (usually in meq/100 g of dry soil) of all exchangeable cations adsorbed. Thus, CEC is a measure of the reversible adsorptive capacity of a material for cations. At equilibrium, the total adsorbed cation charge equals the total negative charge on the solid, resulting in overall neutrality. The portion of CEC not affected by pH changes is caused by adsorption to permanently charged sites. The portion of CEC that increases with pH is caused by pH-dependent charged sites. Below about pH 5, H^+ ions are strongly bound to oxygen atoms at crystal edges, making these sites unavailable for cation adsorption. As pH increases above 5, H^+ ions are increasingly released into solution, making new sites available for cation adsorption.

Anion-exchange capacity (AEC) arises mainly from protonation of hydroxyl groups on the surface of minerals and organic particles. It is mostly pH dependent.

$$Surface - OH + H^+ \rightarrow surface - OH_2^+$$

RULES OF THUMB (see FIGURE 10.5)

- pH-dependent CEC does not change much as pH increases up to about pH 5.
- Above pH 5, CEC increases rapidly with pH.
- AEC increases as pH decreases. Gibbsite, kaolinite, goethite, and allophane clays exhibit small AECs. As pH rises above 3, AEC begins to decrease.

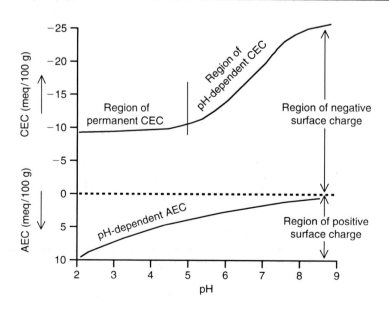

FIGURE 10.5 Ion-exchange capacity dependence on pH.

10.5.3   EXCHANGEABLE BASES: PERCENT BASE SATURATION

The primary exchangeable bases (exchangeable metal cations) are Na^+, Ca^{2+}, Mg^{2+}, and K^+. They usually occupy the majority of CEC sites in natural environments. The remaining CEC sites are occupied mainly by exchangeable H^+. The surface concentration of H^+ is pH dependent. Soil with exchangeable H^+ behaves like an acid, releasing H^+ into solution as a function of pH.

Percent base saturation is defined as the percent of primary exchangeable base cations relative to the total adsorbed cation concentration, at, or near, pH 7 ($CEC_{pH=7}$):

$$\text{Percent base saturation} = (\text{primary exchangeable bases } CEC_{pH=7}) \times 100$$

If there is no pH-dependent surface charge (i.e., only permanent surface charge), $CEC_{pH=7}$ will equal total exchangeable cations (including H^+) at any pH.

EXAMPLE 3

Suppose a clay is measured to have the following cations (in meq/100 g):

$$Ca^{2+} = 16.2, Mg^{2+} = 4.4, K^+ = 0.1, Na^+ = 1.6, \text{ and } H^+ = 10.2$$

What is its percent base saturation?

Answer:

At the pH of the CEC measurement, the total $CEC = 16.2 + 4.4 + 0.1 + 1.6 + 10.2 = 32.5$ meq/100 g. The 22.3 meq/100 g of base exchange capacity due to Ca^{2+}, Mg^{2+}, K^+, and Na^+ represents 68.6% of the total CEC. The 10.2 meq/100 g of exchangeable H^+ is 31.4% of the total CEC. Assume that there is no pH-dependent surface charge. Then

$$CEC_{pH=7} = 32.5 \text{ meq}/100 \text{ g, and percent base saturation} = 68.6\%$$

Percent base saturation is related to the soil pH as follows:

- The higher the percent base saturation, the higher will be the pH (more sites have been vacated by H^+ and occupied by metal cations).
- The lower the percent base saturation, the lower the pH (more sites are occupied by H^+ and unavailable to metal cations).

Leaching of soils reduces base saturation but does not change CEC. Therefore, soil leaching tends to increase soil acidity.

RULES OF THUMB

1. Soil pH is correlated with the percentage of base saturation.
2. The higher the base saturation, the higher the soil pH. Nearly 90%–100% base saturation indicates a soil pH around 7 or higher.
3. Low base saturation (<90%) indicates acidic soils.
4. A 5% decrease in base saturation causes a decrease in pH of about 0.1 unit.

EXAMPLE 4

Estimate the pH of the soil of Example 3.

Answer:

The percent base saturation was measured to be 68.6%. At pH 7, the percent base saturation is about 100% (Rule of Thumb 2). Therefore, 68.6% represents a decrease of

$$100 - 68.6 = 31.4\% \text{ from pH } 7$$

A 5% decrease in base saturation causes about 0.1 decrease in pH (Rule of Thumb 4). Therefore, $31.4/5 \approx 6$ units of 5% decrease in base saturation, or 6×0.1 units decrease in pH. Therefore, $6 \times 0.1 = 0.6$ pH units decrease. Soil pH is around $7 - 0.6 = 6.4$.

If a soil with a low base saturation is made into a slurry with water, the water pH will be acidic, around 4–5 or lower. Because a low base saturation means that many sites are occupied by H^+, acidic water results from exchangeable H^+ being released from the soil into solution. Similarly, passing acidic water through soil causes H^+ to exchange with soil-bound metal cations. The most weakly bound metals are displaced first and, if the pH is low enough, more strongly bound metals are displaced. At first, K^+ and Na^+ are removed, then Ca^{2+} and Mg^{2+} go into solution. Trivalent ions, such as Al^{3+}, are the last to leave the soil and go into solution.

RULE OF THUMB

The presence of an elevated concentration of Al^{3+} (>0.1 mg/L) is a characteristic of acid waters (pH < 5).

10.5.4   CEC in Clays and Organic Matter

10.5.4.1   CEC in Clays

Clays adsorb cations from solution until the fixed total charge is reached that represents their CEC. The total charge of adsorbed cations equals the total negative surface charge on the clay.

A clay particle may be regarded as a system composed of two parts:

- A relatively large, insoluble negatively charged particle.
- A loosely held swarm of adsorbed exchangeable cations.

CEC depends on the clay crystalline layer structure. Clays have internal surfaces as well as external, because of their layered structure. In some clays, the layers are held together more strongly than in others. Water can penetrate into weakly bound layers and force them apart, causing the clay to swell.

Different clays can differ markedly in their layer structures. Montmorillonite is a strongly swelling clay. Kaolinite does not swell significantly. Clays that swell in water have higher CECs than nonswelling clays. Swelling separates the layers and increases the surface area available for ion exchange. Surface charge on the interlayer surfaces attracts additional cations. Between the layers, adsorbed cations with large hydrated diameters (such as Na^+) exert separating forces that can greatly increase clay expansion. This decreases clay permeability to water (see Section 10.2).

RULE OF THUMB

A clay CEC > 10–15 meq/100 g indicates some degree of layer expansion by water swelling.

10.5.4.2   CEC in Organic Matter

Just 1% of humic organic material in a mineral soil contributes a CEC of around 2 meq/100 g of soil, which is about 4 times the CEC of an equal weight of clay.*

RULE OF THUMB

A rough estimate of the CEC of a soil can be made as follows:

1. Estimate or determine the percentages by weight of silicate clay and organic matter.
2. Multiply clay percentage by 0.5 to get the clay contribution.
3. Multiply the organic percentage by 2 to get the humus contribution.
4. Add the results to get the total CEC in meq/100 g.

EXAMPLE 5

Suppose a soil contained 2% organic matter and 16% clay. The soil CEC is estimated to be around $(2 \times 2) + (16 \times 0.5) = 12$ meq/100 g.

10.5.5   RATES OF CATION EXCHANGE

The rates of exchange depend on the clay type. Kaolinite does not swell in water and there is no inner-layer access. Exchange reactions in kaolinites are almost instantaneous because they occur only on the outer surface. Illites swell slightly. Exchange reactions in illites can take several hours because a small part of the exchange occurs between inner crystal unit layers, to which cation diffusion is slow. Montmorillonite expands considerably and most of the exposed surfaces are on the inner layers. Montmorillonites take still longer to reach ion-exchange equilibrium.

---

* Clay is much denser than humic matter. An equal weight of clay represents many fewer milliequivalents of CEC.

10.6 INDICATORS OF FECAL CONTAMINATION: COLIFORM AND STREPTOCOCCI BACTERIA

10.6.1 BACKGROUND

Detecting and preventing fecal contamination is of prime importance for all drinking water systems and recreation water managers. Fecal wastes may contain enteric pathogens (disease-causing organisms from the intestines of warm-blooded animals) such as viruses, bacteria, and protozoans (which include *Cryptosporidium* and *Giardia*). Fecal-contaminated water is a common cause of gastrointestinal illness, including diarrhea, dysentery, ulcers, fatigue, and cramps. It also may carry pathogens that cause a host of other serious diseases such as cholera, typhoid fever, hepatitis A, meningitis, and myocarditis.

Testing water directly for individual pathogenic organisms is, so far, impractical for several reasons:

- There are so many different kinds of pathogens that a comprehensive analysis would be very expensive and time-consuming, whereas time is of the essence for pathogen detection.
- Pathogens can be dangerous at small concentrations, which require large sample volumes for analysis. This adds to the time and cost of analysis.
- Reliable analytical methods for several important pathogens are difficult or not even available. Also, not all waterborne pathogenic microorganisms are known.
- A satisfactory alternative is available, namely, the identification of "indicator" species that are easy to measure and are always present with enteric pathogens.

Hence, awareness of possible contamination by enteric pathogens is based on detecting the more easily identified "indicator" species, whose presence indicates that fecal contamination may have occurred.

The five indicator species most commonly used today are total coliforms, fecal coliforms, *Escherichia coli* (*E. coli*), fecal streptococci, and enterococci. All are bacteria normally present in the intestines and feces of warm-blooded animals, including humans. All but *E. coli* consist of groups of bacterial species that are similar in shape, habitat, and behavior. *E. coli* is a single species within the fecal coliform group. Some strains of *E. coli* are pathogenic but the other indicators are usually not pathogens and do not pose a danger to humans or animals. However, if any of the indicators are present in water, the accompanying presence of a dangerous population of enteric pathogens is a possibility.

All the indicator species are easier to measure than most pathogens, but harder to kill. Therefore, treatment that satisfactorily destroys the indicator species may be assumed to have also destroyed enteric pathogens that were present. For example, in waste water disinfection, it is assumed that a decrease in fecal coliforms to <200 fc/100 mL (fecal coliforms per 100 mL of sample) will have eliminated the great majority of pathogens.

Total coliforms and fecal coliforms are the "old reliable" indicators of fecal contamination, used since the 1920s to protect public health. However, both have limitations that stimulate regulators to continue seeking improved methods.

10.6.2 Total Coliforms

Total coliform bacteria, which include fecal coliforms and *E. coli*, are widespread in nature. In addition to their animal intestine habitat, they occur naturally in plant material and soil. Therefore, their presence does not necessarily indicate fecal contamination. Total coliforms are not recommended as indicators of recreational water contamination, where they are usually present from soil and plant contact. Total coliforms are the standard test for contamination of finished drinking water, where contamination of a water supply or distribution system by fecal, plant, or soil sources is not acceptable. Federal drinking water standards are based on total coliform bacteria. EPA's maximum contaminant level (MCL) for drinking water is 0 total coliforms per 100 mL of water for 95% of samples after treatment (chlorination, ozonation, UV).*

10.6.3 Fecal Coliforms

Fecal coliforms are a more fecal-specific subset of total coliform bacteria. However, even the fecal coliform group contains a genus, *Klebsiella*, with member species not necessarily fecal in origin. *Klebsiella* coliforms are found in large numbers in textile, pulp, and paper mill wastes. Fecal coliforms are widely used to monitor recreational waters and are the only indicator approved for classifying shellfish waters by the U.S. Food and Drug Administration's National Shellfish Sanitation Program. On the basis of statistical data, EPA has recently begun to recommend *E. coli* and enterococci as better indicators of health risk from water contact. However, many states still use fecal coliforms for this purpose, in part so that new data can be directly compared with historical data.

Natural surface waters almost always contain some background level of fecal coliforms, usually less than 15–20 fc/100 mL MPN (most probable number). In sewage entering a waste treatment plant, the fecal coliform count may be over 10 million fc/100 mL. Satisfactory disinfection of secondary effluent from a waste treatment plant is defined by an average fecal coliform count of <200 fc/100 mL. Fecal coliforms are normally absent after wastewater percolates through 5 ft of soil.

State water standards for fecal coliform levels vary, but typical state standards are (geometric mean values):

- Class 1 Primary Contact Recreational: 200 fc/100 mL
- Class 1 Secondary Contact Recreational: 2000 fc/100 mL
- Domestic Water Supply (before treatment): 2000 fc/100 mL

10.6.4 Escherichia coli

E. coli is a single species of fecal coliform bacteria that occurs only in fecal matter from humans and other warm-blooded animals. Some strains of *E. coli* are

---

* The USEPA Total Coliform Rule states: For water systems collecting 40 or more samples per month, no more than 5% can test positive for total coliforms. No more than one sample can be positive for total coliform if fewer than 40 samples per month are collected. Every sample containing total coliforms must be analyzed for fecal coliforms, which must = 0/100 mL. MCLG = 0 for total coliforms, fecal coliforms, and *E. coli*.

disease-causing pathogens. EPA studies (USEPA, 1986) indicate that, in fresh water, *E. coli* correlates better with swimming-related illness than do fecal coliforms. Since 1986, EPA has been recommending that states use *E. coli* as an indicator for fecal-contaminated freshwater recreation areas, instead of fecal coliform. States vary in their adoption of this recommendation.

10.6.5   FECAL STREPTOCOCCI

The normal habitat of fecal streptococci bacteria is the gastrointestinal tract of warm-blooded animals. Because humans differ from most other animals in the relative amounts of coliforms and streptococci normally present in their intestines, it was believed in the past that the ratio of fecal coliforms to fecal streptococci could be used to differentiate human fecal contamination from that of other warm-blooded animals. A ratio >4 was considered indicative of human sources and a ratio <0.7 suggested animal sources. However, this is no longer regarded as a reliable test because of variable survival half-lives in water of different species of streptococci and the effects of wastewater disinfection and different analytical procedures on the measured coliform/streptococcus ratio.

10.6.6   ENTEROCOCCI

Enterococci are a subgroup of fecal streptococci, differentiated by their ability to thrive in saline water over a wide range of temperature. In this respect, they mimic many enteric pathogens better than do the other indicators. EPA recommends enterococci as the best indicator of health risk in salt recreational waters and as a useful indicator in fresh waters as well.

RULES OF THUMB

- To determine whether recreational water or wastewater meets state water quality requirements, find out which indicator (fecal coliform, *E. coli*, or enterococci) your state uses for recreational or wastewaters and measure that one.
- For compliance with federal drinking water standards, measure total coliforms, which include fecal coliform and *E. coli*.
- For a nonregulatory determination of the health risks from recreational water contact, measure *E. coli*, or enterococci.

10.7   MUNICIPAL WASTEWATER REUSE: THE MOVEMENT AND FATE OF MICROBIAL PATHOGENS

10.7.1   PATHOGENS IN TREATED WASTEWATER

Reuse of municipal wastewater requires special care to minimize hazardous exposure of humans and animals to waterborne pathogens. Wastewater that has received secondary treatment generally contains residual active viruses and other pathogens (Rose and Gerba, 1991), which can persist to varying degrees after release to the environment. Removal of microorganisms is accomplished by filtration, adsorption, desiccation, radiation, predation, and exposure to other adverse conditions. Because of their large size, protozoa and helminths are removed primarily by filtration. Bacteria are removed by filtration and adsorption. Fecal coliforms are normally absent after wastewater percolates through 5 ft of soil. Virus removals are not as well documented (USEPA, 1981) but it is generally agreed that viruses are removed from subsurface water flow almost entirely by adsorption to soil particles. As is seen below, removal is not the same as inactivation and is generally not permanent. Viruses adsorbed to soil particles can be released again still in an active state. Once viruses are mobilized in the environment, they may become inactivated by a variety of factors such as higher temperatures, pH, loss of moisture, exposure to sunlight, inorganic cations and anions, and the presence of antagonistic soil microflora.

Viral persistence in the environment can be prolonged by low temperatures, adsorption to particulate surfaces, and moist conditions (Bitton, 1975). Under appropriate conditions, viruses can remain infective for several months in wastewater sludge and environmental waters (Hurst, 1989). In a study of virus occurrences in treated wastewaters in Arizona and Florida, average virus levels ranged from 13 to 130 pfu/100 L (plaque-forming units per 100 L) (Rose and Gerba, 1991). In most of the viral-positive samples, viral effluent quality significantly exceeded the standard for unrestricted irrigation of 1 pfu/40 L. The infections dose for viruses is reported to be 1–10 viral units (USEPA, 1992a). The authors performed a risk analysis, which found that viral effluent quality in Arizona and Florida would potentially produce two infections per 1000 exposed persons.

A review of the general literature shows that the persistence of pathogens associated with the use of municipal wastewater for aquifer recharge is very site-specific. In particular, the behavior of viruses, which are more difficult to remove or inactivate than bacteria or parasites, depends strongly on environmental conditions and the types of viruses present (Bitton, 1980). Some studies find that viruses are completely inactivated quickly after introduction to the soil/groundwater environment, while other studies find that they can persist for long periods of time (over 1 year; USEPA, 1981; Pinholster, 1995) and over long travel distances (100 m; USEPA, 1981).

There is no question that viruses have a potential for high mobility and persistence in groundwater. A National Research Council Report (NRC, 1994) emphasizes that there are significant uncertainties associated with predicting the transport and fate of viruses in recharged aquifers, and that these uncertainties make it difficult to determine the levels of risk from any infectious agents still contained in the disinfected wastewater. The NRC report endorses groundwater recharge practices in general but cautions that current recharge technologies are "especially well-suited to nonpotable uses such as landscape irrigation" and that "potable reuse should be considered only when better quality sources are unavailable." This report states further that, "water quality monitoring and operations management should be more stringent for recharge systems intended for potable reuse."

The State of California has been in the forefront of wastewater recycling applications because of chronic water shortages and the threat of saltwater incursions into freshwater aquifers. The California water reuse regulations pay particular attention to enteric viruses (viruses that are shed in fecal matter) because of the possibility of contracting disease with relatively low doses and the difficulty of routine examination of wastewater for their presence. California requires essentially virus-free effluent via a "full treatment" process for wastewater reuse applications with high potential for human exposure (Asano et al., 1992).

10.7.2 Transport and Inactivation of Viruses in Soils and Groundwater

Viruses are the smallest wastewater pathogens, consisting of a nucleic acid genome enclosed in a protective protein coat called a capsid. A virus capsid contains many ionizable proteins that are subject to protonation and deprotonation reactions in water, depending on the pH and ionic strength of the water. At low pH, virus particles tend to carry a positive charge because of attached H^+ ions. As the pH rises, the positive charge on a virus particle decreases, then passes through zero at the isoelectric point, and becomes negative due to increasing numbers of attached OH^- ions. As a result, viruses can have the ion-exchange characteristics of either cations or anions, depending on the pH.

In groundwater and soils, viruses move as colloidal particles. Due to their small size, it is believed that viruses are not significantly removed from groundwater by membrane filters coarser than RO or NF. Viruses become attached to soil particles mainly by sorption forces arising from electrostatic interactions, London forces, hydrophobic forces, covalent bonding, and hydrogen bonding (USEPA, 1981; Aronheim, 1992). As a result, the extent to which viruses are sorbed to soils depends strongly on pH, temperature, ionic strength, and flow velocity of the water, as well as the mineral-organic composition and particle size distribution of the soil and the particular type of virus. Clay soils are more retentive than sandy soils and finely divided soils retard virus mobility more than coarser soils (USEPA, 1981). Two recent summaries of the behavior of viruses and other pathogens in soils and groundwater are found in Gerba and Smith (2005) and Chu et al. (2003).

The isoelectric point of enteric viruses is usually below pH 5, so that in most soils, enteroviruses carry a negative charge, as do most soils. In general, virus adsorption to soil is enhanced at lower pH values (pH < 7), where soil and virus charges are opposite, and reduced at higher pH values (pH > 7) (Wallis et al., 1972). If chemical conditions change or the flow velocity is increased, either locally by microscopic changes or overall by macroscopic changes, adsorbed virus particles can be detached from soil surfaces and returned to suspension in the flow. Waters with high TDS concentrations favor adsorption to soils because electrostatic repulsion is minimized in waters with high ionic strength. A rain event can dilute TDS levels near the surface and cause a burst of released viruses. The same "burst" effect can occur with a release of higher pH water that locally raises the water column pH from 7.2 to 8 or 9 (USEPA, 1981). For these reasons, virus adsorption to soils cannot be considered a process of absolute immobilization of the viruses from the water. Infective viruses are capable of release from soil particles after immobilization for long periods of time. Any environmental change that reduces their attraction to soil particles will result in their further movement with groundwater.

The presence of organic matter, such as humic and fulvic acids, in soils has been shown to inhibit the adsorption of viruses to soil surfaces by competing with viruses for adsorption sites (Gerba, 1975; Lo and Sproul, 1977; Burge and Enkiri, 1978; Scheurman et al., 1979; Sobsey and Hickey, 1985). In one study (Scheurman et al., 1979), the presence of organic substances in an aqueous environment reduced the retention of viruses in a soil column from greater than 99% to less than 1.5%.

Adsorption to soil particles may prolong viral lifetimes in aqueous environments (Bitton et al., 1979; Foster et al., 1980; Sobsey et al., 1980; Stotsky et al., 1980; USEPA, 1981). Viruses bound to solids are as infectious to humans and animals as the free viruses (Hurst et al., 1980; USEPA, 1981). Virus survival in soil depends on the nature of the soil, temperature, pH, moisture, and the presence of antagonistic soil microflora.

In one study using f2 bacteriophage and poliovirus type 1, 60%–90% of the viruses were inactivated at 20°C within 7 days after the initial release to the soil (Lefler and Kott, 1974). But, after the first 7 days, the inactivation rate slowed and polioviruses could still be detected at 91 days; f2 viruses survived longer than 175 days. At lower temperatures, up to 20% of the polioviruses survived longer than 175 days. Other studies indicate that virus lifetimes may range from 7 days to 6 months in soils and from 2 days to more than 6 months in groundwater (USEPA, 1981).

10.8 OIL AND GREASE

10.8.1 OIL AND GREASE ANALYSIS

When you analyze for oil and grease (O&G), you are actually measuring a group of substances that have similar solubility characteristics in a designated solvent. "Oil and grease" is defined as any substance recovered from an acidified sample by extracting it into a designated solvent, and which is not volatilized during the analysis. The extraction process is called "liquid–liquid extraction (LLE)." After the sample is extracted, the extract may be measured either gravimetrically (drying and weighing) or by infrared spectroscopy.

The solvents used have the ability to dissolve not only O&G but also other organic substances. Some non-O&G materials commonly included in the determination of "O&G" are certain sulfur compounds, chlorophyll, and some organic dyes. There is no known solvent that will dissolve selectively only O&G. Some heavier residuals of petroleum (coal tar, used motor oil, etc.) may contain significant amounts of material that do not extract into the solvent. The method is entirely empirical and duplicate results with a high degree of precision can be obtained only by strict adherence to all details of the analytical procedure.

The designated solvent changed from petroleum ether and *normal*-hexane in 1965 to Freon-113 (trichlorotrifluoroethane) in 1976. While Freon-113 was still the prescribed solvent, an alternative solvent (80% *n*-hexane and 20% MTBE) was developed for gravimetric methods, to help phase-out the use of freons. Then, on May 14, 1999, EPA approved the new Method 1664 for analyzing O&G, which must replace all other methods. It differs mainly in details of the procedure and in the

QA/QC requirements. Method 1664 uses pure *n*-hexane with either LLE or solid phase extraction (SPE). Because *n*-hexane is less dense than water, LLE will not work well with all sample matrices.

10.8.1.1 Silica Gel Treatment

The "total oil and grease" designation includes oils and fats of biological origin (animal and vegetable fats and oils) as well as mineral oils (petroleum products). Silica gel has the ability to adsorb certain organic compounds known as "polar compounds." Since petroleum hydrocarbons are mostly nonpolar, and most hydrocarbons of biological origin are polar, silica gel is used to separate these different types of hydrocarbons. If a solution of nonpolar and polar hydrocarbons is mixed with silica gel, the polar hydrocarbons, such as fatty acids, are removed selectively from the solution. The materials not eliminated by silica gel adsorption are designated as "petroleum hydrocarbons" by this test. Normally, total O&G is measured first and then petroleum hydrocarbons are determined after the animal and vegetable fats are removed by silica gel adsorption. The difference between total O&G and petroleum hydrocarbons is designated as biologically derived hydrocarbons.

10.9 QUALITY ASSURANCE AND QUALITY CONTROL IN ENVIRONMENTAL SAMPLING

Quality assurance and quality control (QA/QC) are the set of principles and practices, which, if strictly followed in a sampling and analysis program, will produce data of known and defensible quality. The effectiveness of any monitoring effort depends on its QA/QC program.

A well-designed QA/QC program reduces, as far as possible, the potential for undetected errors appearing in the final results. It has many components, ranging from competency certification of personnel to replicate sampling. The overall goal is to minimize or correct all the errors that might occur, by introducing explicit steps into the sampling and analysis protocol for identifying, measuring, and controlling these errors.

10.9.1 QA/QC Has Different Field and Laboratory Components

Field QA/QC is intended to collect in the field and deliver to the laboratory, a sample where analyte properties have not changed significantly from their values in the environmental waters being tested. An obvious, but not trivial, part of field QA/QC is properly identifying all samples so that measured analyte concentrations can be assigned to a known location and time with confidence.

Laboratory QA/QC is designed to produce a quantitative measure of certain properties of the sample with known limits of uncertainty. There are some overlapping QA/QC practices, such as including field duplicates and blind known samples with a field sample set to test a laboratory's performance.

This section deals only with field QA/QC protocols. Laboratory practices are generally out of the control of a sampling program manager, who ideally will have sufficient experience with the QA/QC standards of the laboratory being used to have confidence in the final data. However, laboratory performance can change and it is prudent to remain alert to possible laboratory errors. Techniques for checking laboratory performance are included below.

10.9.2   ESSENTIAL COMPONENTS OF FIELD QA/QC

10.9.2.1   Sample Collection

Proper sampling is a separate science in its own right. The choice of automated versus hand sampling equipment, details of how to get a representative sample into the container, and decisions of whether grab, composite, or time/space-integrated samples are appropriate, is not addressed here.* The importance of obtaining a representative sample for analysis is underscored when one recognizes that variability of the sampled media is the greatest source of variability in analytical results, exceeding by far measurement uncertainties.

However, certain QA/QC practices are common to all sampling procedures:

- *Sample contamination must be avoided*: All parts of the sampling equipment and containers that contact the sample must be scrupulously clean. Sample containers must be inert to the sample and preservatives.
- *Samples must be properly preserved*: In general, all samples should be stored and transported at 4°C (ice temperature) or less. In addition, samples for certain analyses require chemical preservatives, for example, acid addition to less than pH 2 for metals and ammonia, base addition to greater than pH 12 for cyanide, etc.
- *Samples must be unambiguously identified*: If the sample container is not prelabeled, it must be securely labeled with a unique identification at the time of collection or removal from a sampler. The collectors' name, date and time, station designation, form of preservation, and desired analyses should also be on the label. Sample identification must be entered on the report form without delay after labeling the sample container.
- *Samples must be packaged for transportation*: Generally, samples are transported in a cooler. The cooler lid must not accidentally open, container lids must not leak, and containers should be protected against breakage.
- *Chain-of-custody and field reporting documentation must be maintained*: Careful completion of chain-of-custody (COC) and sampling report forms are essential parts of field QA/QC. COC form helps to insure sample integrity from collection to data reporting. Entries on the COC form track the possession and handling of each sample from the time of collection through analysis and final disposition, and demonstrate that all standard steps of sample control have been followed.

---

* Detailed guidance for these procedures can be found in Keith (1992) and in reference works such as *Standard Methods for the Examination of Water and Wastewater* (1995).

The sample report form is needed for interpretation of the reported data and must be designed carefully to require an entry for all relevant information. The report form will always include the name or initials of the sampler and have labeled places to enter sampling location, sample identification, date, and sampling time for each sample. It might also include space for other relevant information such as weather conditions, observations concerning the sampled waters (such as odor, color, flow, etc.), and sampling procedure (grab, time-integrated, etc.).

Most commercial laboratories provide coolers, ice packs, clean sample containers with preservatives appropriate for requested analyses, container labels, and COC forms and seals as part of their analytical services.

10.9.3 FIELD SAMPLE SET

A field sample set consists of the environmental samples and several kinds of quality control samples. The project manager must determine which kinds of quality control samples should be collected, based on the purposes of the sampling program.

10.9.3.1 Quality Control Samples

Some field samples should be collected for quality control purposes. Sampling procedures can contribute to both systematic and random errors. Many environmental waters are inherently not uniformly mixed. Field blanks, spikes, and duplicates will help to estimate errors caused by sampling bias (usually contamination and/or non-representative samples) and calculate sampling precision. Field QC samples must be handled exactly the same way as the environmental samples, using identical sampling devices, sampling protocol, storage containers, preservation methods, storage times, and transportation methods. For a correct interpretation, quality control samples must be subject to the same holding time criteria as the environmental samples.

10.9.3.2 Blank Sample Requirements

There are several types of blank samples, each serving a distinct QC purpose. Wherever a possibility exists for a sample to become contaminated, a blank should be devised to detect and measure the contamination. The most commonly collected blanks include field, trip, and equipment blanks. Although it may be prudent to collect a full set of blanks, most of the blank samples usually do not need to be analyzed. Analysis costs can be reduced if the strategy of blank sampling is understood. First analyze only the field blanks, which are susceptible to the broadest range of contaminant sources. If these indicate no problems, the other, more specific, blanks can be discarded or stored. If a problem is suspected, other blanks can be analyzed for confirmation of the problem and to discover the source. Holding time limits must be observed with blanks as well as with environmental samples.

Field blanks: At least one clean water sample should be exposed to the same sampling conditions as the environmental samples at each sampling site. The analytical laboratory can generally provide analyte-free distilled water for this purpose. Field blanks are transferred from one container to another, passed through automatic equipment, or otherwise exposed to the conditions at the sampling site.

At a minimum, the field blank container is opened at the sampling site and exposed to the air for approximately the same time as the environmental samples. It is then capped, labeled, and sent to the laboratory with other samples.

Field blanks measure incidental or accidental sample contamination throughout all the steps of transportation, sampling, sample preparation at the laboratory, and analysis. They help to assure that artifacts are recognizable and are not mistaken as real data.

Trip blanks: For each type of container and preservative, at least one container of analyte-free water should travel unopened from the laboratory to the sampling sites and back to the laboratory. Trip blanks serve to identify contamination from the container and preservative during transportation, handling, and storage.

Equipment blanks: These are especially important with automated sampling equipment. Equipment blanks, sometimes referred to as "rinsate blanks," document adequate decontamination of the sampling equipment. These blanks are collected by passing analyte-free water through the sampling equipment after decontamination and prior to resampling.

The analytical laboratory will run a similar set of blanks to document contamination and errors arising from handling and analysis procedures in the laboratory.

10.9.3.3   Field Duplicates and Spikes

Field duplicates are two separate environmental samples collected simultaneously at the identical source location and analyzed individually. Field duplicates are sensitive to the total sample variability, i.e., variability from all sampling, storage, transportation, and analytical procedures. Where the goals of the sampling program warrant it, for example, for permit monitoring requirements, at least one field duplicate per day should be collected for each analyte. More than two replicates may be required in cases where it is difficult to obtain representative samples.

Field spikes are environmental samples to which known amounts of the analytes of interest are added. Ampoules containing carefully measured amounts of analytes can be purchased from many chemical supply sources. Field spikes can identify storage, transportation, and matrix effects, such as loss of volatile compounds and analytical interferences caused by certain compounds present in the environmental source. In a spiked sample with no problems, the measured analyte concentration should be equal to the concentration measured in the environmental sample from the same source plus the added spike concentration, within the limits of precision of the method.

10.9.3.4   Understanding Laboratory Reported Results

When a laboratory reports that a target compound was not detected, it does not mean that the compound was not present. It always means that the compound was not present in a concentration above a certain lowest reporting limit (RL). There always is the possibility that the compound was present at a concentration below the RL. The laboratory might even have identified the compound below the RL but not reported it because the concentration could not be quantified within acceptable limits of error.

Reported results of analyte concentrations are never exact. There always is some margin of error. The only kind of measurement that can be exact is a tally of discrete objects, for example, dollars and cents or the number of cars in a parking lot. When measuring a quantity capable of continuous variation, such as mass, length, or the concentration of benzene in a sample, there always is some uncertainty. Like an irrational number, measurements capable of continuous variation can always be expressed to more significant figures, and any answer with a finite number of digits is always an approximation.

In addition, many repeated measurements of the same quantity, even under essentially identical conditions, will always yield results that are scattered randomly about some average value, because of uncontrollable variations in environmental, experimental, and operator behavior. The person who interprets experimental data must always bear in mind that no reported experimental value necessarily represents the true value. In other words, it is never possible to be completely certain of a result. Nevertheless, we still try to answer questions such as, "Does this sample indicate a violation of a discharge limit?" or "Does the data indicate that my remediation activities are beginning to work?" Our answers to such questions must always acknowledge that there is a range of error. The purpose of QA/QC is to assure that the reported value lies within a small enough range of error to be useful data.

Part of laboratory QA/QC involves determining how much scatter in the data is to be expected from different analytical procedures. The precision inherent in a particular procedure is a measure of how much deviation may be expected in repeated measurements of the same sample. Standard deviation is a common way of quantitatively expressing the precision, or reproducibility, of a measurement. A more thorough treatment of the standard deviation can be found in elementary statistics texts or in the latest edition of *Standard Methods for the Examination of Water and Wastewater* (Eaton et al., 1995). The procedure for using the standard deviation to expressing the reliability of measurements is described by EPA in several documents (e.g., USEPA, 1992b; USEPA, 2007). Here, it is only necessary to understand that, for many measurements of the same quantity, one standard deviation above and below the average value of the measurements will include about 68% of all the individual values. For example, suppose the concentration of benzene in a sample is measured repeatedly many times to yield an average value of 12.3 μg/L and a standard deviation of 2.7 μg/L. Then, 68% of all the individual measurements can be expected to lie within the range of 12.3 ± 2.7 μg/L, or between 9.6 and 15.0 μg/L. Two standard deviations will include about 95% of all the values.

The larger the standard deviation, the greater the scatter in the data, i.e., the poorer the precision, or reproducibility, of repeated measurements. The standard deviation and, hence, the measurement reproducibility, depends strongly on the matrix in which the analyte is being measured. In general, stream and groundwater samples can be measured more precisely than wastewater and soil samples. For this reason, it is not possible to make general statements about the reliability of measurements based only on the concentration values. The sample matrix is taken into account by determining the standard deviation.

In Figure 10.6, different kinds of measurement limits are defined in terms of the reported concentration expressed as the number of standard deviations above the instrumental zero. Only the method detection limit (MDL) is rigorously defined (USEPA, 2007). A calculation of the MDL puts it about three standard deviations

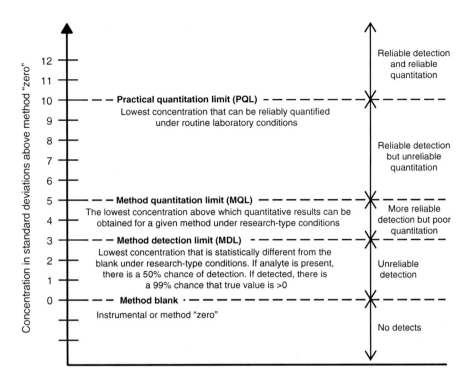

FIGURE 10.6 Statistical measures of data reliability, based on the measured concentration expressed in standard deviations above the instrumental zero. Standard deviation is determined by repeated measurements of a laboratory control sample. Other names for practical quantitation limit (PQL) are EQL (estimated quantitation limit), LLQ (lower limit of quantitation), and LOQ (limit of quantitation). Other names for the method quantitation limit (MQL) are LOQ and ML (method limit).

above the instrumental zero. However, every analysis has a unique set of uncontrollable variables that will cause the measured value to differ from some "true" value in an unknown way. The realities of time and budget limits normally limit sampling programs to rely on fewer sample analyses than would be statistically ideal. For this reason, EPA uses a concept called "quantitation limits" to provide safety margins in their regulations. Although the nomenclature for different quantitation limits has varied over time, their function has remained the same, to establish a legally defensible value that can be used to limit the concentration of a pollutant in the environment.

The practical quantitation limit (PQL) is an estimation of the lowest concentration above which reliable quantitative results can be obtained under routine laboratory conditions, and is arrived at by evaluating the performance of different laboratories. If performance data are not available, the PQL can be estimated from the MDL. EPA believes that setting the PQL at 5–10 times the MDL is generally a fair expectation for routine operation at most qualified government and commercial laboratories (USEPA, 1994). Many laboratories determine their own "routine performance" PQLs and often use 3 times the MDL as the lowest level for a PQL. The method quantitation limit (MQL) is an estimation of the lowest concentration above which reliable quantitative results can be obtained under research laboratory conditions. The MQL can be measured but is often estimated to be about five standard deviations above the instrumental zero. In addition, many laboratories include an RL in their analytical reports, which represents the concentration above which that laboratory has confidence in their quantitative values. Depending on the laboratory, the RL for an analyte may be anywhere between its MDL and its PQL.

RULE OF THUMB

When an "official" PQL has not been designated for a particular analysis, an unofficial PQL of 10 standard deviations above the instrument 0 (or 3.3 times the MDL) is often used.

10.10   CASE STUDY: WATER QUALITY PROFILE OF GROUNDWATER IN COAL-BED METHANE FORMATIONS

This case study combines many of the concepts treated so far in this book.

Drilling for oil and gas often raises concerns about potential detrimental impacts that the drilling and production activities might have on surface waters and water wells in the vicinity. An energy company planning to explore coal-bed methane in a western state undertook a baseline water quality study of water wells and surface waters in the proposed exploration region. The plan was to collect historical and current water quality data into a database, and then compare water samples from the geologic formations where methane was found with other samples from surface water and groundwater sources in the area. The assumption was that waters with similar chemical profiles might be hydraulically connected and that dissimilar profiles would indicate that the water sources were not strongly connected hydraulically. Waters not hydraulically connected to the coal-bed methane formations were deemed unlikely to be affected by drilling and production activities.

This approach is based on the recognition that the water quality of wells and springs is influenced by the hydrogeology of their water sources. TDS concentration in groundwater tends to increase with the length of time that the water is in contact with minerals in the geologic formations. TDS concentrations in wells and springs located in groundwater recharge areas, close to where precipitation first enters an aquifer, are expected to be comparatively low because this groundwater has been in the formation a relatively short time. Likewise, a well or spring located in a groundwater discharge area is likely to have higher TDS concentrations since this water has had a longer residence time in the subsurface. Overall, groundwater flow in the energy company's study area was generally toward deeper parts of the geologic formations. Thus, TDS concentrations should increase with increasing well depth, as was generally observed.

Because the database included water quality measurements from approximately 500 individual water sources, a visual method utilizing Stiff diagrams (see Figure 10.7) proved convenient for comparing chemical profiles. In a Stiff diagram, each ion concentration (in milliequivalents per liter) is plotted on a separate horizontal axis, extending out on each side from a central vertical axis representing zero concentration. In Figure 10.7, anions are plotted to the right of zero and cations to

Stiff diagrams	Comments
(a) Somrst B, 12/12/80 Na — Cl Ca — HCO$_3$ Mg — SO$_4$ 71 35.5 35.5 71 (meq/l)	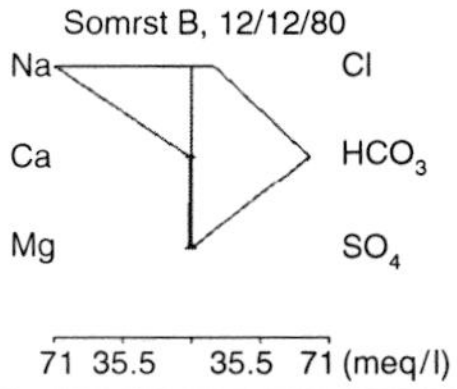 Sodium–bicarbonate type water from a well completed in the coal-bed methane producing formation. This Stiff pattern, with very little Ca, Mg, SO$_4$, and low Cl, is typical for wells completed in the deep coal-bed methane formation far from recharge zones (TDS = 2159 mg/L).
(b) CoorsNichols 1-23 Na — Cl Ca — HCO$_3$ Mg — SO$_4$ 530 265 265 530 (meq/l)	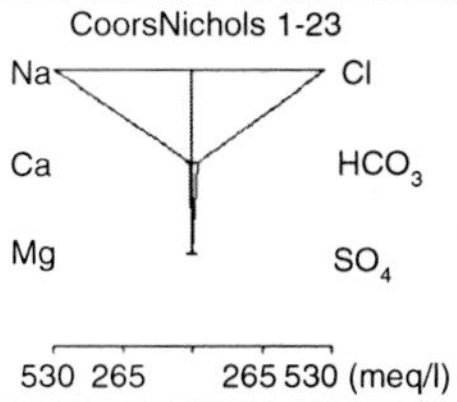 Sodium–chloride type water from a well that passed through the coal-bed methane formation into a formation of marine origin. Note that the meq/L scale is much larger than for (a) above; other ions present are masked by the large amounts of sodium and chloride. This Stiff pattern, with very high TDS arising mainly from sodium and chloride, is typical of water from formations of marine origin (TDS = 30,400 mg/L).
(c) DC-CK1 Na — Cl Ca — HCO$_3$ Mg — SO$_4$ 1.1 0.55 0.55 1.1 (meq/l)	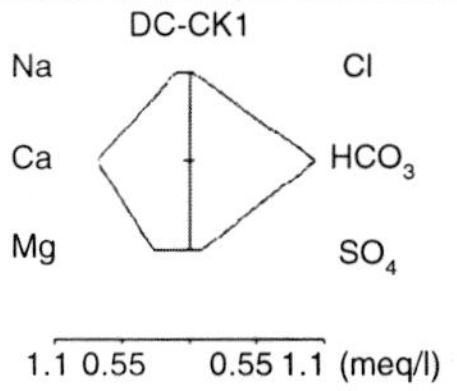 Calcium–bicarbonate type water from a surface stream in the area. The low TDS, low Na, low Cl, but high Ca and Mg indicates that it is not strongly connected to water from the coal-bed methane producing formation. This Stiff pattern is typical for water originating from surface runoff and surface recharge in a shallow unconfined aquifer (TDS = 82 mg/L).
(d) W.Elk Spring G28A Na — Cl Ca — HCO$_3$ Mg — SO$_4$ 7 3.5 3.5 7 (meq/l)	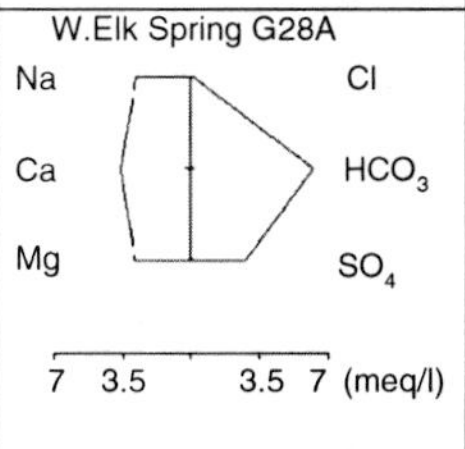 Calcium–magnesium–sodium–bicarbonate–sulfate type water from a spring in the study area. The significant levels of sodium, sulfate, and TDS indicate that it is fed, in part, by discharge from deep formations, possibly the coal-bed methane producing formation. The presence of sulfate indicates a lack of methane at this location (see discussion below). This Stiff pattern is typical for water resulting from mixing of surface water with water from deeper formations (TDS = 427 mg/L).
(e) MVwell 126′ Na — Cl Ca — HCO$_3$ Mg — SO$_4$ 2.7 1.35 1.35 2.7 (meq/l)	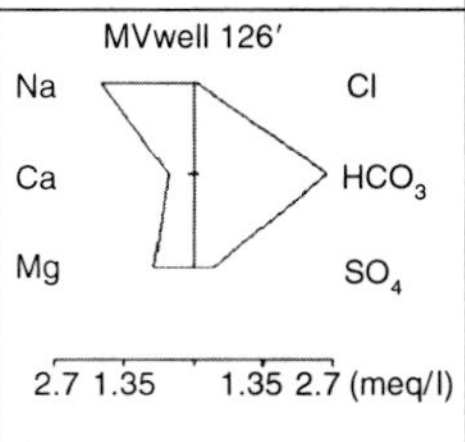 Sodium–magnesium–bicarbonate type water from a well completed in the coal-bed methane producing formation, but about 7 miles south of the well tested in (a). The relatively low TDS and higher calcium and magnesium levels indicate a substantial contribution from surface recharge water. This stiff pattern indicates that this well receives water both from the deep coal-bed methane producing formation and from surface recharge (TDS = 120 mg/L).

FIGURE 10.7 Stiff diagrams of representative water sources.

the left. The plotted points are connected as in Figure 10.7 to give an irregular polygon shape that is distinctive for different water types. It is obviously important to always arrange the ion plots in the same order.

For these studies, the Stiff diagrams plotted the sample concentrations of six major ions (sodium, calcium, magnesium, chloride, bicarbonate, and sulfate) as shown in Figure 10.7. These major ions normally comprise over 90% of all dissolved solids in natural waters and, therefore, their total concentration may be taken to closely represent the TDS concentration of the sample. The shape of a Stiff diagram allows a rapid identification of the water quality profile, for example, predominantly sodium bicarbonate versus NaCl. The total area of the polygon gives an approximate indication of the TDS concentration.

The study found that water from exploration wells drilled into the formations where coal-bed methane was expected had a distinct chemical profile compared to streams, springs, and water supply wells in the study area. From this, the energy company concluded that drilling gas wells into the coal-bed methane formations was not likely to affect most existing water supplies because there appeared to be no significant hydraulic connections between them.

Figure 10.7 shows characteristic Stiff diagrams for different water sources obtained from this study.

10.10.1 Geochemical Explanation for the Stiff Patterns

When precipitation flows over and under the earth's surface, it reacts with the minerals that it contacts, accumulating a characteristic dissolved ion content. The chemical makeup of a collected water sample reflects the history of its prior flow path, particularly the minerals it has contacted, its contact time with these minerals, and the temperature, pH, and redox potential along its flow path. The major dissolved ions in groundwater usually are those used in the Stiff diagrams of Figure 10.7, Na^+, Ca^{2+}, Mg^{2+}, Cl^-, CO_3^{2+}, and SO_4^{2+}. As groundwater moves deeper toward bedrock, geochemical processes change its chemical profile, generally in a predictable way. Typically, when comparing water from coal-bed methane producing formations with shallow water in the recharge zone:

- Bicarbonate anion concentration is increased.
- Calcium and magnesium cation concentrations are reduced.
- Sodium cation concentration may increase.
- Sulfate anion concentration is reduced.
- Total dissolved solids concentration is increased.
- Chloride concentrations depend mainly on whether groundwater comes in contact with formations of marine origin and are not changed much by geochemical processes.

As a result, water from a formation where coal-bed methane is produced is usually a sodium bicarbonate type water with a Stiff diagram similar to Figure 10.7a. Water from a surface stream, or water well recharged mainly by surface water, is usually more like Figure 10.7c and d, with less sodium and more calcium, magnesium, and sulfate.

10.10.1.1 Bicarbonate Anion Increase

There are three main geochemical processes producing bicarbonate enrichment as groundwater moves farther from its recharge areas and deeper into the subsurface.

- Bicarbonate anion is enriched when acidic groundwater (Sections 3.4.2, 3.4.3, and 5.4.1) dissolves bicarbonate minerals.
- In surface water and shallow groundwater, oxidative decay and biodegradation of organic matter produce CO_2 as a reaction product. Hydrolysis of CO_2 produces bicarbonate at normal environmental pHs (Section 3.4.2).
- In deeper groundwaters, under reducing conditions, sulfate reduction and methanogenesis (Sections 8.5.7 and 8.5.8) generate CO_2, which hydrolyzes to bicarbonate.

10.10.1.2 Calcium and Magnesium Cation Decrease

There are two main geochemical processes causing depletion of calcium and magnesium as groundwater moves farther from its recharge areas and deeper into the subsurface.

- The most important depletion process is the precipitation of calcium and magnesium carbonates (calcite and dolomite) that occurs because of the higher concentrations of bicarbonate that develop along the flow path. The solubility of calcium and magnesium carbonates decreases with increasing bicarbonate concentration (Stumm and Morgan, 1981).
- Ion exchange with sodium on clays can also deplete calcium and magnesium concentrations. Ion-exchangeable clays along the flow path generally contain a large fraction of sorbed Na^+. The divalent cations Ca^{2+} and Mg^{2+} sorb more strongly to clays than monovalent Na^+ and, therefore, will exchange with Na^+ on the clay surface, releasing Na^+ into solution.

10.10.1.3 Sodium Cation May Increase

There are three main geochemical processes causing sodium to increase as groundwater moves farther from its recharge areas and deeper into the subsurface.

1. In recharge areas, infiltrating groundwater dissolves soil salts, often accumulating relatively high concentrations of calcium, magnesium, and sodium cations. Processes described above, subsequently deplete calcium and magnesium, but not sodium. There are no important sodium precipitation reactions that control sodium concentrations.
2. Where groundwater contacts clay minerals, sodium is displaced from clay surfaces by ion exchange with calcium and magnesium and becomes dissolved into the water.
3. Groundwater in contact with formations of marine origin can accumulate extremely high concentrations of sodium, greater than 100,000 mg/L.

10.10.1.4 Sulfate Anion Decrease

There are two important geochemical processes causing depletion of sulfate as
groundwater moves farther from its recharge areas and deeper into the subsurface
(see "Sulfate" Appendix A, and Section 8.5.7 for additional details).

- Sulfate salts are abundant in soils where oxidizing conditions exist. Infiltrating water can accumulate sulfate concentrations greater than 1500
  mg/L. However, as water moves deeper to where reducing conditions
  prevail, sulfate is lost by being reduced to sulfide, forming hydrogen sulfide
  gas or precipitating as low-solubility metal sulfides.
- Reduction of sulfate to sulfide is greatly accelerated by microorganisms and
  sulfate-reducing bacteria are ubiquitous in the subsurface (Sections 4.4,
  8.5.7, and 8.6). Methane formation in coal beds begins under reducing
  conditions after sulfate and other electron acceptors have been depleted
  (Section 8.5.8). Therefore, methanogenesis can only occur after sulfate is
  bioreduced to a low concentration. Thus, biogenic coal-bed methane development is always accompanied by a nearly total depletion of sulfate by
  bacterial reduction to sulfide.

The unique chemical profile of groundwater associated with coal-bed methane
formations has been suggested as a diagnostic tool for identifying promising sites
for coal-bed methane extraction (Voast, 2003).

REFERENCES

Aronheim, J.S., 1992, Virus transport in groundwater: Modeling of bacteriophage PRD1
    transport through one-dimensional columns and a two-dimensional aquifer tank,
    Masters thesis, Department of Civil, Environmental, and Architectural Engineering,
    University of Colorado, Boulder, CO.
Asano, T., Leong, L.Y.C., Rigby, M.G., and Sakaji, R.H., 1992, Evaluation of the California
    wastewater reclamation criteria using enteric virus monitoring data, *Water Sci. Technol.*,
    26 (7–8), 1513–1524.
Bellar, T.A., Lichtenberg, J.J., and Kroner, R.C., 1974, The occurrence of organohalides in
    chlorinated drinking water, *J. AWWA*, 66 (12), 703.
Bitton, G., 1975, Adsorption of viruses onto surfaces in soil and water, *Water Res.*, 9, 473.
Bitton, G., 1980, *Introduction to Environmental Virology*, John Wiley, New York.
Bitton, G., Davidson, J.M., and Farra, S.R., 1979, On the value of soil columns for assessing
    the transport pattern of viruses through soils: A critical outlook, *Water, Air, Soil Pollut.*,
    12, 449–457.
Bull, R.J. and Kopfler, F.C., 1991, *Health Effects of Disinfectants and Disinfection
    By-Products*, 90577, AWWA Research Foundation, Denver, CO.
Burge, W.D. and Enkiri, N.K., 1978, Adsorption kinetics of bacteriophage $\phi\chi$-174 on soil,
    J. Environ. Qual., 7 (4), 536–541.
Chang, N.-Y., Pearson, W., Chang, J.I.J., Gross, A., Meyer, M., Jolly, M., Vang, B., and
    Samour, H., 1994, Final report on environmentally sensitive sanding and deicing
    practices, ESSD Research Group, Department of Civil Engineering, University of
    Colorado at Denver, Colorado Transportation Institute and Colorado Department of
    Transportation CDOT-CTI-95–5, November 25.

Chu, Y., Yan, J., Thomas, B., and Marylynn, V.Y., 2003, Effect of soil properties on saturated and unsaturated virus transport through columns, *J. Environ. Qual.*, 32, 2017–2025.

Eaton, A.D., Clesceri, L.S., and Greenberg, A.E., Eds., 1995, *Standard Methods for the Examination of Water and Wastewater*, 19th edition, American Public Health Association, American Water Works Association, and Water Environment Federation, Washington, DC.

FAO, 1986, Water Quality for Agriculture, FAO Irrigation and Drainage Paper No. 29, Rev. 1, Food and Agriculture Organization of the United Nations.

Foster, D.M., Emerson, M.A., Buck, C.F., Walsh, D.S., and Sproul, O.J., 1980, Ozone inactivation of cell- and fecal-associated viruses and bacteria, *J. Water Pollut. Control Fed.*, 52, 2174.

Gerba, C.P., 1975, Fate of wastewater bacteria and viruses in soil, *J. Irrig. Drainage Division*, 101 (IR3), 157–173.

Gerba, C.P. and Smith, Jr. J.E., 2005, Sources of pathogenic microorganisms and their fate during land application of wastes, *J. Environ. Qual.*, 34 (1), 42–48.

Hurst, C.J., Gerba, C.P., and Cech, I., 1980, Effects of environmental variables and soil characteristics on virus survival in soil, *Appl. Environ. Microbiol.*, 40 (6), 1067–1079.

Hurst, C.J., 1989, Fate of viruses during wastewater sludge treatment processes, *CRC Crit. Rev. Environ. Control*, 18 (4), 317–343.

Keith, L.H., 1992, *Environmental Sampling and Analysis: A Practical Guide*, Lewis, Press, Boca Raton, FL.

Lefler, E. and Kott, Y., 1974, Enteric virus behavior in Sand Dunes, *Israel J. Technol.*, 12, 298–304.

Lo, S.H. and Sproul, O.J., 1977, Poliovirus adsorption from water onto silicate minerals, *Water Res.*, 11, 653–658.

McKee, J.E. and Wolf, H.W., 1963, *Water Quality Criteria, State of California*, 2nd edition, Publication No. 3-A, The Resources Agency of California, State Water Quality Control Board, Sacramento, CA.

Michigan Department of Transportation, 1993, The use of selected deicing materials on Michigan roads: Environmental and economic impacts, Prepared by Public Sector Consultants, Michigan Department of Transportation, Lansing, MI, December.

NRC, 1994, *Groundwater Recharge: Using Waters of Impaired Quality*, National Research Council, Washington, DC.

Pinholster, G., 1995, Drinking recycled wastewater, *Environ. Sci. Technol.*, 29 (4), 174A–179A.

Rocky Mountain Construction, 1995, Anti-icing: A bold new strategy, Strategic Highway Research Program, September 30, p. S-14.

Rook, J., 1974, Formation of haloforms during chlorination of natural waters, *J. Soc. Water Treat. Exam.*, 23, 234.

Rose, J.B. and Gerba, C.P., 1991, Assessing potential health risks from viruses and parasites in reclaimed water in Arizona and Florida, USA, *Water Sci. Technol.*, 23, 2091–2098.

Scheuerman, P.R., Bitton, G., Overman, A.R., and Gifford, G.E., 1979, Retention of viruses by organic soils and sediments, *J. Env. Eng. Div.*, ASCE, EE4, 629–640.

Sobsey, M.D. and Hickey, A.R., 1985, Effects of humic and fulvic acids on poliovirus concentration from water by microporous filtration, *Appl. Environ. Microbiol.*, 49 (2), 259–264.

Sobsey, M.D., Dean, C.H., Knuckles, M.E., and Wagner, R.A., 1980, Interactions and adsorption of enteric viruses in soil materials, *Appl. Environ. Microbiol.*, 40, 92.

Stotsky, G., Schiffenbauer, M., Lipson, S.M., and Yu, B.H., 1980, Surface interactions between viruses and clay minerals and microbes: Mechanisms and implications,

Presented at the International Symposium on Viruses and Wastewater Treatment, University of Surrey, Guilford, United Kingdom.

Stumm, W. and Morgan, J.J., 1981, *Aquatic Chemistry*, 2nd edition, Chapter 5, John Wiley, New York, p. 780.

USEPA, 1979, *National Interim Primary Drinking Water Regulations; Control of Trihalomethanes in Drinking Water*, Fed. Reg., 44:231:68624, November 29.

USEPA, 1981, *Process Design Manual for Land Treatment of Municipal Wastewater*, EPA 625/1–81–013, October.

USEPA, 1986, *Bacteriological Ambient Water Quality Criteria for Marine and Fresh Recreational Waters*, EPA 440/5-84-002, NTIS PB-86-158-045.

USEPA, 1992a, *Manual: Guidelines for Water Reuse*, EPA/625/R-92/004, September.

USEPA, 1992b, *Test Methods for Evaluating Solid Waste*, Vol. 1A, Laboratory Manual Physical/Chemical Methods, SW 846, 3rd edition, July.

USEPA, 1994, *Drinking Water Standard Setting: Question and Answer Primer*, EPA/811-K-94–001, November.

USEPA, 1996, *Drinking Water Regulations and Health Advisories*, EPA 822-B-96–002, October.

USEPA, 1997, *Small Systems Compliance Technology List for the Surface Water Treatment Rule*, EPA 815-R-97–002, August.

USEPA, 1999, *Alternative Disinfectants and Oxidants Guidance Manual*, EPA 815-R-99–014, April.

USEPA, 2007, *Guidelines Establishing Test Procedures for the Analysis of Pollutants*, 40 CFR Part 136, Appendix B, Federal Register, 72 (57), 14220–14233, March 26.

Voast, W.A., 2003, Geochemical signature of formation waters associated with coal-bed methane, *AAPG Bull.*, 87 (4) April, 667–676.

Wallis, C., Henderson, M., and Melnick, J.L., 1972, Enterovirus concentration on cellulose membranes, *Appl. Microbiol.*, 23 (3), 476–480.

Watershed Research, 1995, Rating deicing agents—road salt stands firm, *Tech. note 55*, 1 (4), Summer 1995.

Appendix A:
A Selective Dictionary
of Water Quality Parameters
and Pollutants

A.1 INTRODUCTION

This section is a concise guide to useful information about frequently measured water quality parameters and pollutants, arranged alphabetically. The entries are not intended to be exhaustive, but to offer a brief summary of background data and information. Most of the parameters can have both natural and human origins. Where relevant, other parts of this book are referenced for further information.*

Where CAS identification numbers have been assigned, they are included for each entry. CAS stands for Chemical Abstracts Service Registry, a division of the American Chemical Society, which assigns unique identification numbers to each chemical compound and uses these numbers to facilitate literature and computer database searches for chemical information.

A.1.1 Water Quality Inorganic Parameters: Classified by Abundance

Although water quality parameters are listed alphabetically in the dictionary section, it is sometimes useful to classify them according to their typical abundance in natural waters. This is done in the listing below.

Major chemical parameters are those most often present in natural waters in concentrations greater than 1.0 mg/L. These are the cations calcium, magnesium, potassium, and sodium and the anions bicarbonate/carbonate, chloride, fluoride, nitrate, and sulfate. Silicon is usually present as nonionic species and is reported by analytical laboratories as the equivalent concentration of silica (SiO_2).

Several additional chemical parameters are sometimes included with the major constituents because of their importance in determining water quality and because some of them sometimes attain concentrations comparable to the parameters above. These are aluminum, boron, iron, manganese, nitrogen in forms other than nitrate

---

* Information in Appendix A has been compiled from many different sources, but particularly from EPA Web site pages.

(such as ammonia and nitrite), organic carbon, phosphate, and the dissolved gases oxygen, carbon dioxide, and hydrogen sulfide.

Minor chemical parameters are those most often present in natural waters in concentrations less than 1.0 mg/L. These include the so-called trace elements and naturally occurring radioisotopes: antimony, arsenic, barium, beryllium, bromide, cadmium, cesium, chromium, cobalt, copper, iodine, lead, lithium, mercury, molybdenum, nickel, radium, radon, rubidium, selenium, silver, strontium, thorium, titanium, uranium, vanadium, and zinc.

Physical and chemical properties are parameters that do not identify particular chemical species but are used as indicators of how water quality may affect water uses. These are acidity, alkalinity, hardness, hydrogen ion concentration (measured as pH), redox potential, biochemical oxygen demand (BOD), chemical oxygen demand (COD), color, corrosivity, gross alpha and beta emitters, odor, sodium adsorption ratio (SAR), Langelier index, specific conductance (conductivity), specific gravity, temperature, total dissolved solids (TDS), total suspended solids (TSS), and turbidity.

A.2 ALPHABETICAL LISTING OF CHEMICAL AND PHYSICAL WATER QUALITY PARAMETERS AND POLLUTANTS
USEPA Drinking water standards are updated to august 2006 (USEPA, 2006)

2,4-D (2,4-dichlorophenoxyacetic acid), CAS No. 94-75-7

2,4-D is a colorless, odorless powder used as a herbicide for the selective control of broad-leaf weeds in agriculture, and for control of woody plants along roadsides, railways, and utilities rights of way. It is one of the most widely used herbicides in the world and is commonly used on such crops as wheat and corn, and on pasture and rangelands. It is also used to control broad-leaved aquatic weeds.

Environmental Behavior

2,4-D is rapidly degraded by microbes in soil and water, with a half-life in different soils of 3–22 days. 2,4-D is weakly sorbed by soil with sorption generally increasing with increasing soil organic carbon content. Leaching to groundwater is most likely in coarse-grained sandy soils with low organic content or with very basic soils. In general little runoff occurs with 2,4-D or its amine salts.

Ecological Concerns

2,4-D is moderately toxic to birds and fish. Ester formulations are especially toxic to fish, and, in the case of a spill of a 2,4-D ester, some fish mortality is likely. The presence of emulsifiers with 2,4-D amine enhances the toxicity of the 2,4-D amine to aquatic species. 2,4-D exposures through spray drift and runoff are considered the greatest potential risks to terrestrial plants, mammals, and birds, whereas exposures to 2,4-D through the direct application to water for aquatic weed control present the greatest potential risk to aquatic plants and animals. There is no evidence that bioconcentration of 2,4-D occurs through the food chain.

Health Concerns

Short-term exposure to 2,4-D at levels above the maximum contaminant level (MCL) can irritate the eyes, skin, and breathing passages and cause nervous system damage. Long-term chronic exposure at levels above the MCL can cause damage to the nervous system, kidneys, and liver. Some 2,4-D-based products contain surfactants, which, if inhaled, may cause neurological damage; skin contact may cause dermatitis if prolonged or repeated; and when, ingested or inhaled, may cause respiratory swelling and excess fluid in tissue.

EPA Primary Drinking Water Standard
MCL: 70 μg/L
Maximum contaminant level goal (MCLG): 70 μg/L

Acrylamide (C_3H_5NO), CAS NO. 79-06-1

Acrylamide is an organic solid of white, odorless, flake-like crystals. Its major use is in the production of polyacrylamide, which is used in water treatment, pulp and paper production, and mineral processing. It is also used in the synthesis of dyes, adhesives, contact lenses, soil conditioners, and permanent press fabrics.

Environmental Behavior
Acrylamide does not bind to soil and will move into soil rapidly, but it is degraded by microbes within a few days in soil and water. When released into the soil, acrylamide is expected to leach into groundwater, and biodegrade to a moderate extent.

Ecological Concerns
Acrylamide is not expected to be toxic to aquatic life. The LC50/96 hour values for fish are over 100 mg/L. It is not expected to significantly bioaccumulate.

Health Concerns
Acrylamide is a neurotoxicant and in animal studies has been shown to be a carcinogen, germ cell mutagen, and reproductive toxicant. Short-term exposure at levels above the MCL may cause damage to the nervous system, weakness, and loss of coordination in the legs. Long-term exposure at levels above the MCL may cause damage to the nervous system, paralysis, and cancer.

EPA Primary Drinking Water Standard
MCL: Treatment technique. EPA requires a water supplier to show that when acrylamide is added to water, the amount of uncoagulated acrylamide is less than 0.5 ppb.
MCLG: zero

Alachlor ($C_{14}H_{20}ClNO_2$), CAS No.15972-60-8

Alachlor is a pre- and post-emergence herbicide used to control annual grasses and many broad-leaved weeds in corn and a number of other crops. There are liquid, dry flowable, microencapsulated, and granular formulations. Alachlor is applied by ground, aerial, and chemigation equipment. It can also be mixed with dry bulk fertilizer. It is lost from soil mainly through volatilization, photodegradation, and biodegradation. Alachlor and its degradation products may be found in soil, groundwater, and surface water.

Environmental Behavior
Alachlor is highly mobile and moderately persistent, with a low affinity to adsorb to soils. It dissipates primarily by aerobic soil metabolism processes with a half-life of

2–3 weeks. Even when used according to label directions, it has a high probability of entering groundwater and surface water.

Ecological Concerns
Alachlor is slightly to practically nontoxic to birds, mammals, and honey bees; highly to moderately toxic to freshwater fish; and highly toxic to aquatic plants.

Health Concerns
Alachlor is considered to be a carcinogen and a metabolite of alachlor, 2,6-diethylaniline, has been shown to be mutagenic.

EPA Primary Drinking Water Standard
MCL: 0.002 mg/L
MCLG: zero

Aldicarb ($C_7H_{14}N_2O_2S$), CAS No. 116-06-3
Aldicarb is a broad-spectrum, systemic carbamate insecticide, used to control nematodes in soil and a variety of insects and mites on citrus crops, dry beans, grain, sorghum, ornamentals, pecans, peanuts, potatoes, seed alfalfa, soybeans, sugar beets, sugarcane, sweet potatoes, and tobacco.

Environmental Behavior
Aldicarb and its degradation products are generally mobile in soil. Adsorption in soil is primarily to organic matter, so leaching is most extensive in sandy or sandy loam soils. Aldicarb is very persistent in groundwater, typically degrading to nontoxic products with a half-life between a few weeks to as long as several years. The primary mode of degradation is chemical hydrolysis, with some microbial decay in shallow groundwater. In soils, the primary mode of degradation is oxidation by soil microorganisms and hydrolysis, depending on soil conditions.

Ecological Concerns
Aldicarb is one of the most acutely toxic pesticides in use, due to acetylcholinesterase inhibition. It is metabolized to the sulfoxide and sulfone. Aldicarb sulfoxide is a more potent inhibitor of acetylcholinesterase than aldicarb itself, whereas aldicarb sulfone is considerably less toxic than either aldicarb or the sulfoxide. The weight of evidence indicates that aldicarb, aldicarb sulfoxide, and aldicarb sulfone are not genotoxic or carcinogenic.

Health Concerns
Ingestion of aldicarb can cause dizziness, weakness, diarrhea, nausea, vomiting, abdominal pain, excessive perspiration, blurred vision, headache, convulsions, and temporary paralysis of the extremities. Recovery is rapid, usually within 6 h. Aldicarb is considered an acutely toxic pesticide, with several incidents of accidental or intentional poisoning being reported. Aldicarb is classified as probably not carcinogenic to humans.

EPA Primary Drinking Water Standard
Unregulated at present, but monitoring required. EPA is currently reviewing aldicarb. Canada has set a maximum acceptable concentration (MAC) for aldicarb in

drinking water of 9 μg/L. This guideline applies to the total of aldicarb plus its toxic metabolites, aldicarb sulfoxide and aldicarb sulfone.

Aldrin ($C_{12}H_8Cl_6O$), CAS No. 309-00-2, and Dieldrin ($C_{12}H_8Cl_6O$), CAS No. 60-57-1

Aldrin and dieldrin are both nonsystemic chlorinated pesticides used to control soil-dwelling pests, for wood protection and, in the case of dieldrin, against insects of public health importance. The two compounds are closely related with respect to their toxicology and mode of action. Since the early 1970s, many countries have either severely restricted or banned the use of both compounds, particularly in agriculture. The last known manufacturer of aldrin and dieldrin, Shell International Chemical Co. (United Kingdom), ceased production of the pesticides in 1989.

Environmental Behavior
Aldrin and dieldrin have low water solubilities (0.027 mg/L for aldrin at 27°C and 0.186 mg/L for dieldrin at 20°C). Aldrin is rapidly converted to dieldrin under most environmental conditions and in the body. Dieldrin is a highly persistent organochlorine compound that has low mobility in soil, but readily volatizes to the atmosphere. Long-range atmospheric transport tends to be from warmer to colder regions where the pesticides are sometimes found where they have never been used. The high mobility of aldrin and dieldrin has led to widespread banning of their use because of concern about the difficulty of controlling the movement of these persistent chemicals in the environment.

Health Concerns
Aldrin and dieldrin are highly toxic to humans and animals, affecting the central nervous system and the liver. They are not considered carcinogenic. Exposure to aldrin and dieldrin occurs by oral ingestion, inhalation, and dermal absorption. In the body, aldrin is quickly metabolized to dieldrin. The biological half-life of dieldrin in humans is about 266 days. Signs and symptoms related to ingestion of or dermal contact with toxic doses of aldrin and dieldrin include headache, dizziness, nausea, general malaise, and vomiting, followed by muscle twitching, myoclonic jerks, and convulsions.

EPA Primary Drinking Water Standard
Aldrin and dieldrin are not registered in the United States or Canada and, therefore, should not be available for general use. No EPA MCL is listed. Canada has set a maximum acceptable concentration (MAC) for combined aldrin and dieldrin in drinking water of 0.7 μg/L.

Aluminum (Al), CAS No. 7429-90-5

Aluminum is the most abundant metal and the third most abundant element in the Earth's lithosphere (after oxygen and silicon) and its compounds are often found in natural waters. Aluminum is mobilized naturally in the environment by the weathering of rocks and minerals, particularly bauxite clays. It is a normal constituent of all soils and is found in low concentrations in all plant and animal tissues. Most naturally occurring aluminum compounds have very low solubility between pH 6 and 9. Therefore, dissolved forms rarely occur in natural waters in concentrations greater than about 0.01 mg/L. Concentrations in water greater than this usually indicate the presence of solid forms of aluminum, such as suspended solids and colloids.

The concentration of Al^{3+} in water is controlled by the solubility of aluminum hydroxide, $Al(OH)_3$, which increases by a factor of about 10^3 for every unit decrease in pH. Thus, the concentration of dissolved aluminum, Al^{3+}, is about 3×10^{-5} mg/L at pH 6, 0.03 mg/L at pH 5, and 30 mg/L at pH 4.

Also, at lower water pH (<pH 5) in the presence of clays and organic-rich soils, dissolved aluminum concentrations increase because of the release of Al^{3+} from the soil. Low pH means high H^+ concentrations. At pH < 5, the concentration of H^+ is high enough for H^+ to partially ion-exchange with other, more strongly bound metals at ion-exchange sites on soil particles. Since Al^{3+} is bound more strongly than divalent and monovalent cations, it is among the last cations to be displaced by H^+ and requires a continued low pH to reach elevated dissolved levels.

RULE OF THUMB

The presence of an elevated concentration of Al^{3+}, often exceeding the concentrations of Ca^{2+} and Mg^{2+}, is a common characteristic of acidic waters (pH < 5), including mine drainage and waters affected by acid rain.

Health Concerns
Naturally occurring aluminum has a very low toxicity to humans and animals. Only a few industrially important aluminum compounds, such as the fumigant aluminum phosphide, are considered acutely hazardous. Exposure and ingestion of aluminum and its compounds is usually not harmful. Aluminum compounds are used in water treatment to remove color and turbidity, food packaging, medicines, soaps, dental cements, drug store items such as antacids and antiperspirants, and are present in many foods because they are grown in soils containing aluminum.

Daily exposure to aluminum is inevitable due to its ubiquitous occurrence in nature and its many commercial uses. Estimated human consumption is about 88 mg/person of aluminum per day, mostly from food. Consumption of 2 L of water per day containing 1.5 mg/L of aluminum (well above the secondary drinking water standard, see below) only contributes 3.0 mg of aluminum per day, or less than 4% of the normal daily intake. Aluminum is not believed to be an essential nutrient. There is no human or animal evidence of carcinogenicity. There is some indication that ingestion of large doses of aluminum in medicines may cause skeletal problems.

EPA Primary Drinking Water Standard
MCL: None
MCLG: None
Secondary Standard: 0.05 to 0.2 mg/L

EPA has no primary drinking water standard for aluminum. The EPA secondary drinking water standard (nonenforceable) is expressed as a range: 0.05–0.2 mg/L. EPA recommends that 0.05 mg/L be met where possible, but allows states to determine the required level on a case-by-case basis because water treatment technologies often use aluminum salts to remove color and turbidity and cannot always achieve the lower value. EPA recommends that aluminum in drinking water not exceed 0.2 mg/L because of taste and odor problems. In the presence of microorganisms, aluminum can react with iron, manganese, silica, and organic material to form fine sediments that can appear at the consumer's tap. If dissolved aluminum exceeds 0.1 mg/L, levels of iron normally acceptable may produce discoloration and staining. No lifetime health advisory has been established. A lifetime health advisory is the concentration in drinking water that is not expected to cause any adverse effects over 70 years of exposure, with a margin of safety.

Other Comments
Because of its usually low concentrations, aluminum is normally of no concern in irrigation waters. However, a limit of 5.0 mg/L is recommended (National Academy of Sciences, 1982) where irrigation waters are used regularly. Swimming pools treated with commercial grade aluminum sulfate compounds, known as alum, may cause eye irritation at concentrations greater than 0.1 mg/L.

Ammonia/Ammonium Ion (NH_3/NH_4^+), CAS No. 7664-41-7
(See Chapter 3 for a more detailed discussion.)

In the biological decay of nitrogenous organic compounds, ammonia is the first nitrogenous product not containing carbon. It arises from the waste products and decay after death of plants, animals, and other life forms. Thus, ammonia is present naturally in surface and groundwaters. It also is discharged in many waste streams, particularly from municipal waste treatment. Unpolluted waters have very low ammonia concentrations, generally less than 0.2 mg/L as nitrogen.

Continued oxidation of ammonia leads sequentially to nitrite and nitrate. Ammonia gas is very soluble in water, where it reacts as a base, raising the pH and forming an ammonium cation and a hydroxyl anion:

$$NH_3 + H_2O \leftrightarrow NH_4^+ + OH^- \qquad (3.22)$$

The unionized form (NH_3) is of greatest environmental concern because of its greater toxicity to aquatic life. However, because NH_4^+ readily converts to NH_3 by its pH dependent equilibrium, Equation 3.22, total ammonia is normally regulated in discharges. Concentrations of unionized (NH_3) or total ($NH_3 + NH_4^+$) ammonia are often reported in terms of the nitrogen content only, e.g., $NH_3 = 10$ mg/L-N, or NH_3–N $= 10$ mg/L. This means that the sample contains unionized ammonia and the nitrogen portion of the unionized ammonia weighs 10 mg/L of sample. The weight of the hydrogen in the ammonia molecules is ignored. To convert milligrams per liter of NH_3 to milligrams per liter of NH_3–N, multiply by 0.822.

The equilibrium between the unionized form (NH_3) and the ionized form (NH_4^+) depends on pH, temperature, and, to a much lesser degree, on ionic strength (salinity or concentration of total dissolved solids), see Chapter 3.

- At 15°C and pH > 9.6, the fraction of NH_3 is greater than 0.5.
- At 15°C and pH < 9.6, the fraction of NH_4^+ is greater than 0.5.
- A temperature increase shifts the equilibrium of Equation 3.22 to the left, increasing the NH_3 concentration.
- A temperature decrease shifts the equilibrium of Equation 3.22 to the right, increasing the NH_4^+ concentration.
- An increase in ionic strength shifts the equilibrium of Equation 3.22 to the right, increasing the NH_4^+ concentration slightly. In waters with very high total dissolved solids (>10,000 mg/L), there will be a small but measurable decrease in the percentage of NH_3.
- Because pH and temperature can vary considerably along a stream or within a lake, the fraction of total ammonia that is unionized is also variable at different locations. Therefore, the amount of total ammonia is usually of regulatory concern, rather than only the unionized form.

Health Concerns
Total ammonia ($NH_3 + NH_4^+$) in drinking water is more an esthetic than a health concern. The odor and taste of ammonia makes drinking water unpalatable at concentrations well below the appearance of any toxic effects to humans. The main health concern with ammonia is its potential oxidation to nitrite (NO_2^-) and nitrate (NO_3^-). Ingested nitrate and nitrite react with iron in blood hemoglobin to cause a blood oxygen deficiency disease called methemoglobinemia, which is especially dangerous in infants (blue baby syndrome) because of their small total blood volume. There is no human or animal evidence of carcinogenicity.

EPA Primary Drinking Water Standard
EPA has no primary or secondary drinking water standards for ammonia. However, the presence of NH_3 greater than 0.1 mg/L may raise the suspicion of recent pollution. The lifetime health advisory, the concentration in drinking water that is not expected to cause any adverse effects over 70 years of exposure, with a margin of safety, is 30 mg/L.

Some states have adopted ammonia limits for water that will receive treatment to produce drinking water. For example, Colorado's ammonia standard for water classified as domestic water supply is 0.05 mg/L–N, 30 day average for total ammonia ($NH_3 + NH_4^+$).

RULES OF THUMB

1. Only NH_3, the unionized form, has significant toxicity for aquatic life.

RULES OF THUMB (Continued)

2. To convert milligrams per liter of unionized or total ammonia to milligrams per liter as nitrogen, multiply by 0.822. Example: 17.4 mg/L $NH_3 = 0.822 \times 17.4 = 14.3$ mg/L–N.

3. Since pH 9.6 is higher than the pH of most natural waters, NH_3–N in natural waters usually is mostly in the less toxic ionized ammonium form (NH_4^+).

4. In high pH waters (pH > 9), the NH_3 fraction can reach levels toxic to aquatic life.

5. The ionized form is not volatile and cannot be removed by air stripping. The unionized form, NH_3, is volatile and can be removed by air stripping.

Antimony (Sb), CAS No. 1440-36-0

Antimony is a metalloid (having properties intermediate between metals and non-metals) in the same chemical group (group 5A) as arsenic, with which it has some chemical similarities, including toxicity. However, it is only about one-tenth as abundant in the earth's crust and soils. The symbol Sb for the element is from stibium, the Latin name for antimony. In the environmental literature, antimony is often included with the metals because it is usually analyzed, along with other metals, by inductively coupled plasma (ICP) or atomic absorption (AA) techniques. Common sources of antimony in drinking water are discharges from petroleum refineries, fire retardants, ceramics, electronics, and solder. It is also found in batteries, pigments, ceramics, and glass.

Antimony is usually adsorbed strongly to iron, manganese, and aluminum compounds in soils and sediments. Soil concentrations normally range between 1 and 9 mg/L. The amount commonly dissolved in rivers is small, less than 0.005 mg/L. There is no evidence of bioconcentration of most antimony compounds.

Health Concerns

Antimony is used in medicines for treating parasite infections. It is present in meats, vegetables, and seafood in an average concentration of about 0.2–1.1 ppb (μg/L). An average person ingests about 5 μg of antimony every day in food and drink. Short-term exposures above the MCL may cause nausea, vomiting, and diarrhea. Potential health effects from long-term exposure above the MCL are an increase in blood cholesterol and a decrease in blood glucose. There is insufficient evidence to state whether antimony has the potential to cause cancer.

EPA Primary Drinking Water Standards
MCLG: 0.006 mg/L
MCL: 0.006 mg/L

Other Comments
Treatment/Best available technologies: Coagulation and filtration, reverse osmosis.

Arsenic (As), CAS No. 7440-38-2
Chemically, arsenic is classified as a metalloid, having properties intermediate between metals and nonmetals. In the environmental literature, it is often included with the metals because it is usually analyzed, along with other metals, by inductively coupled plasma (ICP) or atomic absorption (AA) techniques. Inorganic arsenic occurs naturally in many minerals, especially in ores of copper and lead. Smelting of these ores introduces arsenic to the atmosphere as dust particles.

In minerals, arsenic is combined mostly with oxygen, chlorine, and sulfur. Inorganic arsenic compounds are used mainly as wood preservatives, insecticides, and herbicides. Organic forms of arsenic found in plants and animals are combined with carbon and hydrogen. Organic arsenic is generally less toxic than inorganic arsenic. Arsenic is not abundant, with an average concentration in the lithosphere of about 1.5 mg/kg (ppm). Background levels in soils typically range from 1 to 95 mg/kg. Average levels in U.S. soils are around 5–7 mg/kg. It is widely distributed and is found naturally in many foods at levels of 20–140 ppb, exposing most Americans to a constant low exposure, perhaps around 50 μg/day. Normal human blood contains 0.2–1.0 mg/L of arsenic; however, there is no evidence that arsenic is an essential nutrient.

Many arsenic compounds are water-soluble and may be found in groundwater, especially in the western United States. The average concentration for U.S. surface water is around 3 ppb. Groundwater levels average about 1–2 ppb, except in some western states where groundwater is in contact with volcanic rock and sulfide minerals high in arsenic. In western mining regions, arsenic levels as high as 48,000 ppb have been observed. Many persons dependent on well water in the West ingest higher than average levels of inorganic arsenic through their drinking water supplies.

Health Concerns
High levels (>60 ppm) of arsenic in food or water can be fatal. Arsenic damages tissues in the nervous system, stomach, intestine, and skin. Breathing high levels can irritate lungs and throat. Lower levels can cause nausea, diarrhea, irregular heartbeat, blood vessel damage, reduction of red and white blood cells, and tingling sensations in hands and feet. Long-term exposure to inorganic arsenic may cause darkening of the skin and the appearance of small warts on the palms, soles, and torso.

Inorganic arsenic was recognized as a possible carcinogen as early as 1879, when it was suggested that high rates of lung cancer in German miners might have been caused by inhaled arsenic. Arsenic is currently considered a carcinogen. Breathing inorganic arsenic increases the risk of lung cancer and ingesting inorganic arsenic increases the risk of skin cancer and tumors of the bladder, kidney, liver, and lung.

A crisis of well-water contamination by arsenic was discovered in Bangladesh in 1992. The crisis was created by a well-intended effort by the United Nation Children's Fund (UNICEF) to provide Bangladesh with reliable water sources free of cholera and dysentery organisms. Millions of water wells were installed and the water was tested for microbial contaminants, but not for arsenic and other toxic metals. It is now estimated that 85% of Bangladesh's geographical area contains wells contaminated with inorganic arsenic. Tens of thousands of people now exhibit signs of arsenic poisoning. The World Bank, United Nations, and other sources have begun a multimillion-dollar Bangladesh Arsenic Mitigation Water Supply Project to supply uncontaminated water to Bangladesh's 85,000 villages.

EPA Primary Drinking Water Standard
MCL: 0.010 mg/L
MCLG: zero

Other Comments
A maximum concentration of 0.1 mg/L is recommended for irrigation water and for protection of aquatic plants.

Treatment/Best available technologies: Iron coprecipitation, activated alumina or carbon sorption, ion-exchange, reverse osmosis.

Asbestos, CAS No. 1332-21-4
Asbestos is a generic term for different naturally formed fibrous silicate minerals that are classified into two groups, serpentine and amphibole, based on structure. Six minerals have been characterized as asbestos: chrysotile, crosidolite, anthophyllite, tremolite, actinolite, and andamosite. The most common form is chrysotile, which is a member of the serpentine group; the others belong to the amphibole group. These different forms of asbestos are composed of 40%–60% silica, the remainder being oxides of iron, magnesium, and other metals. The EPA banned most uses of asbestos in the United States on July 12, 1989, because of potential adverse health effects in exposed persons.

Although asbestos may be introduced into the environment by the dissolution of asbestos-containing minerals and from industrial effluents, the primary source is through the wear or breakdown of asbestos-containing materials. Because asbestos fibers are resistant to heat and most chemicals, they have been mined for use in over 3000 different products in the United States, such as roofing materials, brake linings asbestos-reinforced pipe, packing seals, gaskets, fire-resistant textiles, and floor tiles. The EPA banned most uses of asbestos in the United States on July 12, 1989, because of potential adverse health effects in exposed persons. The remaining, currently allowed uses of asbestos include battery separators, sealant tape, asbestos thread, packing materials, and certain industrial uses of gaskets.

Typical background levels in lakes and streams range from 1 to 10 million fibers per liter. Asbestos is nonvolatile, insoluble, nonbiodegradable, and does not tend to adsorb to stream sediments. Asbestos fibers do not chemically decompose to other compounds in the environment and, therefore, can remain in the environment for decades or longer. Small asbestos fibers and fiber-containing particles may be carried long distances by water currents before settling out; larger fibers and particles tend to settle more quickly. Asbestos fibers do not pass through soils to groundwater.

There are no data regarding the bioaccumulation of asbestos in aquatic organisms, but asbestos is not expected to bioaccumulate. Ordinary sand filtration removes about 90% of the fibers.

Health Concerns
Asbestos is not known to cause any health problems when people are exposed to it at levels above the MCL for relatively short periods of time. Long-term inhalation has the potential to cause cancer of the lung and other internal organs. Long-term ingestion above the MCL increases the risk of developing benign intestinal polyps.

EPA Primary Drinking Water Standard
MCL: 7 million fibers per liter (MFL)
MCLG: 7 million fibers per liter (MFL)

Other Comments
Treatment/Best available technologies: Coagulation and filtration, direct and diatomite filtration, corrosion control.

Atrazine ($C_8H_{14}ClN_5$), CAS No.: 1912-24-9
Atrazine is a selective triazine herbicide used to control broad-leaf and grassy weeds in corn, sorghum, sugarcane, pineapple, Christmas trees, and other crops, and in conifer reforestation plantings. It is also used as a nonselective herbicide on non-cropped industrial lands and on fallow lands.

Environmental Behavior
Atrazine is highly persistent in soil. Its degradation mechanisms in soil and water are hydrolysis, followed by biodegradation by soil microorganisms. Hydrolysis is rapid in acidic or basic environments, but is slower around neutral pH. Sunlight and evaporation do not affect its removal rate. Atrazine can persist for longer than 1 year under dry or cold conditions. It is moderately to highly mobile in soils with low clay or organic matter content. Because it does not adsorb strongly to soil particles and has a lengthy half-life (60 to >100 days), it has a high potential for groundwater contamination despite being only moderately soluble in water. It is frequently detected in drinking water wells.

Ecological Concerns
Atrazine is practically nontoxic to birds, slightly toxic to fish and other aquatic life, and nontoxic to bees. It has a low level of bioaccumulation in fish.

Health Concerns
Atrazine is slightly to moderately toxic to humans and other animals. It can be absorbed orally, dermally, and by inhalation. The most likely source is through drinking water. Short exposures at levels above the MCL may cause nausea and dizziness.

EPA Primary Drinking Water Standard
MCL: 0.003 mg/L
MCLG: 0.003 mg/L

Barium (Ba), CAS No. 7440-39-3
Barium is the sixth most abundant element in the lithosphere, averaging about 500 mg/kg. It exists mainly as the sulfate ($BaSO_4$, barite) and, to a lesser extent, the

carbonate ($BaCO_3$, witherite). Traces of barium are found in most soils, natural waters, and foods. Although most groundwaters contain only a trace of barium, some geothermal groundwaters may contain as much as 10 mg/L.

Barium is released to water and soil in the disposal of drilling wastes, from copper smelting, and industrial waste streams. It is not very mobile in most soil systems. In water, the more toxic soluble salts are likely to precipitate as the less toxic insoluble sulfate and carbonate compounds. Background levels for soil range from 100 to 3000 ppm. Barium occurs naturally in almost all surface waters examined in concentrations of 2–340 μg/L, with an average of 43 μg/L. In surface water and most groundwater, only traces of the element are present. However, some wells may contain barium levels 10 times higher than the drinking water standard. Marine animals concentrate the element 7–100 times, and marine plants 1000 times from seawater. Soybeans and tomatoes also accumulate soil barium 2–20 times.

Health Concerns
There is no evidence that barium is an essential nutrient. All soluble barium salts are considered toxic. Short-term exposure at levels above the MCL may cause gastrointestinal disturbances, muscular weakness, and liver, kidney, heart, and spleen damage. Long-term exposure above the MCL may cause hypertension. There is no evidence that barium can cause cancer. No health advisories have been established for short-term exposures.

EPA Primary Drinking Water Standards
MCLG: 2 mg/L
MCL: 2 mg/L

Other Comments
Treatment/Best available technologies: Ion exchange, reverse osmosis, lime softening, electrodialysis.

Benzene (C_6H_6), CAS No. 71-43-2

Benzene is the simplest aromatic hydrocarbon. Its molecule consists of a single planar, conjugated ring with six carbon atoms arranged in a regular hexagon. See Chapters 5 and 6 for its structural formula. Pure benzene is soluble in water to about 1700 mg/L and is miscible with many organic solvents. Benzene is a highly flammable, volatile, clear, colorless aromatic liquid. A major use of benzene is as a building block for making plastics, rubber, resins, and synthetic fabrics like nylon and polyester. Other important uses include gasoline additive, solvent in printing, paints, dry cleaning, etc. In the United States, gasolines are regulated to 1% benzene on a regional average. Canadian and European gasoline specifications contain a similar 1% limit on benzene content. Emissions from gasoline vehicles constitute the main source of benzene in the environment.

Environmental Behavior
Benzene is released to air primarily by vaporization and combustion emissions associated with its use in gasoline. Other sources are vapors from its production and use in manufacturing other chemicals. In addition, benzene may be in industrial effluents discharged into water and accidental releases from petroleum refining and

distribution industries. Benzene released to soil will either evaporate very quickly or leach to groundwater. It can be biodegraded by soil and groundwater microbes. Benzene released to surface water should mostly evaporate within a few hours to a few days, depending on quantity, temperature, water turbulence, etc. Although benzene does not degrade by hydrolysis, it may be biodegraded by microbes.

Ecological Concerns
Benzene has high acute and chronic toxicity to aquatic life. It bioaccumulates only slightly in aquatic organisms. It can cause death in plants and roots and membrane damage in leaves of various agricultural crops.

Health Concerns
Most exposure is by inhalation of benzene vapor and by ingestion of contaminated water. Short exposures above the MCL may cause temporary nervous system disorders, immune system depression, and anemia. Exposure to high atmospheric concentrations can affect the nervous system and may cause dizziness, nausea and vomiting, headache and drowsiness and, at concentrations in the order of 20,000 ppm (65,000 mg/m^3), narcosis, coma and, sometimes, death. Long-term exposure above the MCL may cause chromosome aberrations and cancer.

EPA Primary Drinking Water Standard
MCL: 5 μg/L
MCLG: zero

Benzo[a]pyrene (C$_{20}$H$_{12}$), CAS No. 50-32-8

Benzo(a)pyrene, or BaP, is a polycyclic aromatic hydrocarbon (PAH). PAHs have a molecular structure that consists of two or more fused aromatic rings, where adjacent rings share two or more carbon atoms. PAHs are not produced or used commercially but are formed as a result of incomplete combustion of organic materials. If BaP is found in a sample, it is likely that other PAHs are present. PAHs, including benzo(a) pyrene, are found in exhaust from motor vehicles and other gasoline and diesel engines; emissions from coal-, oil-, and wood-burning stoves and furnaces; cigarette smoke; general soot and smoke of industrial, municipal, and domestic origin; cooked foods, especially charcoal-broiled; and in incinerators, coke ovens, and asphalt processing. BaP is a solid, melting at 179°C, freely soluble in aromatic hydrocarbon solvents but only slightly in water.

Environmental Behavior
The main natural sources of BaP are forest fires and erupting volcanoes. Anthropogenic sources include the combustion of fossil fuels, coke oven emissions, and vehicle exhausts. In surface waters, direct deposition from the atmosphere appears to be the major source of BaP. Benzo(a)pyrene is moderately persistent in the environment. It readily binds to soils and does not readily leach to groundwater, though it has been detected in some groundwater. If released to water, it sorbs very strongly to sediments and particulate matter. In most waters and in sediments it resists breakdown by microbes or reactive chemicals, but it may evaporate or be degraded by sunlight. In water supply systems, it tends to sorb to any particulate

matter and be removed by filtration before reaching the tap. In tap water, its source is mainly from PAH-containing materials in water storage and distribution systems.

Ecological Concerns
Benzo(a)pyrene is expected to bioconcentrate in aquatic organisms that cannot metabolize it, including plankton, oysters, and some fish.

Health Concerns
Short-term exposure at levels above the MCL can suppress the immune system and cause red blood cell damage, leading to anemia. Long-term exposure at levels above the MCL can cause developmental and reproductive effects and cancer.

EPA Primary Drinking Water Standard
MCL: 0.2 μg/L
MCLG: zero

Beryllium (Be), CAS No. 7440-41-7

Background
Beryllium is a metal found in natural deposits as ores containing other elements, and in some precious stones such as emeralds and aquamarine. Beryllium is not likely to be found in natural waters above trace levels due to the insolubility of oxides and hydroxides at normal environmental pHs. It has been reported to occur in U.S. drinking water at 0.01–0.7 μg/L.

A major use of beryllium is as an alloy hardener. Its greatest use is in making metal alloys for nuclear reactors and the aerospace industry. It is also used as an alloy and oxide in electrical equipment and electronic components, in military vehicle armor. The chloride is used as a catalyst and intermediate in chemical manufacture. The oxide is used in glass and ceramic manufacture. Beryllium enters the environment principally as dust from burning coal and oil and from the slag and ash dumps of coal combustion. Some tobacco leaves contain significant levels of beryllium, which can enter the lungs of those exposed to tobacco smoke. It is also found in discharges from other industrial and municipal operations. Rocket exhausts contain beryllium oxide, fluoride, and chloride.

Very little is known about what happens to beryllium compounds when released to the environment. Beryllium compounds of very low water solubility appear to predominate in soils. Leaching and transport through soils to ground water is unlikely to be of concern. Erosion or runoff of beryllium compounds into surface waters is not likely and it appears unlikely to leach to ground water when released to land. Erosion and bulk transport of soil may carry beryllium sorbed to soils into surface waters, but most likely in particulate rather than dissolved form.

Health Concerns
Beryllium is more toxic when inhaled as fine particles than when ingested orally. Short-term air exposure can cause inflammation (chemical pneumonitis) of the lungs when inhaled. Some people develop a sensitivity, or allergy, to inhaled beryllium, leading to chronic beryllium disease. Long-term ingestion in water above the MCL

may lead to intestinal lesions. There is some evidence that beryllium may cause cancer from lifetime exposures at levels above the MCL.

EPA Primary Drinking Water Standard
MCL: 4.0 μg/L
MCLG: 4.0 μg/L

Other Comments
Treatment/Best available technologies: Activated alumina, coagulation and filtration, ion exchange, lime softening, reverse osmosis

Boron (B), CAS No. 7440-42-8

Boron is usually found in nature as the hydrated sodium borate salt kernite ($Na_2B_4O_7 \cdot 4H_2O$) or the calcium borate salt colemanite ($Ca_2B_6O_{11} \cdot 5H_2O$). Most environmentally important boron compounds are highly water-soluble. Natural weathering of boron-containing minerals is a major source of boron in certain geographical locations. In the United States, the minerals richest in boron are found in the Mojave Desert region of California, where concentrations above 300 mg/L have been observed in boron-rich lakes. In other U.S. surface waters, an average boron level is around 100 μg/L, but concentrations vary widely (from around 0.02 to 0.3 mg/L), depending on local geologic and industrial conditions. Background soil levels in the United States range up to 300 mg/kg, with an average of around 26 mg/kg.

Sodium tetraborate (kernite) is also known as borax and finds use as an additive in detergents and other cleaning agents. A major use for boron is the manufacture of borosilicate glass which, because of its low coefficient of thermal expansion, is used in ovenware, laboratory glassware, piping, and sealed-beam headlights. Boric acid (H_3BO_3) is used as a weak antiseptic and eyewash and as a "natural" insecticide. Other uses for boron compounds include fire retardants, leather tanning, pulp and paper whitening agents, and high-energy rocket fuels. Elemental boron is used for neutron absorption in nuclear reactors and in alloys with copper, aluminum, and steel. For these reasons, boron is common in sewage and industrial wastes. Effluent from municipal sewage treatment plants may contain up to 7 mg/L of boron, with an average (in California) of 1 mg/L.

Boron is essential to plant growth in very small amounts but may become toxic at higher amounts. For boron-sensitive plants, the toxic level may be as low as 1 mg/L. The optimum concentration depends on the plant type, but is generally around 0.3–0.4 mg/L to a maximum level of 0.75 mg/L in soil and irrigation water is generally accepted as protective for sensitive plants under long-term irrigation.

Boron is not known to be an essential nutrient for animals or humans. Boron mobility in water is greatest at pH < 7.5. Adsorption to soils and sediments is the main mechanism for removal from environmental waters. Sorption to oxide and hydroxide solids, particularly aluminum species, is enhanced above pH 7.5 and in the presence of Ca and Mg. There is no evidence that boron is bioconcentrated significantly by aquatic organisms and naturally occurring levels of boron do not appear to have an adverse effect on aquatic life. It is sometimes suggested that boron concentrations in discharges to fresh waters be limited to 10 mg/L.

Health Concerns
Moderately high doses of boron compounds appear to have little detrimental health effects. The lethal dose of boric acid for adults varies from 15 to 20 g. Chronic ingestion may cause dry skin, skin eruptions, and gastric disturbances.

EPA Primary Drinking Water Standard
In general, boron in drinking water is not regarded as hazardous to human health and there are no primary or secondary drinking-water standards.

Other Comments
Treatment/Best available technologies: Because most boron compounds are highly water-soluble, boron is not significantly removed by conventional wastewater treatment. Boron may be coprecipitated with aluminum, silicon, or iron solids.

Cadmium (Cd), CAS No. 7440-43-9

Cadmium is usually present in all soils and rocks. It occurs naturally in zinc, lead, and copper ores, in coal and other fossil fuels and shales. It often is released during volcanic action. These deposits can serve as sources to ground and surface waters, especially when in contact with soft, acidic waters. The adsorption of cadmium onto soils and silicon or aluminum oxides is strongly pH-dependent, increasing as conditions become more alkaline. When the pH is below 6–7, cadmium is desorbed from these materials. The oxide and sulfide are relatively insoluble whereas the chloride and sulfate salts are soluble. Soluble cadmium compounds have the potential to leach through soils to groundwater.

Average concentrations in U.S. waters is about 0.001 mg/L. Cadmium concentrations in bed sediments are generally at least 10 times higher than in overlying water. Cadmium for industrial use is extracted during the production of other metals, chiefly zinc, lead, and copper. It is used for batteries, alloys, pigments, metal protective coatings, and as a stabilizer in plastics. It enters the environment mostly from industrial and domestic wastes, especially those associated with nonferrous mining, smelting, and municipal waste dumps. Because cadmium is chemically similar to zinc, an essential nutrient for plants and animals, it is readily assimilated into the food chain. Plants absorb cadmium from irrigation water and low levels are present in all foods, highest in shellfish, liver, and kidney meats. Smoking can double the average daily intake, one cigarette typically contains 1–2 μg of cadmium. The recommended upper limit in irrigation water is 0.01 mg/L.

Health Concerns
Cadmium is acutely toxic; a lethal dose is about 1 g. Acute exposure can cause nausea, vomiting, diarrhea, muscle cramps, salivation, sensory disturbances, liver injury, convulsions, shock, and renal failure. It is eliminated from the body slowly and can bioaccumulate over many years of low exposure. Long-term exposure to low levels of cadmium in air, food, and water leads to a build-up of cadmium in the kidneys and possible kidney disease. Other potential long-term effects are fragile bones and damage to lungs, liver, and blood. There is inadequate evidence to state whether or not cadmium has the potential to cause cancer from lifetime exposures in drinking water.

EPA Primary Drinking Water Standard
MCL: 5.0 µg/L
MCLG: 5.0 µg/L

Other Comments
Treatment/Best available technologies: Coagulation and filtration, ion exchange, lime softening, reverse osmosis.

Calcium (Ca), CAS No. 7440-70-2

Calcium cations (Ca^{2+}) and calcium salts are among the most commonly encountered substances in water, arising mostly from dissolution of minerals. Calcium often is the most abundant cation in river water. Among the most common calcium minerals are the two crystalline forms of calcium carbonate—calcite and aragonite ($CaCO_3$, limestone is primarily calcite), calcium sulfate (the dehydrated form, $CaSO_4$, is anhydrite; the hydrated form, $CaSO_4 \cdot 2H_2O$, is gypsum), calcium magnesium carbonate ($CaMg(CO_3)_2$, dolomite), and, less often, calcium fluoride (CaF_2, fluorite). Water hardness is caused by the presence of dissolved calcium, magnesium, and sometimes iron (Fe^{2+}), all of which form insoluble precipitates with soap and are prone to precipitating in water pipes and fixtures as carbonates, see Chapter 3. Limestone ($CaCO_3$), lime (CaO), and hydrated lime ($Ca(OH)_2$) are heavily used in the treatment of wastewater and water supplies to raise the pH and precipitate metal pollutants. Very low concentrations of calcium can enhance the deleterious effects of sodium in irrigation water by increasing the value of the sodium absorption ratio (SAR).

Health Concerns
Calcium is an essential nutrient for plants and animals, essential for bone, nervous system, and cell development. Recommended daily intakes for adults are between 800 and 1200 mg/day. Most of this is obtained in food; drinking water typically accounts for 50–300 mg/day, depending on the water hardness and assuming ingestion of 2 L/day. Calcium in food and water is essentially nontoxic. A number of studies suggest that water hardness protects against cardiovascular disease. One possible adverse effect from ingestion of high concentrations of calcium for long periods of time may be a greater risk of kidney stones. The presence of calcium in water decreases the toxicity of many metals to aquatic life. Stream standards for these metals are expressed as a function of hardness and pH. Thus, the presence of calcium in water is beneficial and no limits on calcium have been established for protection of human or aquatic health.

EPA Primary Drinking Water Standard
There are no upper limits for calcium concentrations. To the contrary, calcium in water is usually regarded as beneficial.

Carbofuran ($C_{12}H_{15}NO_3$), CAS No: 1563-66-2

Carbofuran is a white crystalline solid in the carbamate family, with a slightly phenolic odor. It is a broad spectrum insecticide that is sprayed directly onto soil and plants just after emergence to control beetles, nematodes, and rootworm.

The greatest use of carbofuran is on alfalfa and rice, with turf and grapes making up most of the remainder. Earlier uses were primarily on corn crops. Carbofuran is allowed for use on only a few U.S. crops, and will soon be banned from use on corn and sorghum in California. Carbofuran is a restricted use pesticide. On August 3, 2006, the U.S. Environmental Protection Agency (EPA) proposed to cancel the registration of most uses of carbofuran.

Environmental Behavior
Carbofuran is soluble in water and is moderately persistent in soil. Its half-life is 30–120 days. Carbofuran enters surface water as a result of runoff from treated fields and enters groundwater by leaching of treated crops. If released to soil, degradation occurs by chemical hydrolysis and biodegradation. The persistence of carbofuran in the soil increases as the clay and organic matter content of the soil increase, and as the pH and moisture content of soil decrease. Chemical hydrolysis occurs more rapidly in alkaline soil as compared to neutral or acidic soils. Carbofuran is likely to leach to groundwater in soils with low organic content. Volatilization from soil is not expected to be significant, although some evaporation from plants may occur. If released to water, carbofuran degrades by hydrolysis under alkaline conditions and by biodegradation. Aquatic volatilization, adsorption, and bioconcentration are not expected to be important.

Ecological Concerns
Carbofuran is very highly toxic to freshwater and estuarine/marine fish and to birds and mammals. It is reported to have been responsible for more bird deaths than any other pesticide. Carbofuran is also highly toxic to freshwater and estuarine/marine invertebrates. Carbofuran is toxic to bees except in the granular formulation. It is not expected to accumulate in aquatic organisms.

Health Concerns
Occupational and general population exposure to carbofuran may occur by inhalation and dermal routes, particularly in the vicinity of aerial spraying of carbofuran as an insecticide. As with other N-methyl carbamate pesticides, the critical effect of carbofuran is cholinesterase inhibition, that is, it can overstimulate the nervous system causing nausea, dizziness, and confusion. As with other carbamate compounds, carbofuran's cholinesterase-inhibiting effect is short-term and reversible. Very high exposures (e.g., accidents or major spills) may cause respiratory paralysis and death. Long-term exposure at concentrations above the MCL can cause damage to the nervous and reproductive systems. Carbofuran is classified as "Not Likely" to be a human carcinogen.

EPA Primary Drinking Water Standard
MCL: 40 μg/L
MCLG: 40 μg/L

Chloride (Cl⁻), CAS No. 7440-39-3

Chlorides are widely distributed in nature, usually in the form of sodium, potassium, and calcium salts (NaCl, KCl, and CaCl$_2$), although many minerals contain small amounts of chloride as an impurity. Chloride in natural waters arises from weathering of chloride minerals, salting of roads for snow and ice control, seawater intrusion in coastal regions, irrigation drainage, ancient groundwater brines, geothermal waters, and industrial wastewater.

Concentrations in unpolluted surface waters and nongeothermal groundwaters are generally low, usually below 100 mg/L. Thus, chloride concentrations in the absence of pollution are normally less than those of sulfate or bicarbonate.

Environmental Behavior
Chloride ion is extremely mobile; all chloride salts are very soluble except for lead chloride ($PbCl_2$), silver chloride (AgCl), and mercury chlorides (Hg_2Cl_2, $HgCl_2$). Chloride is not sorbed to soils and generally moves with water with little or no retardation. Consequently, it eventually moves to closed basins (as the Great Salt Lake in Utah) or to the oceans.

Chloride ion is almost chemically and biologically inert when compared with the other major environmental ions. Under environmental conditions, chloride ions do not significantly enter into redox reactions, form no important solute complexes with other ions except at very high chloride ion concentrations (tens of thousands of milligrams per liter), form few metal precipitates (except with silver, mercury, and lead), are not significantly adsorbed to mineral surfaces, participate in few important biological processes, and have extremely low toxicities for mammalian and aquatic species.

Chloride circulates through the hydrologic cycle mainly by physical processes. Its lack of environmental reactivity is attested to by the common use of chloride as a conservative tracer for groundwater movement and by its absence from USEPA Drinking Water, CERCLA, and RCRA priority pollutant lists. Chloride in its most common form, as sodium chloride, is a "generally regarded as safe" (GRAS) substance. Chloride ion has no federal or state health-based treatment standards. The main environmental problems associated with chloride are reactivity with concrete, metal corrosion, adverse taste effects in drinking water, and toxicity to irrigated crops (less than 100 mg/L chloride recommended for most crops not classified as chloride sensitive).

Ecological Concerns
To protect freshwater aquatic organisms and their uses, chloride levels should not exceed the acute and chronic levels given below more than once in 3 years (USEPA 1988)

860 mg/L for the criteria maximum concentration (CMC, 1 h average acute)
230 mg/L for the criteria continuous concentration (CCC, 4 day average chronic).

Health Concerns
Chloride is the most abundant anion in the human body and is essential to normal electrolyte balance of body fluids. For adults, a daily dietary intake of about 9 mg of chloride per kilogram of body weight is considered essential for good health. Chlorides in water are more of a taste than a health concern, although high concentrations may be harmful to people with heart or kidney problems.

EPA Primary Drinking Water Standard
MCL: none
MCLG: none

There are no primary drinking water standards for chloride. The EPA secondary standard for chloride is 250 mg/L, based on adverse effect on taste.

Other Comments
Treatment/Best available technologies: Conventional watertreatment does not remove chloride ion. Reverse osmosis or nanofiltration is required.

Chromium (Cr), CAS No. 7440-47-3
Chromium occurs in minerals mostly as chrome iron ore, or chromite ($FeCr_2O_2$), in which it is present as Cr(III), with oxidation number +3. Chromium in soils occurs mostly as insoluble chromium oxide (CrO_3), where it is present as Cr(VI) with an oxidation number of +6. In natural waters, dissolved chromium exists as either Cr^{3+} cations or in anions such as chromate (CrO_4^{2-}) and dichromate ($Cr_2O_7^{2-}$), where it is hexavalent with oxidation number +6. Though widely distributed in soils and plants, it generally is present at low concentrations in natural waters. Background levels in water typically range between 0.2 and 20 µg/L, with an average of 1 µg/L.

As a positively charged ion, trivalent chromium (Cr^{3+}) readily sorbs to negatively charged soils and minerals. Unsorbed Cr^{3+} forms insoluble colloidal hydroxides in the pH range of natural surface waters (6.5–9). Thus, it is unlikely that dissolved trivalent chromium will be present in surface waters at levels of concern. Trivalent chromium is also not likely to migrate to groundwater, most of it being retained in the upper 5–10 cm of soil.

The hexavalent form of chromium, existing in negatively charged complexes, is not sorbed to any extent by soil or particulate matter and is much more mobile than Cr(III). However, Cr(VI) is a strong oxidant and reacts readily with any oxidizable organic material present, with the resultant formation of Cr(III). In the absence of organic matter, Cr(VI) can be stable for long periods of time, particularly under aerobic conditions. Under anaerobic conditions, Cr(VI) is quickly reduced to low mobility Cr(III). Thus, most of the chromium in surface waters will be present in particulate form as suspended and bed sediments.

Chromium has many industrial uses. Some major applications are in metal alloys, protective coatings on metal, magnetic tapes, paint pigments, cement, paper, rubber, and composition floor covering.

The main natural environmental source is weathering of rocks and soil. Major anthropomorphic sources include metal alloy production, metal plating, cement manufacturing, and incineration of municipal refuse and sewage sludge.

Health Concerns
Trivalent chromium is an essential trace nutrient, and plays a role in prevention of diabetes and atherosclerosis. Trivalent chromium is essentially nontoxic; the harmful effects of chromium to human health are caused by hexavalent chromium. Since oxidants such as chlorine or ozone readily oxidize trivalent chromium to the toxic hexavalent form, water quality limits are usually written for total chromium concentrations.

EPA has found chromium to potentially cause the following health effects from acute exposures at levels above the MCL: skin irritation or ulceration. Long term exposures to chromium at levels above the MCL have the potential to cause dermatitis and damage to liver, kidney, circulatory and nerve tissues. There is no evidence that chromium in drinking water has the potential to cause cancer from lifetime exposures in drinking water.

EPA Primary Drinking Water Standard
MCL: 0.1 mg/L (total Cr)
MCLG: 0.1 mg/L (total Cr)
These standards are based on the total concentration of the trivalent and hexavalent forms of dissolved chromium (Cr^{3+} and Cr^{6+}).

Other Comments
Treatment/Best available technologies: Coagulation and filtration, ion exchange, reverse osmosis, lime softening (for Cr(III) only)

Copper (Cu), CAS No. 7440-50-8
In nature, copper sometimes occurs as the pure metal, but more often, in the form of mineral ores that contain 2% or less of the metal. The most common copper-bearing ores are sulfides, arsenites, chlorides, and carbonates. Chalcopyrite ($CuFeS_2$) is the most abundant of the copper ores, accounting for about 50% of the world's copper deposits. The weathering of copper deposits is the main natural source of copper in the aquatic environment, but dissolved copper rarely occurs in unpolluted source water above 10 μg/L, limited by the solubility of copper hydroxide ($Cu(OH)_2$), coprecipitation with less soluble metal hydroxides and adsorption. In some cases, copper salts may be added to reservoirs for the control of algae. Copper concentrations in acid mine drainage may reach several hundred mg/L but, if the pH is raised to 7 or higher, most of the copper will precipitate. Smelting operations and municipal incineration may also introduce copper into surface waters.

Copper occurs in drinking water primarily due to corrosion of copper pipes and fittings, which are widely used for interior plumbing of residences and other buildings. This is the reason for an EPA action level based on samples taken from distribution system taps, rather than an MCL. All water is corrosive toward copper to some degree, even water termed noncorrosive or water treated to make it less corrosive. Corrosivity toward copper metal increases with decreasing pH, especially below pH 6.5.

Health Concerns
Copper is an essential nutrient, but at high doses it has been shown to cause stomach and intestinal distress, liver and kidney damage, and anemia. Persons with Wilson's disease may be at a higher risk of health effects due to copper than the general public. There is inadequate evidence to state whether or not copper has the potential to cause cancer from a lifetime exposure in drinking water.

EPA Primary Drinking Water Standard
MCL: Because plumbing in homes and commercial buildings is the main source of copper in drinking water supplies, EPA has established a tap water action level rather than an MCL.
Action level: >1.3 mg/L in 10% or more of tap water samples.
MCLG: 1.3 mg/L

Other Comments
Treatment/Best available technologies: For treating source water: Ion exchange, lime softening, reverse osmosis, coagulation and filtration. For corrosion control: pH and alkalinity adjustment, calcium adjustment, silica- or phosphate-based corrosion inhibition.

Cyanide (CN^-), CAS No. 57-12-5

Hydrogen Cyanide (HCN), CAS No. 74-90-8
Cyanide is a product of natural animal and vegetative decay processes and also is a component in many industrial waste streams. It is used extensively in mining to separate metals, particularly gold, from ores. In water, an equilibrium exists between the ionized (CN^-) and unionized (HCN) forms, the fraction of each depending on pH (Equation A.1 and Figure A.1).

$$CN^- + H^+ \leftrightarrow HCN \tag{A.1}$$

Below pH $= 9$, the predominant form is HCN. HCN is more toxic than CN^- and is the dominant form in most natural waters. HCN is volatile whereas CN^- is nonvolatile.

The most common industrially used form, hydrogen cyanide, is used in the production of nylon and other synthetic fibers and resins. Some cyanide compounds are used as herbicides. The major sources of cyanide releases to water are discharges from metal finishing industries, iron and steel mills, and organic chemical industries. Disposal of cyanide wastes in landfills is a major source of releases to soil.

Cyanides are generally not persistent when released to water or soil, and are not likely to accumulate in aquatic life. They rapidly evaporate and are broken down by microbes. They do not bind to soils and may leach to groundwater.

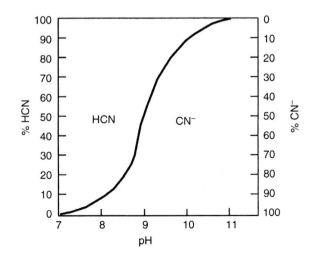

FIGURE A.1 Distribution of cyanide between the HCN and CN^- forms as a function of pH.

Cyanide-containing herbicides, such as Tabun, have moderate potential for leaching, but again are readily biodegraded so they are not expected to bioconcentrate.

Soluble cyanide compounds, such as hydrogen and potassium cyanide, have low adsorption to soils with high pH, high carbonate, and low clay content. However, at pH less than 9.2, most free cyanide converts to hydrogen cyanide, which is highly volatile. Soluble cyanides are not expected to bioconcentrate.

Insoluble cyanide compounds such as the copper and silver salts adsorb to soils and sediments, and generally have the potential to bioconcentrate. Insoluble forms do not biodegrade to hydrogen cyanide.

Health Concerns
Short-term exposure to cyanide compounds above the MCL may cause rapid breathing, tremors, and other neurological effects. Long-term exposure at levels above the MCL may cause weight loss, thyroid effects, and nerve damage. There is inadequate evidence for carcinogenicity from lifetime exposures in drinking water.

EPA Primary Drinking Water Standard
MCL: 0.2 mg/L as free cyanide
MCLG: 0.2 mg/L as free cyanide

Other Comments
Treatment/Best available technologies: Ion exchange, reverse osmosis, chlorination.

Fluoride (F^-), CAS No. 16984-48-8

Most soils and rocks contain trace amounts of fluoride. Much higher concentrations are found in areas of active or dormant volcanic activity. Common fluoride-containing minerals include fluorite (or fluorspar, CaF_2), cryolite (Na_3AlF_6), and fluorapatite ($Ca_5F(PO_4)_3$). Weathering of minerals is the main source of fluoride in unpolluted waters, where concentrations are generally less than 1 mg/L but may sometimes exceed 50 mg/L. Where dissolved calcium is present, the formation of fluorite may limit fluoride concentrations. High fluoride concentrations are more likely in thermal waters and water with low calcium concentrations. Groundwater usually contains higher concentrations than surface water and groundwater concentrations as high as 10 mg/L are common.

The formation of fluoride complexes may be important in solubilizing beryllium, aluminum, tin, and iron in natural waters. Addition of fluoride to drinking water and toothpaste for reducing dental caries, and its subsequent discharge in sewage, also contributes to aquatic fluoride. Discharges from the aluminum, steel, and phosphate production are important industrial sources of fluoride in water.

Health Concerns
Small amounts of fluoride appear to be an essential nutrient. People in the United States ingest about 2 mg/day in water and food. A concentration of about 1 mg/L in drinking water effectively reduces dental caries without harmful effects on health. Dental fluorosis can result from exposure to concentrations above 2 mg/L in children up to about 8 years of age. In its mild form, fluorosis is characterized by white opaque mottled areas on tooth surfaces. Severe fluorosis causes brown to black stains

and pitting. Although the matter is controversial, EPA has determined that dental fluorosis is a cosmetic and not a toxic or an adverse health effect. Water hardness limits fluoride toxicity to humans and fish. The severity of fluorosis decreases in harder drinking water. Crippling skeletal fluorosis in adults requires the consumption of about 20 mg or more of fluoride per day over a 20 year period. No cases of crippling skeletal fluorosis have been observed in the United States from the long term consumption of 2 L/day of water containing 4 mg/L of fluoride. EPA has concluded that 0.12 mg/kg/day of fluoride is protective of crippling skeletal fluorosis. Fluoride therapy, where 20 mg/day is ingested for medical purposes, is sometimes used to strengthen bone, particularly spinal bones.

EPA Primary Drinking Water Standard
MCL: 4.0 mg/L
MCLG: 4.0 mg/L

Other Comments
Treatment/Best available technologies: Anion exchange, nanofiltration, reverse osmosis.

Hydrogen Sulfide (H_2S), CAS No. 7783-06-4

(See Chapters 3 and 8 for more detailed discussions.)

Natural waters acquire sulfur compounds mainly from geochemical weathering of sulfide and sulfate minerals, fertilizers, decomposition of organic matter, and atmospheric deposition from industrial fuel combustion. During the decomposition of organic matter, sulfur is released largely as hydrogen sulfide (H_2S), which oxidizes rapidly to sulfate under aerobic conditions. Therefore, under aerobic conditions in aquatic systems, sulfate is the predominant form of sulfur and concentrations of hydrogen sulfide are very low.

Under anaerobic conditions, sulfate and organic sulfur compounds are reduced to the sulfide anion (S^{2-}) by bacterial reduction. The presence of sulfate-reducing bacteria in drinking water distribution systems can be a major cause of taste and odor problems. Sulfide anion is commonly found under aquatic anaerobic conditions wherever sulfur is present, e.g., domestic and industrial wastewater and sludges, the hypolimnion of stratified lakes, and the bottoms of wetlands. Sulfide anion reacts with water to form hydrogen sulfide (H_2S). H_2S is a flammable, poisonous gas with a characteristic strong odor of rotten eggs. Hydrogen sulfide and the sulfide salts of the alkali and alkaline metals (groups 1A and 2A of the periodic table) are soluble in water. Soluble sulfide salts dissociate in water, forming the sulfide anion, which then reacts with water to form H_2S.

If transition metal cations are present, particularly iron and manganese, metal sulfides of low solubility are formed and precipitated at neutral to alkaline pH values. Thus, anaerobic zones of lakes and wetlands where high levels of dissolved metals are present along with sulfide, are likely to contain metal sulfides in the sediments and very little dissolved sulfide anion. Only after most of the dissolved metals have been precipitated can H_2S accumulate in the water. Black sediments of eutrophic lakes and wetlands consist largely of precipitated iron and manganese sulfides, S^{2-} dissolved in interstitial water, acid-soluble sulfide compounds, elemental sulfur, organic sulfur, and sulfates. H_2S also is added in large quantities to the atmosphere from volcanic gases, industrial sources, and biochemical activity in water and soil.

The human nose is very sensitive to the rotten-egg odor of hydrogen sulfide. Most people can detect the smell of H_2S in water containing as little as 1 μg/L. Typical concentrations of H_2S in unpolluted surface water are <0.25 μg/L. $H_2S >$ 2.0 μg/L constitutes a chronic hazard to aquatic life. Groundwater usually contains little or no sulfide, because contact with metal bearing minerals results in the formation of metal sulfides with low solubility. Note, however, that sulfides may be found in wells, arising from sulfate-reducing bacteria present in the well. Some brines, especially those associated with petroleum deposits, may contain several hundred milligrams per liter of dissolved hydrogen sulfide.

Health Concerns
Hydrogen sulfide is acutely toxic to humans. It is a leading cause of death in the workplace, most often by accidental inhalation of high concentrations (>1000 ppm). However, EPA has no drinking water standards for hydrogen sulfide because its disagreeable taste and odor make water unpalatable at concentrations much lower than the toxic levels. A guideline value is that the presence of H_2S be not detectable by consumers.

EPA Primary Drinking Water Standard
EPA has no drinking water standards for hydrogen sulfide, see Health Concerns above.

Other Comments
Treatment/Best available technologies: See Chapter 3.

Iron (Fe), CAS No. 7439-89-6

Iron is naturally released into waters by weathering of pyritic ores containing iron sulfide (FeS_2) and other iron-bearing minerals in igneous, sedimentary, and metamorphic rocks. It also results from many human sources: mineral processing, coke and coal burning, acid-mine drainage, iron and steel industry wastes, and corrosion of iron and steel. Because iron is an essential nutrient for animals and plants, it is present in organic matter in soil and in sewage and landfill leachate. Many microorganisms use iron as an energy source and play an important part in iron oxidation and reduction processes.

In the aquatic environment, iron is present in two oxidation states, ferrous (Fe^{2+}) and ferric (Fe^{3+}). The reduced ferrous state is highly soluble in the pH range of unpolluted surface waters, whereas the oxidized ferric state is associated with compounds of low solubility at pH values above 5. For example, Fe^{3+} reacts with water to form low solubility iron oxyhydroxides, which form yellow to red-brown precipitates often seen on rocks and sediments in surface waters with high iron concentrations. Iron oxyhydroxides often form as colloidal suspensions of gels or flocs. These have large surface areas and a strong absorptive capacity for other dissolved ionic species. Coprecipitation with iron at elevated pH has been developed as a treatment process for removing other dissolved metals. For example, removal of dissolved zinc by lime precipitation to a concentration below 0.1 mg/L generally requires coprecipitation with iron.

Since the ferrous state is easily oxidized to the ferric state (a process often enhanced by aerobic iron bacteria, which leave slimy deposits of ferric iron), dissolved iron is mainly found under reducing conditions, in groundwater or anaerobic surface waters. In well-aerated water above pH 5, dissolved iron concentrations are generally less than 30 μg/L, whereas groundwater concentrations may be as high as 50 mg/L. When concentrations in the milligram per liter are reported for aerated surface waters, most of the iron is generally associated with sediments. When sampling groundwater for dissolved iron, it is important to purge the well adequately, filter the water immediately onsite to remove iron sediments, acidify the filtrate to prevent further precipitation, and store in a container without any air headspace. It is not uncommon for dissolved Fe^{2+} in groundwater to enter a well and become oxidized to Fe^{3+} after exposure to oxygen in the well or the distribution system. The result is rust-colored water from precipitated ferric iron compounds and potential staining of plumbing fixtures and laundry. Such water often has objectionable taste and eventually may clog plumbing fixtures, reducing the flow of water.

Health Concerns
Iron is an essential nutrient in animal and plant metabolism. It is not normally considered a toxic substance. It is not regulated in drinking water except as a secondary standard for aesthetic reasons. Adults require between 10 and 20 mg of iron per day. Excessive iron ingestion may result in hemochromatosis, a condition of tissue damage from iron accumulation. This condition rarely occurs from dietary intake alone, but has resulted from prolonged consumption of acidic foods cooked in iron utensils and from the ingestion of large quantities of iron tablets.

Iron can be toxic to freshwater aquatic life above 1 mg/L and may interfere with fish uptake of oxygen through their gills above 0.3 mg/L.

EPA Primary Drinking Water Standard
EPA has no primary drinking water standard for iron. The EPA secondary drinking water standard (nonenforceable) is 0.3 mg/L as total iron.

Other Comments
Treatment/Best available technologies: Lime precipitation, aeration, cation exchange, microfiltration, reverse osmosis.

Lead (Pb), CAS No. 7439-92-1

Lead minerals are found mostly in igneous, metamorphic, and sedimentary rocks. The most abundant lead mineral is galena (PbS). Oxide, carbonate, and sulfate minerals of lead are lanarkite (PbO), cerrusite ($PbCO_3$), and anglesite ($Pb(SO_4)$), respectively. Commercial ores have concentrations of lead in the range 30–80 g/kg. Metallic lead and the common lead minerals have very low solubility. Most aquatic environmental lead (perhaps 85%) is associated with sediments, the rest being in dissolved form.

Although some lead enters the environment from natural sources by weathering of minerals, particularly galena, anthropogenic sources are about 100 times greater. Mining, milling, and smelting of lead and metals associated with lead, such as zinc, copper, silver, arsenic, and antimony, are major sources, as are combustion of fossil fuels and municipal sewage. Commercial products that are major sources of lead pollution include acid-storage batteries, electroplating, construction materials, ceramics and dyes, radiation shielding, ammunition, paints, glassware, solder, piping, cable sheathing, roofing and, up to about 1980, gasoline additives such as tetramethyl- and tetraethyllead.

In areas away from mining and smelters, the use of leaded gasoline exceeded all other sources between about 1940 and 1980. In 1970, new regulations in the Clean Air Act led to a reduction in lead additives to gasoline as well as tighter restrictions on industrial emissions. By 1985, these measures had resulted in an overall decrease in lead emissions of around 20%. Organic lead additives to automotive gasoline were completely eliminated in 1996, but soils and water bodies still carry the lead legacy from earlier years.

Levels of dissolved lead in natural surface waters are generally low. Lead sulfides, sulfates, oxides, carbonates, and hydroxides are almost insoluble. Because of their greater abundance, the carbonates and hydroxides are considered to impose an upper limit on the concentrations of lead that can occur in lakes, rivers, and groundwaters. The global mean lead concentration in lakes and rivers is estimated to be between 1.0 and 10.0 $\mu g/L$.

Pb^{2+} is the stable ionic species in most of the natural environment. Sorption is the dominant mechanism controlling the distribution of lead in the aquatic environment, where it forms complexes with organic ligands to yield soluble, colloidal, and particulate compounds that sorb to humic materials. At the low lead concentrations typically found in the aquatic environment, most of the lead in the dissolved phase is likely to be in the form of organic ligand complexes. In the presence of clay suspensions at pH 5–7, most lead is precipitated and sorbed as sparingly soluble hydroxides. Soluble lead is removed from natural waters mainly by association with sediments and suspended particulates. Lead solubility is very low (<1 $\mu g/L$ at pH 8.5–11) in water containing carbon dioxide and sulfate. At constant pH, the solubility of lead decreases with increasing alkalinity. Lead is bioaccumulated by aquatic organisms, including benthic bacteria, freshwater plants, invertebrates, and fish.

Lead in drinking water results primarily from corrosion of materials located throughout the distribution system containing lead and copper and from lead and copper plumbing materials used to plumb publicly and privately owned structures connected to the distribution system. Very little lead enters public distribution systems in water from treatment plants. Most public water systems serve at least some buildings with lead solder or lead service lines.

All water is corrosive to metal plumbing materials to some degree, even water termed noncorrosive or water treated to make it less corrosive. The corrosivity of water to lead is influenced by water pH, total alkalinity, dissolved inorganic carbonate, calcium, and hardness. Galvanic corrosion of lead into water also occurs with lead-soldered copper pipes, due to differences in the electrochemical potential of the two metals. Grounding of household electrical systems to plumbing can accelerate galvanic corrosion.

Lead is not very mobile under normal environmental conditions. It generally is retained in the upper 2–5 cm of soil, especially soils with at least 5% organic matter or a pH 5 or above. Leaching is not important under normal conditions. It is expected to slowly undergo speciation to the more insoluble sulfate, sulfide, oxide, and phosphate salts.

Metallic lead can be dissolved by pure water in the presence of oxygen, but if the water contains carbonates and silicates, protective films are formed preventing further attack. The solubility of Pb is 10 μg/L above pH 8, whereas near pH 6.5 the solubility can exceed 100 μg/L. Lead is effectively removed from the water column by adsorption to organic matter and clay minerals, precipitation as insoluble salts, and reaction with hydrous iron and manganese oxides. Under appropriate conditions, dissolution due to anaerobic microbial action may be significant in subsurface environments. In an oxidizing environment, the least soluble common forms of lead are the carbonate, hydroxide, and hydroxycarbonate. In reducing conditions where sulfur is present, lead sulfide (PbS) is formed as an insoluble solid.

Health Concerns
Short-term exposure to lead at relatively low concentrations can cause interference with red blood cell chemistry, delays in normal physical and mental development in babies and young children, slight deficits in the attention span, hearing, and learning abilities of children, and slight increases in the blood pressure of some adults. It appears that some of these effects, particularly changes in the levels of certain blood enzymes and in aspects of children's neurobehavioral development, may occur at blood lead levels so low as to be essentially without a threshold. Long-term exposure to lead has been linked to cerebrovascular and kidney disease in humans. Lead has the potential to cause cancer from a lifetime exposure at levels above the action level.

EPA Primary Drinking Water Standard
MCL: Because plumbing in homes and commercial buildings is the main source of lead in drinking water supplies, EPA has established a tap water action level rather than an MCL.
Action Level: > 0.015 mg/L in more than 10% of tap water samples.
MCLG: zero

Other Comments
Treatment/Best available technologies: Ion exchange, lime softening, reverse osmosis, coagulation and filtration
Corrosion control: pH and alkalinity adjustment, calcium adjustment, silica- or phosphate-based corrosion inhibition

Magnesium (Mg), CAS No. 7439-95-4
Magnesium is used in the textile, tanning, and paper industries. Lightweight alloys of magnesium are used extensively in molds, die castings, extrusions, rolled sheets and plate forgings, mechanical handling equipment, portable tools, luggage, and general household goods. The carbonates, chlorides, hydroxides, oxides, and sulfates of

magnesium are used in the production of magnesium metal, refractories, fertilizers, ceramics, explosives, and medicinals.

Magnesium is abundant in the earth's crust and is a common constituent of natural water. Along with calcium, it is one of the main contributors to water hardness. The aqueous chemistry of magnesium is similar to that of calcium in the formation of carbonates and oxides. Magnesium compounds are, in general, more soluble than their calcium counterparts. As a result, large amounts of magnesium are rarely precipitated. Magnesium carbonates and hydroxides precipitate at high pH (>10). Magnesium concentrations can be extremely high in certain closed saline lakes. Natural sources contribute more magnesium to the environment than do all anthropogenic sources. Magnesium is commonly found in magnesite, dolomite, olivine, serpentine, talc, and asbestos minerals. The principal sources of magnesium in natural water are ferromagnesium minerals in igneous rocks and magnesium carbonates in sedimentary rocks. Water in watersheds with magnesium-containing rocks may contain magnesium in the concentration range of 1–100 mg/L. The sulfates and chlorides of magnesium are very soluble, and water in contact with such deposits may contain several hundred milligrams of magnesium per liter.

Health Concerns
Magnesium is an essential nutrient for plants and animals, essential for bone and cell development. It accumulates in calcareous tissues, and is found in edible vegetables (700–5600 mg/kg), marine algae (6400–20,000 mg/kg), marine fish (1200 mg/kg), and mammalian muscle (900 mg/kg) and bone (700–1800 mg/kg). Magnesium is one of the principal cations of soft tissue. It is an essential part of the chlorophyll molecule. Recommended daily intake for adults is 400–450 mg/day, of which drinking water can supply from 12 to 250 mg/day, depending on the magnesium concentration and assuming ingestion of 2 L/day. Magnesium salts are used medicinally as cathartics and anticonvulsants. In general, the presence of magnesium in water is beneficial and no limits on magnesium have been established for protection of human or aquatic health.

EPA Primary Drinking Water Standard
There are no primary or secondary drinking water standards for magnesium. Magnesium in drinking water may provide nutritional benefits for persons with magnesium deficient diets.

Manganese (Mn), CAS No. 7439-96-5
Manganese is an abundant, widely distributed metal. It does not occur in nature as the elemental metal, but is found in various salts and minerals, frequently along with iron compounds. Soils, sediments, and metamorphic and sedimentary rocks are significant natural sources of manganese. The most important manganese mineral is pyrolusite (MnO_2). Other manganese minerals are manganese carbonate ($MnCO_3$, rhodocrosite) and manganese silicate ($MnSiO_3$, rhodonite). Ferromanganese minerals, such as biotite mica ($K(Mg,Fe)_3(AlSi_3O_{10})(OH)_2$) and amphibole ($(Mg, Fe)_7Si_8O_{22}(OH)_2$), contain large amounts of manganese. The weathering of manganese deposits contributes small amounts of manganese to natural waters.

Manganese, its alloys, and manganese compounds are commonly used in the steel industry for manufacturing metal alloys and dry-cell batteries, and in the chemical industry for making paints, varnishes, inks, dyes, glass, ceramics, matches, fireworks, and fertilizers. The iron and steel industry and acid mine drainage release a large portion of the manganese found in the environment. Iron and steel plants also release manganese into the atmosphere, from which it is redistributed by atmospheric deposition.

Manganese seldom reaches concentrations of 1.0 mg/L in natural surface waters, and is usually present in quantities of 0.2 mg/L or less. Concentrations higher than 0.2 mg/L may occur in groundwaters and deep stratified lakes and reservoirs under reducing conditions. Subsurface and acid mine waters may contain 10 mg/L. Manganese is similar to iron in its chemical behavior, and is frequently found in association with iron. In the absence of dissolved oxygen, manganese normally is in the reduced manganous (Mn^{2+}) form, but it is readily oxidized to the manganic (Mn^{4+}) form. Permanganates (Mn^{7+}) are not persistent because they are strong oxidizers and are rapidly reduced in the process of oxidizing organic materials. Nitrate, sulfate, and chloride salts of manganese are quite soluble in water, whereas oxides, carbonates, phosphates, sulfides, and hydroxides are only sparingly soluble. In natural waters, a substantial fraction of manganese is present in suspended form. In surface waters, divalent manganese (Mn^{2+}) is rapidly oxidized to insoluble manganese dioxide (MnO_2), which then precipitates as a black solid often observed as black stains on rocks. In drinking water distribution systems, precipitation of MnO_2 may cause unsightly black staining of fixtures and laundry.

Health Concerns
Manganese is an essential trace element for microorganisms, plants, and animals, and, hence, is contained in all or nearly all organisms. Manganese is a ubiquitous element that is essential for normal physiologic functioning in all animal species. The total body load of manganese in an average adult is about 12 mg. Health problems in humans may arise from both deficient and excess intakes of manganese. Thus any quantitative risk assessment for manganese must consider that, although manganese is an essential nutrient, excessive intake causes toxic symptoms. An average dietary intake in the United States ranges between 2 and 10 mg/day, with an average around 4 mg/day. Grains and cereals are the richest dietary sources of manganese, followed by fruits and vegetables. Meat, fish, and poultry contain little manganese. Drinking water supplies almost always contain less than the secondary standard of 0.05 mg/L and drinking water generally contributes no more than about 0.07 mg/day to an adult diet. A maximum adult dietary intake of 20 mg/day is recommended to avoid manganese toxicity. Manganese is not considered to be a cancer risk.

EPA Primary Drinking Water Standard
EPA has no primary drinking water standard for manganese. The EPA secondary drinking water standard (nonenforceable) is 0.05 mg/L.

Other Comments
Treatment/Best available technologies: Lime precipitation, aeration, cation exchange, microfiltration, reverse osmosis.

Mercury (Hg), CAS No. 7439-97-6

Mercury is a liquid metal found in natural deposits of ores containing other elements. Mercury deposits occur in all types of rocks: igneous, sedimentary, and metamorphic. Although cinnabar (HgS) is the most common mercury ore, mercury is present in more than 30 common ore and gangue minerals. Mercury exists in the environment as the elemental metal, as monovalent and divalent salts, and as organic mercury compounds, the most important of which are methyl mercury ($HgCH_3^+$) and dimethyl mercury ($Hg(CH_3)_2$). Methyl and dimethyl mercury are formed from inorganic mercury by microorganisms found in bottom sediments and sewage sludge. There are other microorganisms that can de-methylate mercury back to the inorganic form.

Mercury is noteworthy among environmental metal pollutants by virtue of its volatility and the ease by which inorganic mercury can be converted to organic forms by microbial processes. Its volatility accounts for the fact that mercury is present in the atmosphere as metallic mercury and as volatilized organic mercury compounds. Terrestrial environments appear to be major sources of atmospheric mercury, with contributions from evapotranspiration of leaves, decaying vegetation, and degassing of soils. The major source of mercury movement in the environment is the natural degassing of the earth's crust, which may introduce between 25,000 and 150,000 tons of mercury per year into the atmosphere. It is not unusual for atmospheric concentrations of mercury in an area to be up to four times the level in contaminated soils. Atmospheric mercury can enter terrestrial and aquatic habitats via particle deposition and precipitation. Measuring atmospheric concentrations of mercury at the earth surface or even from aircraft is a form of prospecting for mineral formations of other metals associated with elemental mercury. Inorganic forms of mercury (Hg) can be converted to soluble organic forms by anaerobic microbial action in the biosphere. In the atmosphere, 50% of volatile mercury is metallic mercury (Hg). Twenty-five to fifty percent of Hg in water is organic. Mercury in the environment is deposited and revolatilized many times, with a residence time in the atmosphere of several days. In the volatile phase it can be transported hundreds of kilometers.

Twenty thousand tons of mercury per year also is released into the environment by human activities such as combustion of fossil fuels, operation of metal smelters, cement manufacture, and other industrial releases. Mercury is used in the chloralkali industry, where mercury is used as an electrode to produce chlorine, caustic soda (sodium hydroxide), and hydrogen by electrolysis of molten sodium chloride. It is also used to produce electrical products such as dry-cell batteries, fluorescent light bulbs, switches, and other control equipment. Electrical products account for 50% of mercury used. Aquatic pollution originates in sewage, metal refining operations, chloralkali plant wastes, industrial and domestic products such as thermometers and batteries, and from solid wastes in major urban areas, where electrical mercury switches account for a significant release of mercury to the environment.

In most unpolluted surface waters, mercuric hydroxide ($Hg(OH)_2$) and mercuric chloride ($HgCl_2$) are the predominant mercury species, with concentrations generally less than 0.001 mg/L. In polluted waters, concentrations up to 0.03 mg/L may occur. In aquatic systems, mercury binds to dissolved matter or fine particulates. In freshwater habitats, it is common for mercury compounds to be sorbed to particulate matter and sediments. Sediment binding capacity is related to organic content, and is little affected by pH. Mercury tends to combine with sulfur in anaerobic bottom sediments. Organic methyl mercury bioconcentrates along aquatic food chains to an extent that fish in mildly polluted waters may become unsafe for food use.

Health Concerns
Mercury is highly toxic. Organic alkyl mercury compounds, such as ethylmercuric chloride (C_2H_5HgCl) that used to be used as fungicides, produce illness or death from the ingestion of only a few milligrams. Because inorganic forms of mercury can be converted to very toxic methyl and dimethyl mercury by anaerobic microorganisms, any form of mercury must be considered as potentially hazardous to the environment. Most human mercury exposure is due to consumption of fish. EPA has found that short-term and long-term exposure to mercury at levels in drinking water above the MCL may cause kidney damage. There is inadequate evidence to state whether or not mercury has the potential to cause cancer from lifetime exposures in drinking water.

EPA Primary Drinking Water Standard
MCL: 0.002 mg/L
MCLG: 0.002 mg/L

Other Comments
Treatment/Best available technologies: Granular activated carbon for influent mercury concentrations above 10 μg/L; coagulation and filtration, lime softening, and reverse osmosis for influent mercury concentrations less than 10 μg/L.

Molybdenum (Mo), CAS No. 7439-98-7
Molybdenum is widely distributed in trace amounts in nature, occurring chiefly as insoluble molybdenite (MoS_2) and soluble molybdates (MoO_4^{2-}). Molybdenum is relatively mobile in the environment because soluble compounds predominate at pH > 5. The solubility of molybdenum increases as redox potential is lowered. Below pH 5, adsorption and coprecipitation of the molybdate anion by hydrous oxides of iron and aluminum are effective at removing dissolved molybdenum.

The weathering of igneous and sedimentary rocks (especially shales) is the main natural source of molybdenum to the aquatic environment.

Molybdenum metal is used in the manufacture of special steel alloys and electronic apparatus. Molybdenum salts are used in the manufacture of glass, ceramics, pigments, and fertilizers. The use of fertilizers containing molybdenum is the single most important anthropogenic input to the aquatic environment. Other contributions to the aquatic environment come from mining and milling of molybdenum, the use of molybdenum products, the mining and milling of some uranium and copper ores, and the burning of fossil fuels. Fresh water usually contains less than 1 mg/L molybdenum. Concentrations ranging between 0.03 and 10 μg/L are typical of unpolluted waters. Levels as high as 1500 μg/L have been observed in rivers in industrial areas. The average concentration of molybdenum in finished drinking water is about 1–4 μg/L.

Health Concerns
Molybdenum is an essential trace nutrient for all plants and animals. It is considered nontoxic to humans, but excessive levels (0.14 mg/kg body weight; 10 mg/day for a 70 kg adult) may cause high uric acid levels and an increased chance of gout. The recommended daily intake is 70–250 μg/day for adults. Local concentrations may vary by a factor of 10 or more depending on regional geology, causing both deficient and excessive intake of molybdenum by plants and ruminants. Average adults contain about 5 mg of molybdenum in their body, and ingest about 100–300 μg/day. Twenty enzymes in plants and animals are known to be built around molybdenum, including xanthine oxidase, which helps to produce uric acid, essential for eliminating excess nitrogen from the body.

EPA Primary Drinking Water Standard
There are no primary or secondary drinking water standards for molybdenum.

Nickel (Ni), CAS No. 7440-02-0
Nickel is found in many ores as sulfides, arsenides, antimonides, silicates, and oxides. It's average crustal concentration is about 75 mg/kg. Because nickel is an important industrial metal, industrial waste streams can be a major source of environmental nickel. Inadvertent formation of volatile and poisonous nickel carbonyl (C_4NiO_4) can occur in various industrial processes that use nickel catalysts, such as coal gasification, petroleum refining, and hydrogenation of fats and oils. Nickel oxide is present in residual fuel oil and in atmospheric emissions from nickel refineries. The atmosphere is a major conduit for nickel as particulate matter. Contributions to atmospheric loading come from both natural sources and anthropogenic activity, with input from both stationary and mobile sources. Nickel particulates eventually precipitate from the atmosphere to soils and waters. Soil-borne nickel enters waters with surface runoff or by percolation of dissolved nickel into groundwater.

Nickel is one of the most mobile heavy metals in the aquatic environment. Its concentration in unpolluted water is controlled largely by coprecipitation and sorption with hydrous oxides of iron and manganese. In polluted environments, nickel forms soluble complexes with organic material. In reducing environments where sulfides are present, insoluble nickel sulfide is formed. Average concentrations in U.S. surface waters are typically between 10 and 100 μg/L, with concentrations as high as 11,000 μg/L in streams receiving mine discharges. In surface waters, sediments generally contain more nickel than the overlying water.

Health Concerns
Nickel appears to be an essential trace nutrient in animals and humans, but its exact role is not yet understood. Daily intake of nickel, mainly from food, is around 150 μg/day; the adult requirement appears to be between 5 and 50 μg/day. Although a few nickel compounds, such as nickel carbonyl, are poisonous and carcinogenic, most nickel compounds are nontoxic. The toxicity of dissolved nickel ingested orally is low, comparable to zinc, chromium, and manganese, perhaps because only 2%–3% of ingested nickel is absorbed.

EPA has not found nickel to cause adverse human health effects from short-term exposures at levels above the MCL. Long-term exposure above the MCL can cause decreased body weight, heart and liver damage, and dermatitis. There is no evidence that nickel has the potential to cause cancer from lifetime exposures in drinking water.

EPA Primary Drinking Water Standard
EPA remanded the drinking water standard for nickel in 1995. Prior to 1995, the MCLG and MCL) both were 0.1 mg/L. Currently there are no drinking water standards for nickel.

Other Comments
Treatment/Best available technologies: Ion exchange, lime softening, and reverse osmosis

Nitrate (NO_3^-), CAS No. 14797-55-8

Nitrite (NO_2^-), CAS No. 14797-65-0

Background (See Chapter 3 for a more detailed discussion.)
Nitrate and nitrite anions are highly soluble in water. Due to their high solubility and weak retention by soil, nitrate and nitrite are very mobile, moving through soil at approximately the same rate as water. Thus, nitrate has a high potential to migrate to groundwater. Because they are not volatile, nitrate and nitrite are likely to remain in water until consumed by plants or other organisms.

Nitrate is the oxidized form and nitrite is the reduced form. Aerated surface waters will contain mainly nitrate and groundwaters, with lower levels of dissolved oxygen, will contain mostly nitrite. They readily convert between the oxidized and reduced forms depending on the redox potential. Nitrite in groundwater is converted to nitrate when brought to the surface or exposed to air in wells. Nitrate in surface water is converted to nitrite when it percolates through soil to oxygen-depleted groundwater.

The main inorganic sources of contamination of drinking water by nitrate are potassium nitrate and ammonium nitrate. Both salts are used mainly as fertilizers. Ammonium nitrate is also used in explosives and blasting agents. Because nitrogenous materials in natural waters tend to be converted to nitrate, all environmental nitrogen compounds, particularly organic nitrogen and ammonia, should be considered as potential nitrate sources. Primary sources of organic nitrates include human sewage and livestock manure, especially from feedlots.

Health Concerns
Nitrate is a normal dietary component. A typical adult ingests around 75 mg/day, mostly from the natural nitrate content of vegetables, particularly beets, celery, lettuce, and spinach. Short-term exposure to levels of nitrate in drinking water higher than the MCL can cause serious illness or death, particularly in infants. Nitrate is converted to nitrite in the body, and nitrite oxidizes Fe^{2+} in blood hemoglobin to Fe^{3+}, rendering the blood unable to transport oxygen, a condition called methemoglobinemia. Infants are much more sensitive than adults to this problem because of their small total blood supply. Symptoms include shortness of breath and blueness of the skin. This can be an acute condition in which health deteriorates rapidly over a period of days.

Long-term exposure to levels of nitrate or nitrite in excess of the MCL may cause diuresis, increased starchy deposits, and hemorrhaging of the spleen. There is inadequate evidence to state whether or not nitrates or nitrites have the potential to cause cancer from lifetime exposures in drinking water.

EPA Primary Drinking Water Standard
Nitrate (as N) MCL: 10 mg/L
Nitrate (as N) MCLG: 10 mg/L
Nitrite (as N) MCL: 1.0 mg/L
Nitrite (as N) MCLG: 1.0 mg/L
Total (Nitrate + Nitrite, as N) MCL: 10 mg/L
Total (Nitrate + Nitrite, as N) MCLG: 10 mg/L

Perchlorate (ClO_4^-), CAS No.: 014797-73-0

Perchlorate (ClO_4^-) and Perchlorate Salts: 02/18/2005
CASRN 7790-98-9 Ammonium perchlorate
CASRN 7778-74-7 Potassium perchlorate
CASRN 7601-89-0 Sodium perchlorate

Perchlorate is a soluble oxychloro anion most commonly used as a solid salt in the form of ammonium perchlorate, potassium perchlorate, or sodium perchlorate, all of which are highly soluble. Ammonium perchlorate is the most widely used perchlorate compound. In their pure forms, these salts are white or colorless crystals or powders. Perchlorate salts dissolve in water and readily move from surface to groundwater. Perchlorate is known to originate from both natural and man-made sources.

The most common uses for ammonium perchlorate are in explosives, military munitions, and rocket propellants. In addition, perchlorate salts are used in a wide range of nonmilitary applications, including pyrotechnics and fireworks, blasting agents, solid rocket fuel, matches, lubricating oils, nuclear reactors, air bags, and certain types of fertilizers. Improper storage and disposal related to these uses are the most typical route for perchlorate to enter into the environment.

Environmental Behavior
Perchlorate is extremely soluble in water and, because it adheres poorly to mineral surfaces and organic material, perchlorate is very mobile in aquifer systems. It is relatively inert in typical groundwater and surface water conditions and can persist in the environment for many decades.

Ecological Concerns
Available information about ecological effects of the perchlorate anion is very limited, but is receiving increased attention. The current judgment is that perchlorate may have deleterious effects on other species throughout the environment. Field studies have demonstrated an overall lack of bioconcentration. Detectable concentrations of perchlorate are found only in a limited number of terrestrial mammals, birds, fish, amphibians, and insects exposed to elevated levels of perchlorate in the environment.

Health Concerns
Perchlorate interferes with thyroid functions because it is similar in size to the iodide anion and can mistakenly be taken up in place of iodide by the thyroid gland. Even at low concentrations, perchlorate impairs normal thyroid function and may contribute to thyroid cancer, although perchlorate has not positively been linked to cancer in humans. Recent studies also show possible adverse effects on normal growth and development of both fetuses and children. Perchlorate has been detected in some food products but the primary route of exposure is through the consumption of water containing perchlorate.

EPA Primary Drinking Water Standard
No MCL or MCLG has been set for perchlorate, but in February 2005 EPA established an official reference dose (RfD) of 0.0007 mg/kg/day of perchlorate. A reference dose is a scientific estimate of a daily exposure level that is not expected to cause adverse health effects in humans. The 0.0007 mg/kg/D RfD equates to a drinking water equivalent level (DWEL) of 24.5 μg/L, which is the concentration in drinking water that supplies the RfD quantity.

Because margins of safety are built into the RfD and the DWEL, exposures above the DWEL are not necessarily considered unsafe. Note that DWELs are not enforceable standards. USEPA is in the process of establishing an MCL for perchlorate. Meanwhile, states have been left to set their own limits, such as 14 μg/L in Arizona and 22 μg/L in Texas.

Polychlorinated Biphenyls
(There are 209 different PCBs, each with its own CAS identification number. See Chapter 7 for more information about PCBs.)

Polychlorinated biphenyls (PCBs) are a family of stable man-made organic compounds produced commercially by direct chlorination of biphenyl. PCBs have very high chemical, thermal, and biological stability; low water solubility; low vapor pressure; high dielectric constant; and high flame resistance. PCBs were manufactured and sold under various trade names (Aroclor, Pyranol, Phenoclor, Pyralene, Clophen, and Kaneclor) as complex mixtures differing in their average chlorination level. PCB mixtures range from oily liquids to waxy solids. Due to their nonflammability, chemical stability, high boiling point, and electrical insulating properties, PCBs were used in hundreds of industrial and commercial applications including electrical, heat transfer, and hydraulic equipment; as plasticizers in paints, plastics, and rubber products; in pigments, dyes, and carbonless copy paper; and in many other applications. More than 1.5 billion pounds of PCBs were manufactured in the United States until production was stopped in 1977.

Environmental Behavior
PCBs are very stable and do not degrade readily in the environment. They even survive ordinary incineration and can escape as vapors up the smokestack. The wide use of PCBs has resulted in their common presence in soil, water, and air. PCB dispersion from source regions to global distribution occurs mainly through atmospheric transport and subsequent deposition. They sorb strongly to soil and sediments. In aquatic systems, sediments are an important reservoir.

Ecological Concerns

Environmental contamination was first reported in 1966, when high levels of PCBs were found in fish, due to bioaccumulation. Fish consumption remains the major route of exposure to PCBs and there are health consequences associated with these exposures.

PCBs do not easily biodegrade. Although they are no longer manufactured, they still leak from old electrical devices, including power transformers, capacitors, television sets, and fluorescent lights, and can be released from hazardous waste sites and historic and illegal refuse dumps. They also persist in fatty foods, such as certain fish, meat, and dairy products.

Health Concerns

The toxicity of PCBs is a complicated issue since each congener, of which there are 209, differs in its toxicity. Furthermore, commercial PCBs are blends of the different pure congeners. In general, PCBs cause a wide variety of health effects, often at very low exposure levels. PCBs alter major systems in the body (immune, hormone, nervous, and enzyme systems) and, therefore, they can affect most of the body organs and functions. All PCBs are listed by EPA as known carcinogens and priority pollutants. When they are incinerated, they can produce dioxins, which are rated by EPA among the most toxic substances.

EPA Primary Drinking Water Standard
MCL: 0.5 μg/L
MCLG: zero

Selenium (Se), CAS No. 7782-49-2

Selenium is widely distributed in the earth's crust at concentrations averaging 0.09 mg/kg. It occurs in igneous rocks, with sulfides in volcanic sulfur deposits, in hydrothermal deposits, and in porphyry copper deposits. The major source of selenium in the environment is the weathering of rocks and soils. In addition, volcanic activity contributes to its natural occurrence in waters in trace amounts. Volcanic activity is an important source of selenium in regions with high soil concentrations.

Most selenium for industrial and commercial purposes is produced from electrolytic copper-refining shines and from flue dusts from copper and lead smelters. Anthropogenic sources of selenium in water bodies include effluents from copper and lead refineries, municipal sewage, and fallout of emissions from fossil fuel combustion. Selenium in surface waters can range between 0.1 and 2700 μg/L, with most values between 0.2 and 20 μg/L.

Selenate is more mobile under oxidizing conditions than under reducing conditions and can be reduced by bacteria in anaerobic environments to form methylated selenium compounds, which are volatile. Dissolved selenium exists mostly as the selenite (SeO_3^{2-}) and selenate (SeO_4^{2-}) anions. Ferric selenite, however, is insoluble and offers a treatment for removing dissolved selenium. Alkaline and oxidizing conditions favor the formation of soluble selenates, which also are the biologically available forms for plants and animals. Acidic and reducing conditions readily reduce selenates and selenites to insoluble elemental selenium, which precipitates from the water column.

Health Concerns
Selenium is a nutritionally essential trace element for all vertebrates and most plants. Human blood contains about 0.2 mg/L, 1000 times higher than typical surface waters, demonstrating that selenium is bioaccumulated. The adult daily requirement is between 20 and 200 µg/day. Drinking water seldom contains more than a few micrograms per liter of selenium, not enough for it to be a significant dietary source. Depending on the food eaten, daily intake is between 6 and 200, with levels near 150 µg/day being typical. Cereals, nuts, and seafood are good sources of selenium. Ingestion of quantities above the recommended maximum daily intake of 450 µg/day increases the risk of selenium poisoning, the most obvious symptom of which is bad breath and body odor caused by volatile methyl selenium produced by the body to eliminate excess selenium.

EPA has found that short-term exposure to selenium at levels above the MCL may cause hair and fingernail changes, damage to the peripheral nervous system, fatigue, and irritability. Long-term exposure above the MCL may cause hair and fingernail loss, damage to kidney and liver tissue, and damage to the nervous and circulatory systems. There is no evidence that selenium has the potential to cause cancer from lifetime exposures in drinking water.

EPA Primary Drinking Water Standard
MCL: 0.05 mg/L
MCLG: 0.05 mg/L

Other Comments
Treatment/Best available technologies: Activated alumina, coagulation and filtration, lime softening, reverse osmosis, electrodialysis.

Silver (Ag), CAS No. 7440-22-4
Silver is a white, lustrous, ductile metal that occurs naturally in its pure, elemental form and in ores, mostly as argentite (Ag_2S). Other silver ores include cerargyrite (AgCl), proustite ($3AgS\text{-}As_2S_3$), and pyrargyrite ($(Ag_2S)_3 \cdot Sb_2S_3$). Silver is also found associated with lead, gold, copper, and zinc ores. Silver is among the less common but most widely distributed elements in the earth's crust. It's concentration in normal soil averages around 0.3 mg/kg.

Before the recent wide spread adoption of digital photography, a large portion of silver consumption is for photographic materials. Also, because silver has the highest known electrical and thermal conductivities of all metals, it finds extensive use in electrical and electronic products such as batteries, switch contacts, and conductors. Other major uses include sterling and plated metalwork, jewelry, coins and medallions, brazing alloys and solders, catalysts, mirrors, fungicides, and dental and medical supplies.

Natural processes, such as weathering and volcanic activity, release silver to the environment. Silver has been found associated with sulfides, sulfates, chlorides, and ammonia salts in deposits and discharges of hot springs and volcanic materials. Anthropogenic sources of silver include discharges from landfills and waste lagoons, fallout from incineration and industrial emissions, and direct waste discharge to water. Some home water treatment devices use silver as an antibacterial agent and may represent a contamination source. Surface waters in nonindustrial regions average around 0.2–0.3 μg/L of silver, while streambed sediments range between 140 and 600 μg/kg of silver. In industrial areas, silver concentrations in surface waters may reach 40 μg/L and stream sediment concentrations 1500 μg/kg. Finished drinking water seldom contains more than 1 μg/L of silver.

Metallic silver is stable over much of the pH and redox range found in natural waters, but has very low water solubility. Insoluble silver compounds, such as AgCl, Ag_2S, Ag_2Se, and Ag_3AsS_3, may be present in aquatic systems in colloidal form, adsorbed to various humic substances, or incorporated with sediments. At pH < 7.5 under aerobic conditions, the Ag^+ cation is soluble and mobile. Around pH 7.5–8.0, aquatic Ag^+ reacts with water to form the insoluble oxide. Silver is dispersed through the aquatic environment as dissolved and colloidal species, but it eventually resides in the bottom sediments. Sorption, particularly by manganese dioxide, and precipitation of silver halides, particularly silver chloride, are the main processes that remove dissolved silver from the water column. These processes, along with the low crustal abundance of silver, account for its low observed concentrations in the aqueous phase.

Health Concerns
There is no evidence that silver is an essential nutrient. Metallic silver is not considered toxic to humans, but most of its salts exhibit toxic properties. Silver in all forms is acutely toxic to aquatic life. Large oral doses of silver nitrate can cause severe gastrointestinal irritation and ingestion of 10 g is likely to be fatal. Chronic human exposure to silver in drinking water seems only to cause argyria, a discoloration of the skin resulting from the deposition of metallic silver in tissues. EPA considers this condition to be mainly cosmetic. The minimum adult cumulative dose of silver for inducing argyria is about 1000 mg, an amount likely to be encountered only in industrial environments. The accumulation of 1000 mg of silver over a lifetime (70 years), would require the retention of 40 μg/day. Since around 90% of the silver intake is excreted, the required daily intake for inducing argyria would be more like 400 μg/day. An average daily diet may contain 20–80 μg of silver and drinking 2 L of water may contribute an additional 2 μg. Thus, food and drinking water are not likely to deliver toxic quantities of silver.

EPA Primary Drinking Water Standard
EPA has no primary drinking water standard for silver. The EPA secondary drinking water standard (nonenforceable) is 0.10 mg/L.

Other Comments
Treatment/Best available technologies: Lime softening at pH 11, sand filtration followed by activated carbon, ion exchange, nano- and ultrafiltration.

Sulfate (SO_4^{2-}), CAS No. 14808-79-8
The sulfate anion (SO_4^{2-}) is the stable, oxidized form of sulfur. Sulfate minerals are widely distributed in nature, and most sulfate compounds are readily soluble in water. All sulfate salts are very soluble except for calcium and silver sulfates, which are

moderately soluble, and barium, mercury, lead, and strontium sulfates, which are insoluble.

It is estimated that about one-half of the river sulfate load arises from mineral weathering and volcanism, the other half from biochemical and anthropogenic sources. Industrial discharges are another significant source of sulfates. Mine and tailings drainage, smelter emissions, agricultural runoff from fertilized lands, pulp and paper mills, textile mills, tanneries, sulfuric acid production, and metalworking industries are all sources of sulfate-polluted water. Aluminum sulfate (alum) is used as a sedimentation agent for treating drinking water. Copper sulfate is used for controlling algae in raw and public water supplies.

Air emissions from industrial fuel combustion and the roasting of sulfur-containing ores carry large amounts of sulfur dioxide and sulfur trioxide into the atmosphere, adding sulfates to surface waters through precipitation. Sulfate concentrations normally vary between 10 and 80 mg/L in most surface waters, although they may reach several thousand milligrams per liter near industrial discharges. High sulfate concentrations are also present in areas of acid mine drainage and in well waters and surface waters in arid regions where sulfate minerals are present.

Environmental Behavior
Nearly all natural waters contain sulfate anions. Sulfate is commonly found as a prominent component of unpolluted waters and is included among the six major surface and shallow groundwater ions (Na^+, Ca^+, Mg^+, Cl^-, HCO_3^{2-}, and SO_4^{2-}), second to bicarbonate as the most abundant anion in most freshwaters. Sulfur is an essential plant and animal nutrient, and sulfate is the most common inorganic form of sulfur in aerobic environments. Sulfate water concentrations that are too low have a detrimental effect on both land and aquatic plant growth.

Sulfate is redox sensitive and is bacterially reduced to sulfide ion under anaerobic conditions. Sulfide may be released to the atmosphere as H_2S gas or precipitated as insoluble metal sulfides. Oxidation of sulfides returns sulfur to the sulfate form.

Sulfates may be leached from most sedimentary rocks, including shales, with the most appreciable contributions from such sulfate deposits as gypsum ($CaSO_4 \cdot 2H_2O$) and anhydrite ($CaSO_4$). The oxidation of sulfur-bearing organic materials can contribute sulfates to waters.

Ecological Concerns
Sulfates serve as an oxygen source for bacteria. Under anaerobic conditions, sulfate-reducing bacteria reduce dissolved sulfate to sulfide, which then may be volatilized to the atmosphere as H_2S, precipitated as insoluble salts, or incorporated in living organisms. These processes are common in the anaerobic regions of wetlands and lakes fed by surface and groundwaters with high sulfate levels. Oxidation of sulfides returns the sulfur to the sulfate form.

Sulfate is a major contributor to salinity in many irrigation waters. However, toxicity usually is not an issue, except at very high concentrations, where high sulfate can interfere with uptake of other nutrients. As with boron, sulfate in irrigation water is a plant nutrient, and irrigation water often carries enough sulfate for maximum production of most crops. Seawater contains about 2700 mg/L of sulfate.

Health Concerns

The sulfate anion is generally considered nontoxic to animal, aquatic, and plant life. It is an important source of sulfur, an essential nutrient for plants and animals. Sulfates are used as additives in the food industry, and the average daily intake of sulfate from drinking water, air, and food is approximately 500 mg. As examples, some measured sulfate concentrations in beverages are 100–500 mg/L in drinking water, 500 mg/L in coconut milk, 260 mg/L in beer (bitter), 250 mg/L in tomato juice, and 300 mg/L in red wine (FNB 2004). Available data suggest that people acclimate rapidly to the presence of sulfates in their drinking water.

No upper limit likely to cause detrimental human health effects has been determined for sulfate in drinking water. However, concentrations of 500–750 mg/L may cause a temporary mild laxative effect, although doses of several thousand milligrams per liter generally do not cause any long-term ill effects. Because of the laxative effects resulting from ingestion of drinking water containing high sulfate levels, EPA recommends that health authorities be notified of sources of drinking water that contain sulfate concentrations in excess of 500 mg/L.

The presence of sulfate can adversely affect the taste of drinking water. The lowest taste threshold concentration for sulfate is approximately 250 mg/L as the sodium salt but higher as calcium or magnesium salts (up to 1000 mg/L).

EPA Primary Drinking Water Standard
MCL: none
MCLG: none
EPA has no primary drinking water standard for sulfate. The EPA secondary drinking water standard (nonenforceable) is 250 mg/L, based on aesthetic effects.

Other Comments
Treatment/Best available technologies: Anion exchange, reverse osmosis, nanofiltration

Sulfide (S^{2-}, see Hydrogen Sulfide)

Thallium (Tl), CAS No. 7440-28-0

Thallium is a soft, lead-like metal that is widely distributed in trace amounts. It may be found naturally as the pure metal or associated with potassium and rubidium in copper, gold, zinc, and cadmium ores. Thallium compounds are nonvolatile, although many are water-soluble. Thallium is generally present in trace amounts in fresh water. Unpolluted soil levels range between 0.1 and 0.8 mg/kg, with an average around 0.2 mg/kg.

Thallium compounds are used mainly in electronics industry, and to a limited extent in the manufacture of pharmaceuticals, alloys, and high refractive-index glass. Man-made sources of thallium pollution are gaseous emission from cement factories, coal burning power plants, and metal sewers. Small amounts of thallium in fallout from these sources frequently contaminates nearby farms and gardens, where it is readily absorbed by the plant roots to contaminate food crops. The leaching of thallium from ore processing operations is the major source of elevated thallium concentrations in water. Thallium is a trace metal associated with copper, gold, zinc, and cadmium.

Most thallium is released into the environment by weathering of minerals. Human sources of thallium are wastes from the production of other metals, e.g., from the roasting of pyrite during the production of sulfuric acid, and in mining and smelting operations of copper, gold, zinc, lead, and cadmium. Waste streams of these industries may contain as much as 90 μg/L.

In the aquatic environment, thallium is transported as soluble complexes with humic materials (above pH 7), sorption to clay minerals, and bioaccumulation. In reducing environments, thallium may be precipitated as elemental metal or, in the presence of sulfur, as the insoluble sulfide. In waters of high oxygen content, Tl^+ is the dominant oxidation state, forming soluble chloride, carbonate, and hydroxy salts. Thallium sorption to sediments is pH dependent. Thallium is strongly sorbed by montmorillonite clay at pH 8 but only slightly at pH 4. In a study of heavy metal cycling in a lake in southwestern Michigan, thallium was detected only in the sediments. Since thallium is soluble in most aquatic systems, it is readily available to aquatic organisms and is quickly bioaccumulated by fish and plants.

Health Concerns
Thallium is a toxic metal with no known nutritional value. On the contrary, it is notorious for its use by murderers as a poison; the lethal dose for an adult is around 800 mg. Environmental exposure is mainly through contaminated foods, which are estimated to contain, on average, about 2 ppb of thallium. The adult total body burden of thallium is 0.1–0.5 mg thallium, the greatest part of which is carried by muscle tissue.

Short-term exposures to thallium at levels above the MCL can cause gastro-intestinal irritation, numbness of toes and fingers, the sensation of burning feet, and muscle cramps. Long-term exposure to thallium at levels above the MCL can cause damage to liver, kidney, intestinal and testicular tissues, as well as changes in blood chemistry and hair loss. There is no evidence that thallium has the potential to cause cancer from lifetime exposures in drinking water.

EPA Primary Drinking Water Standard
MCL: 2.0 μg/L
MCLG: 0.5 μg/L

Other Comments
Treatment/Best available technologies: Activated alumina, ion exchange, reverse osmosis, nanofiltration

Vanadium (V), CAS No. 7440-62-2
Vanadium is widely dispersed in the earth's crust at an average concentration of about 150 ppm. Deposits of ore-grade mineable vanadium are rare. There are more than 65 known vanadium-bearing minerals. In the United States, vanadium occurs in uranium-bearing ores and in phosphate shales and rocks in parts of the western states. Vanadium is also present in coal and crude oil. The bulk of commercial vanadium is obtained as a by-product or coproduct from the processing of iron,

titanium, and uranium ores, and to a lesser extent, from phosphate, bauxite, and chromium ores and the ash or coke from burning or refining petroleum. The main use for vanadium is as an alloy additive.

Vanadium enters the aquatic environment mainly by surface erosion and natural seepage from carbon-rich deposits such as tar and oil sands, atmospheric deposition, weathering of vanadium-rich ores and clays, and leaching of coal mine wastes. Vanadium concentrations in fresh water typically range between <0.3 $\mu g/L$ and around 200 $\mu g/L$. Groundwater concentrations of vanadium are typically <1 $\mu g/L$.

Health Concerns
There are no particular health or safety hazards associated with vanadium and its compounds. Dust and fine powders present a moderate fire hazard. Vanadium compounds, of which the most common is vanadium pentoxide, may irritate the conjunctivae and respiratory tract. Toxic effects have been observed from airborne concentrations of vanadium compounds of several milligrams or more per cubic meter of air. OSHA threshold limits in the workplace are 0.5 mg/m^3 for dust and 0.05 mg/m^3 for fumes. Oral toxicity in humans is minimal.

As an environmental pollutant, vanadium is of concern mainly because of its high levels in residual fuel oils and its subsequent contribution to atmospheric particulate levels from the combustion of these fuels in urban areas.

EPA Primary Drinking Water Standard
EPA has no drinking water standards for vanadium.

Zinc (Zn), CAS No. 7440-66-6
Zinc is a common contaminant in surface and groundwaters, storm water runoff, and industrial waste streams. Its average concentration in the earth's crust is around 70 mg/kg. Because it is very chemically reactive, zinc is not found free in nature but is always associated with one of the 55 known zinc minerals (mainly sulfides, oxides, carbonates, and silicates). Ninety percent of industrial zinc production is from the minerals sphalerite (also called zincblende) and wurtzite, which are different crystalline forms of zinc sulfide (ZnS). Zinc minerals tend to be associated with minerals of other metals, particularly lead, copper, cadmium, mercury, and silver. The most important chemical property of zinc is its high reduction potential, which places it above iron in the electromotive series. Because of this, zinc coatings are used to protect iron and steel from corrosion. When iron or steel coated with zinc is exposed to corrosive conditions, zinc displaces iron from solution and prevents dissolution of the iron. In the process of dissolving, zinc reduces dissolved iron, and many other dissolved metals, to the metallic state.

Zinc occurs in natural waters in both suspended and dissolved forms. The dissolved form is the divalent cation, Zn^{2+}. Dissolved zinc is readily sorbed to or occluded in mineral clays and humic colloids. In water of low alkalinity and below pH 7, Zn^{2+} is the dominant form. As pH and alkalinity increase, the potential for sorption and humic complex formation also increase. Above pH 7, sorption to metal oxides, clays, and apatite (calcium phosphate minerals, $Ca_5(F,Cl,OH,0.5CO_3)$ $(PO_4)_3$) can bind over 90% of dissolved zinc. Sorption is generally not significant below pH 6. At low pH or under anaerobic conditions, metal oxides can redissolve and release bound zinc cation into solution. However, if the redox potential rises and aerobic conditions are established, zinc sulfide is oxidized to soluble zinc sulfate ($ZnSO_4$), also releasing Zn^{2+} into solution.

Zinc concentrations in unpolluted surface waters typically range between about 5 and 50 $\mu g/L$. Streams draining mined areas often exceed 100 $\mu g/L$. Industries with waste streams containing significant levels of zinc include steel works with galvanizing operations, zinc and brass metal works, zinc and brass plating, and production of viscose rayon yarn, ground wood pulp, and newsprint paper. Reported concentrations of zinc in industrial waste streams reach 48,000 mg/L. More representative values are between 10 and 200 mg/L.

Health Concerns
Zinc is an essential nutrient and is not toxic to humans. About 1 g/day may be ingested without ill effects. Recommended dietary allowance is 15 mg/day for adults.

EPA Primary Drinking Water Standard
EPA has no primary drinking water standard for zinc. The EPA secondary drinking water standard (nonenforceable) is 5 mg/L, based on a metallic taste detectable by many people above that level.

Other Comments
Treatment/Best available technologies: Chemical precipitation, ion-exchange, eative recovery of salts, reverse osmosis, electrolytic plating.

REFERENCES

FNB, 2004, *Dietary Reference Intakes for Water, Potassium, Sodium, Chloride, and Sulfate*, Panel on dietary reference intakes for electrolytes and water, food and nutrition board, National Academies Press, Washington, DC, 640 p.

National Academy of Sciences, 1982, *Drinking Water and Health*, Vol. 4, National Academy Press, Washington, DC, 299 p.

USEPA, 1988, United States Environmental Protection Agency, Ambient Water Quality Criteria for Chloride—1988. Office of Research and Development, Environmental Research Laboratory, Duluth, MN. EPA 440/5-88-001.

USEPA, 2006, 2006 Edition of the Drinking Water Standards and Health Advisories, EPA 822-R-06-013, Office of Water, U.S. Environmental Protection Agency, Washington, DC.

Answers to Selected Chapter Exercises

CHAPTER 1

1. 475 mL of a water sample was evaporated to determine the amount of dissolved salts contained in it. After evaporation, the dried precipitated salts weighed 1475 mg. What was the concentration in parts per million (ppm) of dissolved salts (also called total dissolved solids [TDS])?

   Answer: 3105 ppm

2. The annual arithmetic mean ambient air quality standard for sulfur dioxide (SO_2) is 0.03 ppmv. What is this standard in $\mu g/m^3$?

   Answer: 1300 $\mu g/m^3$

3. The primary drinking water maximum contaminant level (MCL) for barium (Ba) is 2.0 mg/L. If the sole source of barium is barium sulfate ($BaSO_4$), how much $BaSO_4$ salt is present in 1 L of water that contains 2.0 mg/L of Ba? (*Note*: The moles of Ba in 2.0 mg equals the moles of $BaSO_4$ in 1 L sample.)

   Answer: 3.40 mg

4. Most people can detect the odor of ozone in concentrations as low as 10 ppb. Can they detect the odor of ozone in samples with an ozone level (a) 0.118 ppm; (b) 25 ppm; (c) 0.001 ppm?

   Answer:

   a. 0.118 ppm $= 0.118$ ppm $\times$ 1000 ppb/ppm $= 118$ ppb ozone. Since this is much higher than 10 ppb, you probably can smell the sample.
   b. 25 ppm $= 25$ ppm $\times$ 1000 ppb/ppm $= 25,000$ ppb ozone. This also is much higher than 10 ppb. You probably can smell it.
   c. 0.001 ppm $\times$ 1000 ppb/ppm $= 1$ ppb ozone. You probably cannot smell it.

5. Determine the percentage by volume of the different gases in a mixture containing 0.3 L of O_2, 1.6 L of N_2, and 0.1 L of CO_2.

   Answer:

   When the pressures of the individual gases before mixing and the pressure of the combined gases after mixing are the same, the individual volumes are additive in the mixture and the percentage by volume of any gas in the mixture will be

$$\text{Percentage of given gas} = \frac{\text{volume of given gas}}{\text{total volume of all gases}} \times 100$$

Thus

$$\text{Percentage of O}_2 = \frac{0.3\text{L}}{0.3 + 1.6 + 0.1\text{ L}} \times 100 = 15\% \text{ (normal atmosphere \% } = 21\%)$$

$$\text{Percentage of N}_2 = \frac{1.6\text{L}}{0.3 + 1.6 + 0.1\text{ L}} \times 100 = 80\% \text{ (normal atmosphere \% } = 78\%)$$

$$\text{Percentage of CO}_2 = \frac{0.1\text{L}}{0.3 + 1.6 + 0.1\text{ L}} \times 100 = 5\% \text{ (normal atmosphere \% } = 0.03\%)$$

As a check on the accuracy of the answers, note that all the percentages add up to 100%. When figures are rounded to maintain the correct significant figures, percentages might not always add to exactly 100% because of cumulative rounding errors.

6. What is the significance of the fact that the percentage of oxygen in air is 21% by numbers of molecules and 23% by mass?

Answer:

The mass of O_2 molecules must be greater than the average mass of all other gases in air. The average mass of air molecules depends mainly on the two most abundant gases, nitrogen and oxygen. The predominant gas in air is N_2, with a number (same as volume) percentage of 78% and a molecular mass of 28. Oxygen is heavier, molecular mass of 32, so its mass percentage is greater than its number percentage. The mass percentage of N_2 will be a little lower than its number percentage. If all the gas molecules in air have the same mass, then the percentages by number and mass would be the same.

7. Express the 0.9% argon content of air in parts per million.

Answer:

Percentage is the same as parts per hundred (pph). One percent is the same as 1 part per 100 parts, or $\frac{1\text{ part}}{100\text{ parts}}$ and 1 ppm is the same as $\frac{1\text{ part}}{10^6\text{ parts}}$. To express percent as ppm, you must convert parts per hundred to parts per million. Therefore

$$0.9\% = \frac{0.9\text{ parts}}{100\text{ parts}} \times \frac{10^2 \times 10^4}{10^6} = \frac{0.9 \times 10^4}{10^6} = 9000 \text{ ppm}$$

8. Express 400 ppm of CO_2 in inhaled cigarette smoke as a percentage of the smoke inhaled.

   Answer:

   Since percent is parts per hundred, the question is how many pph will be equal to 400 ppm?

   $$400 \text{ ppm} = \frac{400}{1,000,000} = \frac{0.04}{100} = 0.04\%$$

9. The permissible 1 h average limit for ozone is 0.12 ppm. If Little Rock, Arkansas registers a reading of 0.15 ppm for 1 h, by what percent does Little Rock exceed the limit for atmospheric ozone?

   Answer:

   In this case you want to compare the standard with the amount that the Little Rock measured value exceeds the standard value. The percent exceedance will be the exceedance divided by the standard, times 100. The difference between the Little Rock value and the standard is
   0.15 − 0.12 = 0.03. The percentage that the Little Rock value exceeds the standard is

   $$\frac{0.03}{0.12} \times 100 = 25\% \text{ above the standard}$$

10. A certain water soluble pesticide is fatal to fish at 0.5 mg/L (ppm). Nearly 5 kg of the pesticide is spilled into a stream. The stream flow rate was 10 L/s at 1 km/h. For what approximate distance downstream could fish potentially be killed?

    Answer:

    First, find the volume of water that 5 kg of the pesticide will contaminate to a lethal concentration. Then, find what length of stream contains that concentration.

    1. Total volume of water contaminated to lethal concentration is
       $$\frac{5000 \text{ g}}{0.5 \times 10^{-3} \text{ g/L}} = 1 \times 10^7 \text{ L}$$
    2. Find the time required until the stream water is diluted by new flow to less than lethal concentration.

       $$\text{At a flow of } 10 \text{ L/s:} \frac{10^7 \text{L of contaminated water}}{10 \text{ L/s}} = 10^6 \text{ s}$$

       $$\frac{1 \times 10^6 \text{ s}}{3600 \text{ s/h}} = 278 \text{ h of flow until pesticide is diluted}$$

    Potential distance of fish kill is 278 h × 1 km/h = 278 km or 173 miles.
        This assumes that the pesticide does not move downstream by plug flow but is distributed uniformly throughout the contaminated length of stream.

CHAPTER 2

5. Various compounds and some of their properties are tabulated below.

Compound	Boiling Point (°C)	Water Solubility (g/100 mL)	Dipole Moment (D)
H_2	−253	2×10^{-4}	0
HCl	−84.9	82	1.08
HBr	−67	221	0.82
CO_2	−78	0.15	0
CH_4	−164	2×10^{-3}	0
NH_3	−33.5	90	1.3
H_2O	100	∞	1.85
HF	19.5	∞	1.82
LiF	1676	0.27	6.33

a. Which compounds are nonpolar?

Answer:

All compounds with a dipole moment equal to zero are nonpolar, e.g., H_2, CO_2, and CH_4.

b. Which compounds are polar?

Answer:

All compounds with a dipole moment greater than zero are polar, e.g., HCl, HBr, NH_3, H_2O, HF, and LiF.

c. Make a rough plot of boiling point versus dipole moment. What conclusions may be inferred from the graph?

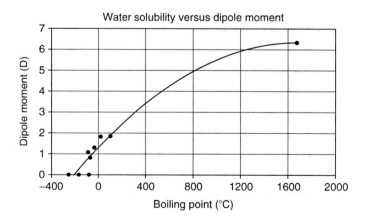

Answer:

The plot shows a general increase of boiling point with dipole moment.

d. Discuss briefly the trends in water solubility. Why is solubility not related to dipole moment in the same manner as boiling point? Zero dipole moment indicates low solubility and higher dipole moments up to at least about $D = 2$ tend to indicate higher solubility. But LiF, with a very high dipole moment of $D = 6.3$, has a low solubility characteristic of compounds with dipole moments around $D = 0$ (see part e. for the reason).

Answer:

While the boiling point of a compound is related only to the strength of the attractive forces between molecules of the pure compound, its water solubility is related to that plus the strength of the attractive forces to water molecules.

e. The ionic compound LiF appears to be unique. Try to suggest a reason for its low solubility.

Answer:

The large dipole moment of LiF indicates that its bonding is ionic. Both the Li^+ and the F^- ions are small, allowing them to come very close together, which results in an unusually strong ionic bond. The very strong ionic attractions within the LiF solid compound are too strong for the attractive forces between LiF and water molecules to pull apart.

6. The chemical structures of several compounds are shown below. Fill in the table with estimated properties based only on their chemical structures. Do not look at reference data for the answers.

Compound	Structure	Physical State at Room Temperature (Gas, Liquid, or Solid)	Water Solubility (Very Low, Low, Moderate, High)	Lipid Solubility (Very Low, Low, Moderate, High)
Ethane	$H_3C{-}CH_3$	Gas	Very low	High
Benzene	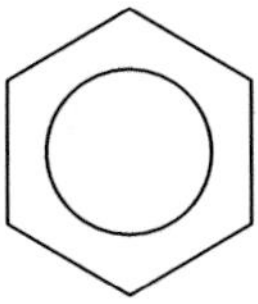	Liquid	Low	High

(*Continued*)

(Continued)

Compound	Structure	Physical State at Room Temperature (Gas, Liquid, or Solid)	Water Solubility (Very Low, Low, Moderate, High)	Lipid Solubility (Very Low, Low, Moderate, High)
Citric acid	(structure)	Solid	High	Low
Anthracene	(structure)	Solid	Very low	High
Glyphosate	(structure)	Solid	High	Low

CHAPTER 3

2 a. The $[H^+]$ of water in a stream is 6.1×10^{-8} mol/L. What is the pH?

   Answer: $pH = 7.21$

   b. The pH of water in a stream is 9.3. What is the hydrogen ion concentration?

   Answer: $[H^+] = 5.0 \times 10^{-10}$ mol/L

3. An engineer requested a water sample analysis that included the following parameters: pH, carbonate ion, and bicarbonate ion. Explain why she probably is wasting money.

   Answer:

   If there is a charge for each analysis, the request for a carbonate analysis is unnecessary because the carbonate concentration can be found from Figure 3.2 and accompanying discussion, when pH and bicarbonate are known.

6. A wastewater treatment plant removed ammonia with a nitrification–denitrification process that also reduced alkalinity. For each gram of NH_3–N removed, the process also removed 7.14 g of alkalinity-$CaCO_3$. If the plant was designed to remove 25 mg/L of NH_3–N and the total throughput was 250,000 gal/day, how many pounds of caustic soda (sodium hydroxide, NaOH) must be added each day to restore the alkalinity to its original value before ammonia removal?

Answer:

According to Figure 3.3, the addition of NaOH, a strong base, represents a vertical increase on the alkalinity versus total carbonate graph. Therefore, the milliequivalents per liter of NaOH needed is equal to the milliequivalents per liter of alkalinity lost.

Alkalinity-$CaCO_3$ lost $= 7.14 \times 25$ mg/L NH_3–N $= 178.5$ mg/L alkalinity-$CaCO_3$

Eq. wt of $CaCO_3 = 50.04$ g/eq (Table 1.1)

Eq/L of alkalinity-$CaCO_3$ lost $= (178.5$ mg/L$)/(50.04$ mg/meq$) = 3.57$ meq/L

Equivalents of NaOH required $=$ eq/L of alkalinity-$CaCO_3$ lost $= 3.57$ meq/L

Eq. wt of NaOH $= 40.0$ g/eq

$$Wt/d\ NaOH = \frac{40.0\ mg/meq \times 3.57\ meq/L}{4.54 \times 10^5\ mg/lb \times 0.264\ gal/L}$$

$$\times\ 2.50 \times 10^5\ gal/day = 298\ lb/day$$

7. A wastewater flow contains 30 g/L total ammonia nitrogen and has a 10 g/L discharge limit. An air-stripping tower is to be used. Its temperature varies from 20°C to 30°C and the pH is normally about 9. At what pH must the stripper be operated?

Answer:

Total NH_3–N $= [NH_3] + [NH_4^+]$. The pH must be adjusted so that enough ammonia is in the volatile form to meet the discharge requirement.

The ratio $\frac{[NH_3]}{[NH_3]+[NH_4^+]}$ must be at least as large as the fraction of NH_3–N to be removed.

Fraction of NH_3–N to be removed is $\frac{20}{30} = 0.67$

Stripping efficiency is increased with increasing pH and higher temperatures. Figure 3.8 shows that to get 0.67 (67%) of the total NH_3–N in the form of volatile NH_3, the pH must be raised to about 10.0, for the worst case of 20°C.

8. A water sample from a lake has a measured alkalinity of 0.8 eq/L.

In the early morning, a monitoring team measures the lake's pH as part of an acid rain study and finds pH $= 6.0$. The survey team returns after lunch to recheck their data. By this time, algae and other aquatic plants have consumed enough dissolved CO_2 to reduce the lake's total carbonate (C_T) to one-half of its morning value.

a. What was the morning value for C_T?

Answer:

Use the Deffeyes diagram, Figure 3.3. A horizontal line from alkalinity $= 0.8$ eq/L intercepts the pH 6.0 line at about $C_{T,\ morning} = 2.5 \times 10^{-3}$ mol/L.

b. What was the pH after lunch?

Answer:

C_T after lunch was $(2.5 \times 10^{-3} \text{ mol/L})/2 = 1.25 \times 10^{-3}$ mol/L. Since no acid was added, the alkalinity is essentially unchanged.
Move vertically up from $C_{T, \text{afternoon}} = 1.25 \times 10^{-3}$ mol/L to alkalinity $= 0.8$ eq/L.
The pH at that point is about 6.6.
Depletion of CO_2 by photosynthesis caused pH to rise from 6.0 to 6.6.

9. A groundwater has the following analysis at pH 7.6:

Analyte	Concentration (mg/L)
Calcium	75
Magnesium	40
Sodium	10
Bicarbonate	300
Chloride	10
Sulfate	112

Calculate alkalinity, total hardness, carbonate (temporary) hardness, and non-carbonate (permanent) hardness.

Answer:

Note that the carbonate concentration will be 0 at pH 7.6 (see the distribution diagram, Figure 3.2).
Alkalinity $= 0.820 \, [HCO_3^-] + 1.668 \, [CO_3^{-2}] = (0.820)(300) + (1.668)(0)$
$\qquad = 246$ mg/L
Total hardness $= 2.497 \, [\text{Ca, mg/L}] + 4.118 \, [\text{Mg, mg/L}]$
$\qquad\qquad = (2.497)(75) + (4.118)(40) = 352$ mg/L
Carbonate hardness $= 0.820 \, [HCO_3^-] + 1.668 \, [CO_3^{-2}]$
$\qquad\qquad = (0.820)(300) + (1.668)(0) = 246$ mg/L
Noncarbonate hardness $=$ total hardness $-$ carbonate hardness
$\qquad\qquad = 352 - 246 = 106$ mg/L

Note: Noncarbonate hardness is the hardness from noncarbonate salts of calcium and magnesium, mainly the salts of sulfate, chloride, and silicate, all expressed as mg/L of $CaCO_3$.

CHAPTER 4

1. Is a solution of ferric sulfate, $Fe_2(SO)_3$, in water expected to be acidic, basic, or neutral? Explain your answer.

Answer:

The solution is expected to be acidic. Ferric sulfate dissolves to form ferric cations and sulfate anions:

$$Fe_2(SO)_3 = 2Fe^{3+} + 3SO_4^{2-}$$

Each ion attracts a hydration shell of water molecules. The triple positive charge of a Fe^{3+} ion can attract the electrons of adjacent water molecules so strongly that the O–H bond in a water molecule is broken, releasing H^+ into the solution and making the solution acidic (see Figure 4.1).

3. A groundwater sample contains dissolved Fe^{2+}. What difficulties might be encountered in using this sample for determining the in situ pH of the groundwater?

Answer:

The pH must be measured without exposing the sample to the atmosphere. Once the sample is exposed to the atmosphere, two sources of error can occur:

1. Carbon dioxide dissolved in the sample will equilibrate with CO_2 in the atmosphere. Typically, groundwater is enriched in CO_2 relative to the atmosphere because it is produced by biodegradation reactions in the subsurface. In this case, CO_2 will be lost from the sample and tend to cause an increase in pH over the in situ value.
2. Oxygen dissolved in the sample will equilibrate with O_2 in the atmosphere. Typically, groundwater is depleted in O_2 relative to the atmosphere because it is consumed by biodegradation reactions in the subsurface. This establishes the anaerobic redox conditions that reduce any iron present to the ferrous Fe^{2+} form. When the sample is exposed to the atmosphere, O_2 will enter the sample causing Fe^{2+} to be oxidized to Fe^{3+}. Fe^{3+} dissolved in water produces acidity as in Example 1, and will tend to lower the pH below the in situ value.
   The net effect on pH will depend on the relative concentrations of CO_2 and Fe^{2+} in the sample.
5. An industrial discharge permit has limits for chromium, copper, and cadmium of 0.01 mg/L each. Assuming that the untreated discharge will always exceed these limits, what difficulties are encountered if you try to comply by using only pH adjustment? Use Figure 4.2, assuming that the only complexing anion present is OH^-. How might you meet the limit for all three metals?

Answer:

The most soluble of the metals is cadmium, which has a solubility for $Cd(OH)_2$ of 0.01 mg/L at a pH of 10.8. For a safety margin, you might want to go to a little higher pH, but not above pH 12 (see why on Figure 4.2). However, pH 10.8 is above the pH for the minimum solubilities of chromium and copper. At pH 10.8, the solubility of $Cu(OH)_2$ is about 0.017 mg/L and that of $Cr(OH)_3$ is about 1.3 mg/L. If you lower the pH to a value where chromium and copper will meet the discharge limit, around pH 8.1, cadmium will not; at $pH = 8.1$ the solubility of $Cd(OH)_2$ is greater than 100 mg/L, off the scale of Figure 4.2. To meet the limits for all 3 metals, first lower the pH to around 11, which lowers the solubility of $Cd(OH)_2$ well below the limit, and remove the precipitated metals by filtering. Then adjust the pH to around 8.1 the meet the limits for chromium and copper.

CHAPTER 5

1. What are soil horizons?

   Answer:

   Stratified soil layers formed by the cumulative effects of many rock-weathering reactions and the actions of bioorganisms on organic matter.

2. The pores of soil contain a significant amount of air. How does the composition of soil air is different from atmospheric air and why is it so?

   Answer:

   It tends to be depleted in oxygen and enhanced in CO_2 because of microbial respiration.

3. Many soils have ion-exchange properties. What does this mean?

   Answer:

   They can immobilize many metals by charge attraction. They contain materials such as clays or humus that have many polar and charged molecular sites to which dissolved ions are attracted and adsorbed. Typically, these sites are occupied by the abundant Na^+ and H_3O^+ ions, which are relatively weakly held. When multiply charged metal cations are available, they attach to the charged sites and displace (exchange) the Na^+ and H_3O^+ ions, which then enter the water as dissolved ions.

4. Discuss how fluctuations in the groundwater level can influence the distribution of LNAPL hydrocarbon contamination in the subsurface.

   Answer:

   Light nonaqueous phase liquids (LNAPL) "floats" on the surface of the groundwater. When the water table elevation fluctuates, it pushes the LNAPL up and down, spreading the LNAPL into a "smear zone," effectively widening the region where LNAPL is immobilized by sorption to soil particles. When the water table moves up, some LNAPL is trapped by capillary action in the saturated zone below the water table and the floating LNAPL layer is thinned. When the water table moves down again, some of the previously trapped LNAPL is released. This released LNAPL moves downward under gravity to the new water table surface and the floating LNAPL layer becomes thicker.

5. A chemical spill has contaminated a lake with carbon tetrachloride (CCl_4). For carbon tetrachloride, Table 5.5 (p. 158) shows that log $K_{ow} = 2.73$ (thus, $K_{ow} = 5.4 \times 10^2$). The maximum concentration of carbon tetrachloride that a certain species of fish can accumulate with a low probability of health risks is reported to be 10 ppm. Assuming that the fatty tissue of these fish behaves similarly to octanol with respect to carbon tetrachloride partitioning, at what value should the water quality standard for carbon tetrachloride be set in this lake to protect this fish?

$$K_{ow} = 5.4 \times 10^2 = \frac{\text{Concentration in octanol}}{\text{Concentration in water}} = \frac{\text{Maximum concentration in fish}}{\text{Maximum concentration in water}}$$

$$\text{Maximum concentration in water} = \frac{10 \text{ ppm}}{5.4 \times 10^2}$$

$$= 1.85 \times 10^{-2} \text{ ppm} = 18.5 \text{ ppb} = 18.5 \text{ μg/L}$$

CHAPTER 9

1. Complete the following nuclear equations:

   (In every case, the unknown is just one kind of particle.)

   a. $^{218}_{84}\text{Po} \rightarrow ^{4}_{-2}\text{He} + ^{214}_{86}\text{Rn}$

   c. $^{1}_{0}\text{n} + ^{235}_{92}\text{U} \rightarrow ^{90}_{38}\text{Sr} + 2^{1}_{0}\text{n} + ^{144}_{54}\text{Xe}$

5c. Each atom of the heavier sample has four more neutrons in the nucleus than do the atoms in the lighter sample.

7. Use Figure 9.2 to write an equation for the probable mode of decay of each of the following radionuclides. Your equation should be in the form of the answers to question 9.1.

   a. $^{38}_{16}\text{S} \rightarrow ^{0}_{-1}\text{e} + ^{38}_{17}\text{Cl}$

9. Radon-222 has a half-life of 3.8 days. The source water for a drinking water treatment plant contains 1200 pCi/L of radon. Because aeration is not an option at this plant, how long must they store the water to meet the Environmental Protection Agency (EPA) guideline of 300 pCi/l?

$$k = \frac{0.693}{3.8\text{d}} = 0.182/\text{day}$$

$$\ln\frac{300 \text{ pCi/L}}{1200 \text{ pCi/L}} = -1.39 = -0.182/\text{day} \times t$$

$$t = 7.62 \text{ days}$$

Index

A

Abiotic and biotic particles, 240
Acetic acid, hydrogen bonding in, 45
Acid-base potential (ABP)
    determination of, 127–128
    of soil, 127
Acid-base reactions, 27, 57–58
Acidity, 66
    pH and carbonate relationship, 71, 73
Acid rock drainage, 124–128
    acid-base potential
        determination of, 127–128
        of soil, 127
    acid formation in, 125–126
    non-iron metal sulfides in, 127
    by pyrite oxidation, 126
Acrylamide, CAS NO. 79-06-1, 367;
        see also Water quality parameters
Activated charcoal, hydrogen sulfide
        sorption to, 96
Active biodegradation, rate of, 252
Advanced oxidation process, 335
AEC, *see* Anion-exchange capacity
A emission, 278
Aerobic biodegradation, 24, 29, 246
Aerobic microbial reactions, 245
Aerobic respiration, 252
Age-dating fuel spills
    BTEX ratios in groundwater and, 211–213
    diesel fuel, 213–215
    gasoline, 210–211
Agricultural water quality, 311–314
Air
    concentration calculation, 13
    pollutant concentrations, 11
Air-stripping ammonia, method for removing
        nitrogen from waste water, 87
Air-water partition coefficient, 145–147;
        see also Henry's law constant
    estimation methods for, 156–161
Alachlor (herbicide), 367–368
Aldicarb (insecticide), 368–369
Aldrin (pesticides), 369
Alkali metals, 109
Alkalinity
    calculation of, 68
    carbonate and pH relationship, 69–74

importance of, 67
water quality
    criteria and standards for, 67
    in terms of, 4
Alkanes, properties of, 47
Aluminum (Al), 119
Aluminum, CAS No. 7429-90-5
    concentrations, 370
    drinking water standard for, 371
    health concerns, 370
    solubility, 369
Ammonia, 30
    chlorine needed to remove, 89–90
    toxicity with pH and temperature, 84
    water quality, 83–85
        criteria and standards for, 85
        in terms of, 4
Ammonia/ammonium ion (NH_3/NH_4^+)
    health concerns, 372
    source of, 371
Ammonia dissolved in water, hydrogen
        bonding in, 45
Ammonium ion, and water quality,
        83–85
Anaerobic biodegradation, 24, 29, 244
Anaerobic electron acceptors,
        252–253
Anaerobic microbial reactions, 245
Anion-exchange capacity, 242
Anthracene, 30
Antimony, health concerns and
        EPA primary drinking
        water standards, 373
Antiskid materials
    environmental concerns of, 318
    sand, 317
AOP, *see* Advanced oxidation process
Aqueous phase contamination in unsaturated
        zone, 230
Aquifer, 230
Arsenic (As), 30, 120, 123
    EPA primary drinking water standard, 375
    health concerns associated with, 374
Asbestos
    health concerns and EPA primary drinking
        water standard, 376
    sources, 375
Atomic mass unit, 272

Solubility Table for Ionic Compounds in Water

Compounds Containing	Soluble	Moderately Soluble	Insoluble
Lithium (Li^+) Sodium (Na^+) Potassium (K^+) Ammonium (NH_4^+)	All		
Chlorides (Cl^-)	All except →	Pb^{2+}	Ag^+, Hg_2^{2+}, Hg^{2+}
Bromides (Br^-)	All except →	Hg^{2+}, Pb^{2+}	Ag^+, Hg_2^{2+}
Iodides (I^-)	All except →		Ag^+, Hg_2^{2+}, Hg^{2+}, Pb^{2+}
Fluorides (F^-)	All except →		Mg^{2+}, Ca^{2+}, Sr^{2+}, Ba^{2+}, Pb^{2+}
Nitrates (NO_3^-) Chlorates (ClO_3^-) Perchlorates (ClO_4^-)	All		
Acetates ($C_2H_3O_2^-$)	All except →	Ag^+, Hg_2^{2+}	
Carbonates (CO_3^{2-}) Phosphates (PO_4^{3-}) Oxalates ($C_2O_4^{2-}$) Chromates (CrO_4^{2-})	Alkali metals (Group 1A cations), NH_4^+		All except ←
Hydroxides (OH^-) Oxides (O^{2-})	Alkali metals (Group 1A cations)	Ca^{2+}, Sr^{2+}, Ba^{2+}	All except ←
Sulfides (S^{2-})	Group 1A and 2A cations and NH_4^+		All except ←
Sulfates	All except →	Ca^{2+}, Ag^+	Sr^{2+}, Ba^{2+}, Pb^{2+}, Hg_2^{2+}, Hg^{2+}
Sulfites (SO_3^{2-})	Group 1A cations, NH_4^+		All except ←